ELECTRONICS FOR NUCLEAR
PARTICLE ANALYSIS

United Kingdom Atomic Energy Authority Research Group

ELECTRONICS FOR NUCLEAR PARTICLE ANALYSIS

EDITED BY L. J. HERBST

Harwell Post-Graduate Series

OXFORD UNIVERSITY PRESS 1970

Oxford University Press, Ely House, London W. 1

GLASGOW NEW YORK TORONTO MELBOURNE WELLINGTON
CAPE TOWN SALISBURY IBADAN NAIROBI DAR ES SALAAM LUSAKA ADDIS ABABA
BOMBAY CALCUTTA MADRAS KARACHI LAHORE DACCA
KUALA LUMPUR SINGAPORE HONG KONG TOKYO

PRINTED IN GREAT BRITAIN BY
SPOTTISWOODE, BALLANTYNE AND CO. LTD.
LONDON AND COLCHESTER

Preface

THIS book is based on a series of lectures delivered for the first time in 1964, and repeated about once a year since, at the Education and Training Centre of the Atomic Energy Research Establishment, Harwell. The decision to publish the lecture notes in book form was made in 1966. By then the occupational background of the participants had widened considerably from that obtaining in 1964. The first course was mainly intended for users of A.E.R.E. Type 2000 modular nucleonic instrumentation in the various divisions of A.E.R.E. and the adjacent Rutherford High Energy Laboratory. The objective, an appreciation in some depth of that instrumentation, has been a major aim of the course and indeed one of the main reasons for its existence initially.

Two factors largely account for the subsequent changes that led to the text in its present form. First, as the 2000 series instrumentation gathered momentum in both the number of different modules and the volume of production, it became used increasingly in other establishments of the U.K.A.E.A. and elsewhere, notably in universities and laboratories engaged in nuclear work of one form or other. Second, the system approach and the circuit techniques covered constitute material of considerable interest to anyone concerned with the design or use of nucleonic instrumentation. The result has been a substantial change in the subject matter now that the course is aimed at and largely attended by members of industrial establishments, government research organizations outside the U.K.A.E.A., and academic staff from universities and polytechnics.

The first major revision of the lecture notes was carried out in accordance with a pattern planned early in 1967. That plan has been adhered to, but there have been further substantial revisions during the last two years to take recent developments into account. The text offered presents, with minor alterations, the notes handed out on the course held in the summer of 1969. The authors were chosen to prepare and deliver the lectures for the first course held in 1964 and have run the course between them ever since. In that capacity they have represented the various teams of the Electronics and Applied Physics Division engaged in the design, development, production, and maintenance of nucleonic instrumentation to whom equally this book is due.

PREFACE

Having given an indication of the readers for whom this book is intended, it may be added that the text is aimed at designers and users of nucleonic instrumentation. The contents reflect this aim by embracing systems, circuit techniques and design, data processing, equipment practice, and reliability in one volume.

Two opening chapters outline system requirements and pulse shaping. Then follow three chapters on semiconductor fundamentals and general circuit practice. These may appear to be superfluous but were included because many experimentalists requiring nucleonic instrumentation have had little, if any, training in circuit design. Next come the chapters on analogue, counting, and timing circuits. Here a problem arose of what to include and what to leave out. On the one hand are numerous established circuits, on the other many new designs that have not yet reached the stage of production. Should the emphasis be laid on the former or the latter or should both be treated as of equal importance? It is a problem to which much thought was given. In the end the criterion was formulated that if a design had been proven in operational equipment it was included, if not it was left out. Underlying this decision was the conviction that only established circuits should be offered to the reader. The present rapid rate of progress means that many designs have only a short working life of a few years, sometimes less. However, the inclusion of circuits not yet beyond the drawing-board stage is doubtful practice. Anyone engaged in translating a design into production will know how easily a promising circuit may fail on the route leading to a fully-engineered model. Data-processing and display is covered in four chapters. One of these is devoted to multi-channel analysis. The other three embrace logic design, logic circuits, and the use of digital computers. The final chapters deal with practical design-aspects and reliability.

There are inevitably areas that overlap, although these have been kept to a minimum. For example, it was felt advisable to outline pulse shaping early on, even though some of that account is covered in far greater depth in the chapter on pulse amplifiers. Another instance is the effect of pulse shape on timing, a matter discussed in the chapters on pulse shaping, pulse-amplitude discriminators, and time-spectrometer circuits. There is also some diversity, rather than standardization, of symbols. Different symbols have been used for logic gates, for various transistor parameters, etc. The reader is likely to encounter this proliferation in other scientific literature. Where such diversity is established, the various accepted representations of a particular parameter or logic symbol have been included, so that the reader may become familiar with them all.

To sum up, the authors have tried to produce a book covering both fundamentals and circuit practice to a degree that will make it a self-contained text for most readers. The designer needs to appreciate system aspects and basic circuit fundamentals. The user will benefit by understanding circuit practice even if he is not engaged in actual design. Both are therefore likely to need the entire text rather than just part of it. The level of the treatment is not aimed at a specifically qualified reader. The standard is, by and large, at degree level, but the treatment is predominantly practical. The subject matter is within the grasp of lesser qualified readers although naturally these will be more extended. The main objective is helping newcomers to nucleonic instrumentation to stand on their own feet.

Thanks are due to the numerous typists and draughtsmen who helped produce the manuscript. Also the help of the publishers at all stages in the production of this book has been most gratifying. It remains for me to end on a personal note by acknowledging the unfailing co-operation I have had from all the authors. In the majority of cases the original contributions were completely re-written and often underwent further major changes at my request to reach their final form. All this was carried out in a spirit that made my task of editing a pleasure.

L. J. HERBST

Electronics and Applied Physics Division
Atomic Energy Research Establishment
Harwell, Berkshire
September 1969

Contents

CONTENTS

CONTENTS

CONTENTS

CONTENTS

List of Authors

	Chapter(s)
G. H. Moss†	1
L. J. Herbst‡	2, 4, 5, and 8
M. O. Deighton	3
A. B. Gillespie	6 and 7
F. H. Hale	9
P. R. Orman	10
F. H. Wells	11
J. M. Richards	12, 13, and 14
W. D. Brownrigg	15
L. A. Kilbey	16

† Now at Ministry of Technology, London
‡ Now at Teesside Polytechnic, Middlesbrough

I System Requirements for Nuclear Particle Analysis

By G. H. MOSS

1.1. Introduction

THIS book is an introduction to pulse circuit design for users of nuclear particle analysis systems, who may or may not have been trained in electronics. The appreciation of the techniques involved is dependent on an understanding of the systems in which they are employed, and this chapter seeks to fill in some of the background before the specific aspects are dealt with, the main aims being:

(1) to present a generalized picture of analysis systems;
(2) to show where the subject of each succeeding chapter fits into this picture;
(3) to discuss some of the factors governing the choice of system characteristics;
(4) to show how analysis systems of differing characteristics are related to one another.

Fig. 1.1 shows how the book covers the subject by reference to a generalized counting system. Some chapters of this book are specialized and relate to no more than one or two stages of the system, while others are much broader

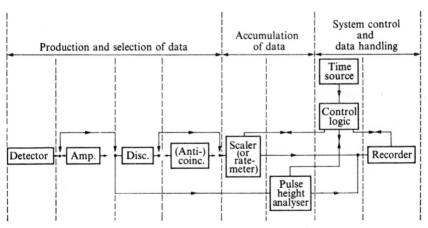

FIG. 1.1. Generalized counting system.

1

in scope. It is hoped that the latter will help to preserve continuity throughout the book.

1.2. General system requirements

Most common counting systems can be derived from the general scheme in Fig. 1.1 by selecting or duplicating the units as necessary. The three main sections are (1) data production and selection, (2) data storage, and (3) data handling. Every system includes these three even if some of the functions are combined in one unit or performed by an operator, and each system has its own special requirements controlling the degree of complexity of each section.

1.2.1. Data production and selection

The first stage in this section is the detector, which produces electrical pulses in response to radiation. It may also perform an initial selection of data since its response can often be made negligible to unwanted types of radiation. This stage must be followed by an amplifier if the pulses are of insufficient amplitude, and also by a discriminator that rejects pulses below a preset level. For special purposes a further stage of data selection is provided by a coincidence or anti-coincidence unit, which accepts or rejects coincident pulses from two or more detection channels.

1.2.2. Data storage

The next general section of a nuclear counting system consists of an accumulating data store, which may be a scaler, pulse-height analyser, or a ratemeter. If this is a pulse-height analyser its function overlaps with that of the first section, since it sorts pulses of varying amplitudes into different sections of the data store.

1.2.3. Data handling

The function of the final stage is to control the system and record the data. Since the data stores in common use have visual displays, this part of the system can be an operator with stop-watch, pencil, and paper, or it can be an automatic system with a stored program whose output is fed to a printer, tape punch, or analogue recorder. The system output is thus permanently recorded ready for inspection or further data processing.

1.2.4. Typical systems

As an example of how widely system requirements vary, three specific cases are shown in Figs. 1.2, 1.3, and 1.4. Fig. 1.2 represents a very simple system such as might be used for counting alpha particles, perhaps for measuring the quantity of plutonium when it can be assumed that no other alpha emitters are present. The operator performs control and recording functions and a minimum of apparatus is used. The system of Fig. 1.3 might be used to count output coincidences from two detectors and also to monitor the pulse output independently. In this case full automatic control would be most essential if the counting periods were to be synchronized with the operation of external apparatus. The third example, Fig. 1.4, can be used

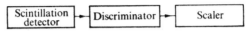

FIG. 1.2. Simple counting system.

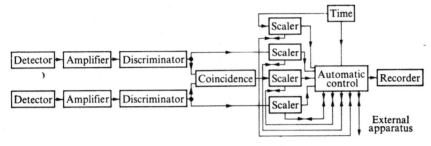

FIG. 1.3. Automatic counting system for coincidence measurements.

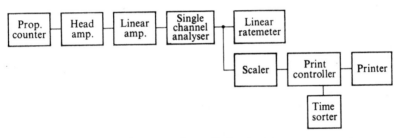

FIG. 1.4. Automatic system for radiation spectra measurements.

for radiation spectra measurements, since the detector and amplifiers are chosen to preserve a linear relationship between radiation energy and output pulse height, while a single-channel pulse-height analyser precedes the scaler. The system includes automatic control, and approximate check of the count rate is continuously available from the ratemeter.

These three systems illustrate fundamentally different types of measurement, viz. count rate, coincidence rate, and energy. It is also clear that where flexibility of use is required a modular range of electronic units is advantageous, since the assembly of a tailor-made system is simple provided the range of available functions is sufficiently wide.

1.3. Nuclear radiation and its detection

Nuclear experiments are observed via the interaction of radiation with matter to produce electrical pulses, and therefore the properties of nuclear radiation, together with some common methods of detection, will be surveyed briefly before the principles of analysis-system design are considered.

1.3.1. *The properties of nuclear radiation*

The term 'nuclear radiation' includes all emissions that result from changes in structure or excitation of the nucleus or electron shells of an atom. Detection of these radiations must depend on a transfer of all or part of their energy to the detector, and the transfer mechanisms vary with the particle under consideration. Table 1.1 lists the properties of some common radiation types and a few of these will be considered as typical examples.

TABLE 1.1

Characteristics of nuclear radiation

Type of particle	Charge	Rest (amu)	Mean life (s)
Neutron (n)	0	1·008982	$1·013 \times 10^3$
Proton (p)	1	1·007593	Stable
Deuteron (d)	1	2·014187	Stable
Alpha particle (α)	2	4·002777	Stable
Electron (e)	−1	0·000549	Stable
Gamma ray (γ)	—	—	Stable
X-Ray (X)	—	—	Stable
Fission fragment (light)	20	95	
Fission fragment (heavy)	22	139	

1.3.1.1. *Proton-like particles.* This heading includes protons, alpha particles, and other nuclei of comparable mass and charge, which may be ejected with energies of up to some 10 MeV during nuclear transformations. When these particles pass through an absorber the principal cause of energy loss is by ionization and excitation, due to interaction of their Coulomb fields with bond electrons. There are many collisions along each track because the electrons are relatively light and only small amounts of energy are lost at a time. The deflexions are therefore negligible, the standard deviation of track length for particles of the same energy is small, and the range is almost linearly related to energy. The actual range is approximately inversely proportional to the absorber density and for protons it may amount to a few cm in air at s.t.p. Because of their double charge and mass, alpha particles have greater ionizing power than protons and so are stopped in a fraction of the distance. Fission fragments create even more intensive ionization but in this case energy loss by nuclear collisions is also important due to the increased nuclear charge.

1.3.1.2. *Electrons*. Because of their small mass, beta rays of a given energy (electrons) have a much higher velocity and penetrating power as compared with protons. However, the average energy loss per collision is large (up to 50 per cent) and deflexions are correspondingly large. Electrons are therefore brought to rest by a relatively small number of collisions, resulting in tortuous paths that are variable in total length for electrons of equal energy. The range also varies greatly and an appreciable proportion of electrons may be scattered back towards the source. Energy loss is still by ionization and excitation but in this case the electromagnetic radiation due to electron deceleration at collisions (Bremsstrahlung) is also important.

1.3.1.3. *Gamma rays and X-rays*. Gamma rays and X-rays are photons of electromagnetic radiation and are absorbed as single events, exciting the nucleus or producing electrons by the photoelectric effect. Photons of large energy can also be absorbed in the production of positive and negative electron-pairs, any excess energy appearing as kinetic energy of the particles.

1.3.1.4. *Neutrons*. Because they have no charge, neutrons cannot ionize atoms directly. They can, nevertheless, be made to induce ionizing radiations, for example, by producing recoil protons or by causing fission when captured by nuclei. It is not always easy to make these processes efficient because the probability of collisions is not high.

1.3.2. *Radiation detectors*

Detectors are used for measurements such as intensity, time, and energy of particles or electromagnetic radiation, and the selection of a type suitable for a particular experiment demands some knowledge of their characteristics. Before some principal detectors are discussed the concept of 'dead time' with consequent counting losses must be introduced, as it affects the choice of detector, its output coupling circuit, and the apparatus that follows.

1.3.2.1. *Counting losses*. If two photons or particles, each of which would cause a detector output on its own, are separated by a shorter period of time than the net resolving time of the whole system, then only one will be counted. Since the emission of nuclear radiation is a random process and each new event must be considered as independent of any other, some proportion of the events will not be counted, however low the mean count rate and however short the resolving time of the system. Accurate radiation

5

measurement therefore depends on the use of detectors and systems capable of a far higher mean count rate than that under measurement. It is possible to correct for the counting losses provided the dead time per pulse can be defined accurately and if the correction factor is not large. This is illustrated in Fig. 1.5. If N_0 is the observed count rate, then the system must have

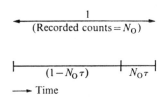

FIG. 1.5. 'Counting loss' principle.

been non-receptive ('dead') for a total period $N_0\tau$, where τ is the dead time per pulse. The actual time the system is receptive (the 'live' time) must therefore be $(1 - N_0\tau)$ and so the 'true' mean count rate N_m is

$$N_m = \frac{N_0}{(1 - N_0\tau)},\qquad(1.1)$$

$$\frac{\text{true rate}}{\text{recorded rate}} = \frac{N_m}{N_0} = \frac{1}{(1 - N_0\tau)}.\qquad(1.2)$$

The simple analysis above is not applicable to all systems. For example, the dead time of pulse-height analysers cannot be regarded as constant. Even in simple systems the formula has to be used with caution. With some detectors, the dead time per pulse is a function of count rate.

1.3.2.2. *Gas-filled chambers.* Gas-filled chambers, in which there is an electric field established between two electrodes, are often used as direct detectors of ionizing radiation, and their characteristics depend on the p.d., type of gas, geometry, etc. A simple ionization chamber, in which all ions produced by the incident radiation travel to the electrodes and where no secondary ionization is generated, will collect a charge proportional to the energy of an incident particle. The output will, however, be small and will require much amplification before the data collection stage. If the voltage gradient is increased until secondary ionization occurs, then it is possible to collect more charge while retaining proportionality provided the 'gas multiplication' is kept to a moderate amount. Such a chamber is known as a proportional counter and, like the ionization chamber, can be used for energy measurements. If the field is increased still further above a critical value, the initial ionization triggers a continuous discharge that is self-sustaining until the potentials are reduced. A chamber operated in this condition is known as a 'Geiger counter' and its output in a suitable circuit may be several volts, the value being almost independent of the initial ionization. The recovery time of such a tube may be hundreds of microseconds so that the usable count rates are limited, but the tube is convenient and available in sealed-off form, usually with a gas mixture that assists quenching of the discharge. Counting efficiency will depend on the probability of absorption of radiation, and will not be high for gamma rays, due to the low density of the gaseous filling. G.M. tubes will also detect all forms of ionizing radiation and neutrons in principle, provided conditions are arranged to favour the production of secondary radiation either by recoil or fission. It should be noted that sealed-off tubes need windows whose thickness normally precludes counting of alpha particles and protons.

1.3.2.3. *Semiconductor detectors.* A slab of intrinsic conductivity semiconductor with p and n contacts (p-i-n) can be looked upon as the solid-state equivalent of the ionization chamber in that the radiation gives rise to electron–hole pairs, which separate and move to the positive and negative electrodes respectively. In both types of detector the rise time of the pulse will depend on the position of ionization within the slab and the time constants of the succeeding circuits must be long in comparison if energy measurements are to be made. Most detectors in this category are formed from silicon, but other materials are of interest, particularly the heavy elements that hold out the promise of great stopping power for γ-radiation. At present the advantages relative to gaseous detectors are better energy resolution, sensitivity (production of electron–hole pairs only requires one-tenth of the energy for a pair of ions in a gas), high density and stopping power, and convenience in use. Some transistor-like structures can also give gain that increases the sensitivity still further (without improving the signal-to-noise ratio). The response of such detectors is dependent on their geometry, although in practice the pulse width is usually of the order of a microsecond.

1.3.2.4. *Scintillation detectors.* A scintillation detector consists of a phosphor that emits a light flash in response to radiation and feeds a photomultiplier for converting this flash into an electrical pulse. This type of detector is often used for energy measurement of γ-rays, due to its linearity, high sensitivity, and fast response, but it can be made suitable for detecting other ionizing radiation. Sodium iodide is often used as the scintillator because the high atomic number of iodide ($Z = 53$) leads to a high conversion efficiency. This crystal is also very transparent to its own radiation and is quite fast, having a luminiscent decay time constant of $2 \cdot 5 \times 10^{-7}$ s. The fluorescence of inorganic materials is usually dependent on imperfections in a crystal lattice so that these substances are only useful in the solid phase.

7

In organic phosphors emission of light is the result of interactions within individual molecules so that these materials can be used in solution or even as a gas. Two such modern materials are stilbene and terphenyl, which have a decay time for the primary component of a few nanoseconds. Use of such detectors makes possible time resolution (using the leading edge of the pulse) of the order of 1 ns and also permits the measurement of very high count rates.

1.3.2.5. *Other detectors.* Bubble chambers, cloud chambers, and nuclear emulsions are detectors that produce a picture of the paths of charged particles, as can arrays of counters to a limited extent when combined with suitable coincidence and anti-coincidence circuitry. Except for the counter arrays these detectors are receptive for relatively long periods and may show many unwanted background tracks. The spark chamber has been developed to overcome this limitation and it consists of a series of parallel plates, alternate members being connected to the positive and negative poles of a high-voltage source. The ionization from a charged particle triggers sparks along its path which expose a photographic emulsion. The supply voltage need only be on for short periods and indeed it is often switched on by the pulses from a pair of detectors observing the particle entering and leaving the chamber.

1.3.2.6. *Coupling circuits.* The circuits used to couple detectors to pulse-counting systems are important, having a marked effect on performance. This happens because electrically a charge-collecting detector has an appreciable self-capacitance and is connected to a polarizing source through a high resistance (Fig. 1.6). During the charge-collecting process the voltage

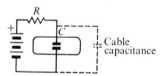

FIG. 1.6. Charge-collecting detector.

on C falls. At the same time the signal charge begins leaking away through R, but if CR is long compared to the duration of the signal, the voltage is inversely proportional to C. Naturally the effective value of C is the total of detector and connecting-cable capacitance. Where a long cable run is unavoidable, a head amplifier acting primarily as an impedance changer is often used so that there is plenty of current available to charge the cable capacitance. The basic sensitivity of the detector is then retained. In cases where head amplifiers cannot be used, recourse may be made to a charge-sensitive amplifier that accepts charge without an appreciable change in voltage so that little of the collected charge is used in changing the potential across C. This sort of amplifier is also helpful in taking advantage of the excellent linearity possible when semiconductor detectors are employed.

If energy measurements are to be made, it is important that the height of each pulse should have no contribution remaining from the preceding one, and it is usual to provide a 'differentiating' (high pass) circuit, the output of which responds to changes. As signal-to-noise ratio is also important, an 'integrating' (low pass) network may also be used. These

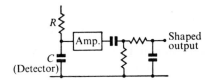

FIG. 1.7. Pulse-shaping circuits.

pulse-shaping circuits are normally placed after the amplifier as in Fig. 1.7 and are usually made adjustable to suit many applications.

1.4. Choice of system characteristics

Each stage of the generalized counting system will now be considered in more detail.

1.4.1. The detector

The following factors and their relation to one another must be considered in choosing a detector: (a) type of radiation to be detected, (b) efficiency of detection, (c) linearity of response to radiation energy, (d) selectivity to wanted and unwanted (background) radiation, (e) resolution, (f) stability, and (g) convenience.

Suppose a detector is required to measure β-activity in urine for radiation protection purposes. In this case both the efficiency and selectivity of the detector are important and special measures may be necessary to reduce the background to below the expected count rate. For operational use, convenience is also important and useful results can be obtained with a G.M. tube of low background type screened with an array of G.M. tubes in anti-coincidence, the whole assembly being enclosed in a substantial lead shield. An alternative scheme might use special gas-flow counters in anti-coincidence which can have a greater efficiency/background ratio.

The example quoted above is one in which disintegration rate is being measured. If energy measurements are being attempted, the prime consideration is the linearity of response, that is the ratio pulse height/energy must be constant, and the stability of this ratio with respect to time and temperature is also important. Depending on the application, scintillation, proportional, semiconductor, or even ionization chamber counters may be used. The choice of detector type may be dictated by considerations other than linearity, for example resolution time for detection of successive events, or background rejection. The choice of a detector is often difficult because of the interrelated factors, particularly for very low or very high count rates or where accurate energy measurements are required.

1.4.2. *Linear pulse-amplifiers*

A linear pulse-amplifier raises the signal level at the detector sufficiently to operate the discriminator and in its design the following factors are important: (a) gain, (b) maximum output, (c) linearity, (d) gain stability, (e) bandwidth or pulse response, and (f) noise.

If pulses are to be sorted into groups, the maximum output amplitude must be such that the group width is large compared to any uncertainty that may exist in the discriminator threshold. In addition, the linearity and gain stability must both be good so that the pulse height/energy relation is maintained constant up to the maximum amplitude required. For applications requiring good resolution, very wide bandwidths are necessary (perhaps to 100 MHz or beyond) but usable bandwidth is also a function of the equivalent input noise, which may set a lower limit to the energy measurable or determine the accuracy of measurement. Depending on the system, gain requirements may vary from unity to 10^7 according to detector type. For example a very high gain is always needed where the detector is an ionization chamber with no gas multiplication, but little or no gain may be required following the P.M. tube of scintillation detector, the tube itself being an efficient pulse amplifier.

1.4.3. *Pulse discrimination*

The basic function of pulse discrimination is to respond only to input pulses above a critical amplitude (bias voltage or current), but other characteristics are also important such as (a) bias stability, (b) non-blocking of input circuit, (c) resolving time, (d) output waveform constancy, and (e) controlled paralysis time.

A simple discriminator can be used to investigate the energy distribution of radiation pulses by measuring the count rate at a number of different bias levels and must therefore have good bias stability. If large count rates are to be measured, the resolving time must be short and independent of pulse height and rate (the input circuit must not block). The output

10

requirements are less stringent. There must be no output, or an output pulse suitable for operating the data store in use.

In addition to satisfying these requirements, many discriminator units include a circuit for giving a pre-set dead time (paralysis) after each input pulse. This facility is essential where the detector in use has a dead time that may vary with its history. The controlled dead time is set larger than the inherent dead time for the detector so that an accurate correction can be applied to the measured count rate. It may also prevent blocking of the data store by closely spaced pulses.

If two simple discriminators are used in such a way that one accepts pulses smaller than A_2, and the other larger than A_1 ($A_2 > A_1$), then the system becomes a single-channel pulse-height analyser selecting pulses from the radiation within a certain energy band.

1.4.4. (Anti-)coincidence unit

Coincidence units detect the presence of pulses in two counting channels within a finite time known as the resolving time. An example of their use is the detection of nuclear events where more than one quantum of radiation is emitted simultaneously or within a short time. For this kind of experiment, very short resolving times may be required and an allowance must be made for cable lengths, etc. The technique is not limited to two channels and has many applications in nuclear physics.

Sometimes a comparatively long resolving time may be used, for example for reducing the background counts in a β-counting assembly using G.M. tubes. In this case an anti-coincidence arrangement is necessary.

1.4.5. Data storage

Data stores can be either analogue or digital. The ratemeter is an example of an analogue store with a short-term memory. The most frequently used digital stores are scalers and pulse-height analysers. Each type has its special uses although in some complicated experiments all three may be in simultaneous use.

1.4.5.1. *Ratemeters.* A ratemeter assesses the mean count rate by means of an integrating circuit (usually capacitive) having a time constant long compared with the mean time between pulses. With suitable circuitry the integrator output as displayed on a meter can be made proportional to the mean rate, in which case it is known as a linear ratemeter. Such an instrument has only moderate accuracy and usually has a number of switched ranges to cover high and low count rates. It can be made an inexpensive and robust instrument for monitoring, alarm, and control applications, where extreme accuracy is not required. Sometimes it is required to measure very different count rates on a single range, for example where the rate is being recorded on a chart without an operator available to change ranges. In this case a logarithmic ratemeter is the natural choice, with constant percentage reading accuracy. A third type is the differential ratemeter, which indicates the difference between count rates from two sources.

1.4.5.2. *Scalers*. Scalers or other digital stores are used in experiments where good statistical accuracy is needed. Electronic techniques are employed for all but the slowest counting rates. The important characteristics of scalers come under the following headings: (a) resolution, (b) capacity, (c) display, (d) read-out, and (e) control facilities.

Electronic scalers of different types have resolving times ranging from 10^{-4}s to below 10 ns for paired pulses and the choice of a suitable scalar is related to detector resolution, the slowest being adequate for use with a G.M. tube while the fastest may be necessary in nuclear physics experiments using scintillation counters. One point liable to be overlooked is that the capacity required (maximum count) is not necessarily related to resolving time since an experiment may produce a few very closely spaced pulses, all of which must be counted. In completely automatic systems, read-out and remote control facilities are essential and scalers of this type should always be used if automatic operation may be required for future experiments.

1.4.5.3. *Multi-channel pulse-height analysers*. It has been shown how some detectors produce voltage pulses proportional to energy and where the distribution of radiation energy is of interest these pulses must be sorted into separate counting channels by height. Pulse-height analysers were developed primarily for this application. Other uses, such as measurement of pulse-time distributions and multi-channel scaling, are comparable in importance. Some applications require additional apparatus, such as a time-to-amplitude converter, or circuitry that will allow the subtraction of a reference energy spectrum from the measured one in order to expose peaks not present in the reference (spectrum stripping). To use a multi-channel scaler, providing up to several hundred separate scalers, is often much more economical than separate scalers could possibly be, but the scaling speed is not great because the pulse-sorting process limits the speed by blocking the inputs not in use for the appropriate dead time. It is sometimes possible to improve the apparent dead time by using some form of derandomization for the input pulses.

1.4.6. *Automatic control*

Automatic system control should be used either when manual control cannot give sufficiently accurate timing of control functions, or where an

12

experiment lasts sufficiently long to make the saving in manpower worth while. Sometimes partial automation is worth considering because full automation may not give a comparable manpower-saving/expense ratio, and this should be looked into at the system design stage.

In digital system control a time source is the first essential, and the type chosen will depend on the accuracy required. Some of the available time sources are (a) pulses derived directly from 50-Hz mains, (b) pulses derived from a mains-driven synchronous motor, (c) spring-driven clock, (d) pendulum clock, (e) tuning fork, and (f) crystal oscillator.

With the exception of the a.c. mains, the timing accuracy of all these sources can be sufficient for most nuclear counting, although the relatively high frequency available from a crystal may dictate its use when short intervals for count control are features of the experiment. The problem with mains timing is that the accuracy is unpredictable, since it depends on changes in the electricity supply loading in relation to the number of power stations operating and those at readiness. However, mains time can always be used where 1 per cent accuracy is ample. Conversely, it must never be used for 0·1 per cent accuracy unless the timed period extends over several hours. These two rules apply in all but exceptional circumstances.

With the advent of cheap packaged computers it has become feasible to provide computer control for moderate size systems and also to provide multi-channel analysis facilities by programming the computer accordingly. In bigger, multiple experiments, a general computer-controlled system may give individual users control and data-reduction facilities on a self-service basis.

The extra investment necessary in giving a system complete automatic control is often large, since special timing circuits and data recorders are necessary in addition to the control circuitry. Automatic control may even require a computer with special interface circuitry, but there are often unexpected gains, the most cogent being repeatability of results.

1.4.7. *Automatic recording*

Automatic measurements may be recorded in analogue form using a chart recorder, or in digital form using printer, punch, or tape recorder. For very low count-rates, or for count-controlled systems where only time is recorded, an electromechanical printing counter can be used, combining the functions of data accumulation and printing. This method can be used if a paralysis time of 100 ms is acceptable, and such a system can be quite cheap. For higher count rates, a separately controlled printer must be used and different types can vary in speed from 7 to 100 characters per second or more.

Whether data is accumulated on punched tape, cards, or magnetic tape

3

will depend on the application, the quantity of data, and computer requirements. Where preliminary data-sorting is essential cards may be the best
choice, but as data-output speeds increase, punched tape and finally magnetic tape must be considered. Typical speed for cards is sixteen characters
per second, while tape punches are available for speeds up to 300 characters
per second. Magnetic tape must be used for higher recording rates.

1.4.8. *Modular systems*

If a laboratory requires a flexible instrumentation system, it is convenient
if its structure is modular as is the A.E.R.E. 2000 system. Within such
a system, the physical dimensions of the units are standardized so that
mechanical compatability should also include defined digital coupling pulses
and d.c. control levels, and the same principles should apply to data highways
that connect with digital recorders, computers, etc.

In conclusion, the interrelation between the various parts of a nuclear
counting system can be complex, requiring much thought in order to choose
the best system, and during the design stage the overall requirements of
the experiment must always be kept in mind. This chapter has indicated
some of the problems, together with the principles that allow of their
solution.

2 *Pulse Shaping*

By L. J. HERBST

2.1. Introduction

THE outputs of the various pulse detectors employed in nuclear particle counting require amplification and shaping prior to processing for presentation to the experimenter. Several such arrangements are outlined in the preceding chapter, which also contains a diagram of a shaping network (Fig. 1.7). The detector output consists of an electronic signal, a charge, which constitutes the input for the amplifier. The amount of charge and the time over which it is delivered depend on the detector and on the nature of the incident particle. Detector charges lie in the range of about 10^{-15} to 10^{-10} coulombs per pulse and the time interval of charge delivery varies from a few nanoseconds to several milliseconds. Fig. 2.1 indicates typical outputs

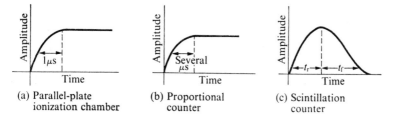

(a) Parallel-plate
ionization chamber

(b) Proportional
counter

(c) Scintillation
counter

FIG. 2.1. Typical detector outputs.

of various detectors. These pulses have a relatively fast initial rise time. Generally only the initial portion of the signal up to the time of reaching peak amplitude contains information for processing. In ionization chambers, scintillation detectors with a decay time of less than 1 ms, and semiconductor radiation detectors, the pulse output can be regarded as a step function of charge. A modified approach is needed for very fast pulses obtained from organic scintillators and fast photomultipliers. The requirements for pulse shaping are outlined in the next section.

2.2. Requirements for pulse shaping

For a number of reasons the shape of the output pulse from the detector is invariably changed considerably by the amplifier chain. The nature of

15

PULSE SHAPING

the shaping depends to a large extent on the application. For this purpose applications can be divided into three broad categories:

(a) counting measurements,
(b) energy measurements,
(c) time-interval measurements.

One of the most important considerations is resolution of adjacent pulses. Nuclear experiments always involve many events with random spacing. It is then essential to distinguish between closely spaced pulses, even when the mean rate of events is quite low. Fig. 2.2 illustrates this point. The

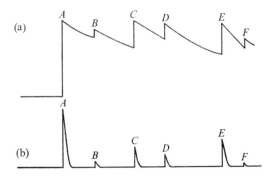

FIG. 2.2. Differentiating of detector output. (a) Detector output. (b) Differentiation of (a).

detector output in Fig. 2.2(a) is shaped into the form shown in Fig. 2.2(b) by differentiation, a technique that will be discussed later. The shaping converts the detector outputs A to F into pulses with zero base line. Timing and amplitude information of the original signals are retained and the restoration to zero base line prevents overloading of the following amplifier stages. The original pulses, which have the pattern of step-function charge outputs mentioned in the previous section, have been narrowed, making it possible to distinguish between closely spaced events such as E and F. The need to narrow pulses in order to retain amplitude information between a closely spaced pair is brought out more clearly in Fig. 2.3 where B is superimposed on A. In the idealized shaping of Fig. 2.3(b) the original amplitudes of A and B are preserved. Without such shaping false amplitude information for B would have been supplied to the amplifier. This phenomenon of pulse pile-up and pulse amplitude distortion, due to the occurrence of a second and further pulses on the trailing edge of the first pulse, can cause trouble and is very much reduced by differentiation, which shortens the resolving time.

A further advantage of differentiation is improvement in the final signal-to-noise ratio. *RC* shaping, involving differentiation and integration by the passage of a signal through networks like those shown on Figs. 2.4 and 2.5,

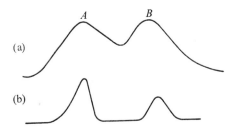

FIG. 2.3. Superimposed signals. Differentiation. (a) Detector output. (b) Output differentiated.

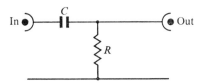

FIG. 2.4. Differentiating network.

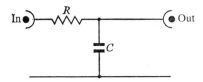

FIG. 2.5. Integrating network.

improves the signal-to-noise ratio by reducing the bandwidth of the transmission system while substantially preserving the peak amplitude of the signal. The matter is dealt with more fully in Chapters 6 and 7.

RC networks exist not only for pulse shaping, but also for a.c. coupling in amplifiers. It is generally not practicable to have the amplifier d.c. coupled throughout. An important consideration of *RC* networks is the influence on pulse transmission when the network exists solely for d.c. isolation between successive stages.

An alternative method of differentiation employs delay-line shaping. Essentially the delay line produces a rectangular pulse of defined width from a step-function input. Fig. 2.6 illustrates the process and brings out the contrast of this technique compared with Fig. 2.4. The latter performs the operation dv/dt under certain conditions. The generally adopted designation

17

'differentiation' for the delay-line shaping in Fig. 2.6 is, strictly speaking, not correct. It is nevertheless adhered to because in nuclear instrumentation delay-line and RC shaping often achieve the same function of narrowing the input pulses. Delay-line shaping is also employed for optimum timing information in coincidence and other timing measurements. The techniques and design considerations are somewhat different from those governing shaping in pulse amplifiers for energy measurements.

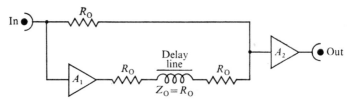

FIG. 2.6. Delay-line shaping. A_1, inverting amplifier unity gain, near zero output resistance. A_2, amplifier, near zero input resistance.

Pulse shape is defined in terms of rise time T_r, delay time T_d, decay (or fall) time T_f, and pulse duration. The definitions are explained with the aid of Fig. 2.7. Amplifier characteristics are often defined by T_r, T_f, and

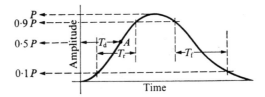

FIG. 2.7. Pulse shape definitions.

T_d of the output pulse for a step-function input. T_d is the time required for the pulse to reach half its maximum amplitude A in Fig. 2.7. It must be distinguished from the time delay through a transmission element. In the latter case the time delay is the delay of the leading edge, more precisely a specific point of the leading edge (generally half the maximum amplitude) in transmission through the element. T_r may be defined as the reciprocal of the slope dV/dt at A. An alternative and generally preferred definition of rise time is the time required for the pulse to rise from 0·1 to 0·9 of its maximum amplitude. The following sections give a general account of RC and delay-line shaping and also of shaping for optimum timing.

2.3. RC shaping

The simple RC network of Fig. 2.8 represents the usual interstage coupling network of an amplifier and is also used for pulse shaping. The

18

generator resistance R_g of the voltage source V_i is frequently ignored compared with R and the reader is more likely to encounter Fig. 2.8(b) which presents that condition. In many cases R_g is sufficiently large to be

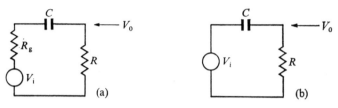

FIG. 2.8. CR differentiating network. (a) Including R_g. (b) Excluding R_g.

significant and its presence must then be allowed for. The response for a sinusoidal input is readily obtained, remembering that the impedance of C equals $1/j2\pi fC$ where f is the frequency. This gives a transmission gain A equal to

$$A = V_o/V_i = 1/(1 + f_1/jf), \qquad (2.1)$$

where $f_1 = (1/2\pi CR)$. Hence V_o leads V_i by a phase angle θ equal to $\tan^{-1}(f_1/f)$. Also, from (2.1),

$$|A| = 1/\{1 + (f_1/f)^2\}^{1/2}. \qquad (2.2)$$

At $f_1 = f$, A is reduced to $(1/2)^{1/2} - 3$ dB, so that f_1 is known as the 3 dB frequency or simply as the 3 dB bandwidth of the circuit. The bandwidth, unless stated otherwise, is understood to apply to the 3 dB point.

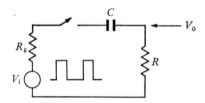

FIG. 2.9. CR differentiation. Step function input.

For Fig. 2.9, ignoring R_g and with time T reckoned from the instant of closing the switch S (Cutler 1960),

$$C\frac{d(V_i - V_o)}{dt} = V_o/R, \qquad (2.3)$$

$$I = -C\,dV_o/dt = V_o/R. \qquad (2.4)$$

Eqns (2.3) and (2.4) with the boundary condition $V_o = V_i$ for $t = 0$ lead to

$$V_o = V_i \exp(-t/RC). \qquad (2.5)$$

If R_g is allowed for, the equation becomes

$$V_i = I(R_g + R) + Q/C, \qquad (2.6)$$

where Q is the charge across C. Differentiating,

$$0 = (R_g + R)\, dI/dt + I/C \qquad (2.7)$$

which gives

$$I = I_o \exp\{-t/(R_g + R)C\} \qquad (2.8)$$

where $I_o = V_i(R + R_g)$, so that

$$V_o = \{(V_i \times R)/(R + R_g)\} \exp\{-t/(R_g + R)C\}. \qquad (2.9)$$

In the subsequent treatment R_g will be ignored because, more often than not, it is much smaller than R. However, its influence will be mentioned again later.

If now the input voltage V_i is not a step function, so that V_i can no longer be treated as a constant $(dV_i/dt = 0)$, eqn (2.3) becomes

$$dV_i/dt = V_o/RC + dV_o/dt. \qquad (2.10)$$

If $(V_o/RC) \gg dV_o/dt$, (2.10) reduces to

$$dV_i/dt = V_o/RC. \qquad (2.11)$$

Fig. 2.9 is thus a differentiating circuit because the output voltage will then be the time differential of the input. Eqn (2.11) is a general expression that can be applied to step-function pulses, pulses with finite rise time, and sinusoids alike. Note the condition for differentiation given in the inequality below (2.10). It is sometimes stated that, provided R is small enough, so

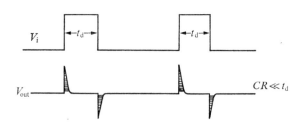

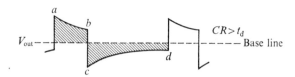

FIG. 2.10. Waveforms for Fig. 2.9.

that V_o is very much less than V_i, the charging current through C will be very nearly $C(dV_i/dt)$, giving an output of V_o equal to $CRdV_i/dt$. A more rigorous assessment of differentiation is necessary and eqn (2.10) indicates that V_o, the RC product, and dV_o/dt are the determining factors. For the reasons just given the RC network of Figs. 2.8 and 2.9 is known as a differentiating network.

An alternative description, high-pass filter, signifies the transmission behaviour outlined in eqns (2.1) and (2.2). Eqn (2.10) gives the necessary condition for transmission without alteration of pulse shape or loss of amplitude, namely

$$V_o/RC \ll dV_o/dt. \tag{2.12}$$

Eqn (2.12) indicates a long time constant for transmission without differentiation. The response of Fig. 2.9 for a step-function input is shown in Fig. 2.10 for various values of CR. Fig. 2.10 gives pulse shaping for various

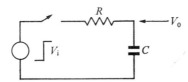

FIG. 2.11. *RC* integrating network transient input.

values of CR. Fig. 2.11 is an RC integrating network. The transmission gain is

$$|A| = V_o/V_i = 1/\{1 + (f/f_2)^2\}^{1/2} \tag{2.13}$$

where $f_2 = 1/2\pi CR$. V_o lags behind V_i by θ where $\theta = \tan^{-1}(f/f_2)$. Fig. 2.11 is a low-pass circuit: the output voltage decreases with increasing frequency.

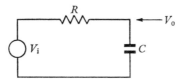

FIG. 2.12. *RC* low-pass network.

For a step-function input applied with switch S in Fig. 2.12 closed at time $T = 0$.

$$V_i = IR + V_o, \tag{2.14}$$

$$I = C(dV_o/dt). \tag{2.15}$$

21

PULSE SHAPING

Combining (2.14) and (2.15),

$$V_i = CR(dV_o/dt) + V_o. \tag{2.16}$$

The solution of (2.16) is

$$V_o = V_i\{1 - \exp(-t/CR)\}. \tag{2.17}$$

If $CR(dV_o/dt) \gg V_o$, (2.16) becomes

$$V_o = (1/CR) \int V_i \, dt, \tag{2.18}$$

showing that the circuit of Fig. 2.12 integrates the input waveform under those conditions. In most applications the radiation detector output pulse

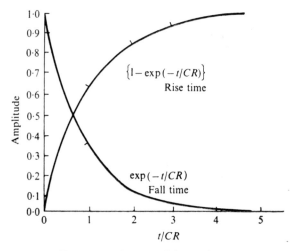

FIG. 2.13. Exponential waveforms.

has an exponential rise of the form given in (2.17) and illustrated in Fig. 2.13. If we substitute an input of the form

$$V_i = \hat{V}\{1 - \exp(-t/\tau)\}, \tag{2.19}$$

where τ is a constant, the solution of (2.10) becomes

$$V_o = \{\hat{V}m/(m-1)\}\{\exp(-t/m\tau) - \exp(-t/\tau)\}, \tag{2.20}$$

where m is (RC/τ). The 10–90 per cent rise times of (2.17) and (2.19) are $2 \cdot 2 \times$ (time constant), i.e. $2 \cdot 2CR$ and $2 \cdot 2\tau$ respectively. A plot of V_o versus t/τ for various values of m is shown in Fig. 2.14. When τ is much less than $RC - (m \gg 1)$, (2.20) approximates to (2.5). The peak value of the output voltage in (2.20) is $\{m/(m-1)\}\hat{V}$, i.e. less than $\hat{V}$. This signifies that the capacitor C has accumulated some charge during the finite rise time of the input pulse.

22

If an input of the form given in (2.17) is applied to the RC network in Fig. 2.12, the solution comes to

$$V_0 = \hat{V}[1 + \{1/(m-1)\}\exp(-t/\tau) - \{m/(m-1)\}\exp(-t/m\tau)] \quad (2.21)$$

for $m \neq 1$. For m equal to 1

$$V_0 = \hat{V}\left\{1 + \left(1 + \frac{t}{\tau}\right)\exp(-t/\tau)\right\}. \quad (2.22)$$

Fig. 2.15 is a plot of (2.21) and shows that the delay, interpreted to be the time taken by the pulse to reach 50 per cent of its peak amplitude, increases

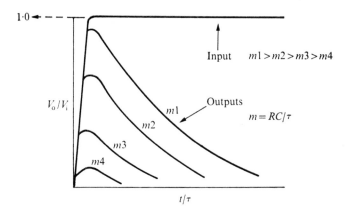

FIG. 2.14. Response of CR differentiating network. Exponential inputs.

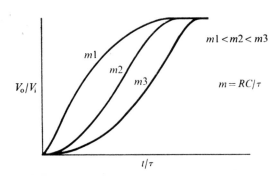

FIG. 2.15. Response of RC integrating network. Exponential inputs.

with m. Integration is used mainly to improve the final signal to noise ratio and is dealt with more fully in Chapter 6.

Eqns (2.20) and (2.21) are significant for another reason: they represent very closely the pulse outputs of many detectors for slow and fast working.

PULSE SHAPING

Even the fast pulse obtained from the scintillators via photomultipliers approximate to the general form of (2.20), namely

$$V_o = K\{\exp(-t/T_1) - \exp(-t/T_2)\}, \qquad (2.23)$$

where K, T_1, and T_2 are constants. The treatment so far has dealt with shaping the leading edge of the input pulse. Consider now the effect of a CR differentiating network like Fig. 2.8 on the pulse shown in Fig. 2.10. The first input pulse is assumed to start from zero base level. If in Fig. 2.10 $CR > t_d$, the response over the interval ab is simply $V_o = V_i \exp(-t/CR)$ so that the amplitude at b equals $V_i \exp(-t_d/CR)$. At b the output must follow the step-function input change exactly, because a capacitor cannot change its charge instantaneously. The amplitude at c is $V_i\{\exp(-t_d/CR)-1\}$ and from c to d the pulse will decay exponentially towards zero at a rate $\exp\{-(t - t_d)/CR\}$, remembering that time is reckoned from a. Hence the peak value of the output has been shifted relative to the input. The capacitor cannot pass direct current so that the shaded areas in Fig. 2.10 are equal. The resultant base-line shift constitutes a major problem in pulse height analysis, where it introduces a source of error. The problem is especially serious in experiments with pulsed accelerators having a low-duty cycle. A.c. coupling is used in nearly all instrumentation, simply because d.c. coupling is not feasible. The error due to base-line shift is difficult to correct, because pulses occur at random and not at regular intervals. It is possible to remove the pulse undershoot due to RC coupling by a pole-zero network of the form shown in Fig. 2.16. The action is illustrated in Fig.

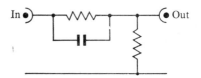

FIG. 2.16. Pole-zero cancelling network.

2.17, which gives waveforms for pulses transmitted through RC and pole-zero coupling networks. The pulses have semi-Gaussian shape corresponding to actual pulses obtained in many experiments. The action of a pole-zero cancelling network is explained more fully in Chapter 6. Here is must be stressed that Fig. 2.16 involves d.c. coupling. Pole-zero cancellation is frequently employed in the major shaping circuits but the need for RC coupling to achieve d.c. isolation remains and the problem of base-line shift remains also.

The inevitable shunt capacitance arising from the self-capacitance of R

24

and circuit strays effects differentiation chiefly by reducing the output voltage. For a step function the reactance of C_2 in Fig. 2.18 will be very much smaller than R and

$$V_o/V_i = \{1/j\omega C_2\}/\{(1/j\omega C_1) + (1/j\omega C_2)\} = C_1/(C_1 + C_2). \quad (2.24)$$

The time constant of the network will also be changed appreciably if C_2 is a significant fraction of C_1. For this reason C_1 should be made relatively large for a given CR product to achieve the desired differentiation. There may still be some voltage attenuation because R is now more likely to be comparable with the generator resistance R_g in Fig. 2.9.

(a)
Base line

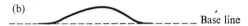

(b)
Base line

FIG. 2.17. Transmission through pole-zero cancelling network. (a) Normal RC coupling. (b) Pole-zero coupling.

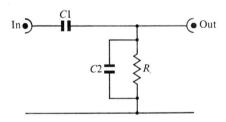

FIG. 2.18. Differentiating circuit with capacitance across R.

The relation between pulse rise time and transmission bandwidth can be deduced from eqns (2.2) and (2.17). The 3 dB bandwidth B of an amplifier with uniform frequency response and an output load consisting of a resistor R in parallel with a capacitor C is

$$B = 1/2\pi CR. \quad (2.25)$$

The 10–90 per cent rise time t_r is, from (2.17),

$$t_r = 2\cdot2CR. \quad (2.26)$$

Combining (2.25) and (2.26),

$$t_r = 0\cdot31/B. \quad (2.27)$$

25

In practice it is found that t_r is given more closely by (Wells and Lewis 1959)

$$t_r = 0.45/B. \tag{2.28}$$

2.4. Delay-line shaping

Delay-line shaping is an alternative to RC differentiation for shortening pulses to a required duration. In its basic form the shaping consists of a delay line of characteristic impedance Z_0, driven by an amplifier terminated with a load R equal to Z_0 (see Fig. 2.19). The line is short-circuited at the

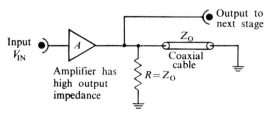

FIG. 2.19. Delay-line shaping giving bipolar pulse.

other end, resulting in anti-phase reflection of the pulses. Pulses are therefore shaped in a manner indicated for an ideal step function in Fig. 2.20. The duration T is determined by the length of line and its delay properties. For the purpose of overlap prevention the method is superior to RC shaping because it gives immediate base-line return at the end of the shaped pulses (edge B in Fig. 2.20). Double delay-line shaping, in which one delay-line

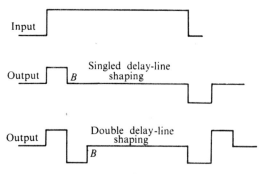

FIG. 2.20. Delay-line shaped pulses.

shaper is followed by another, not necessarily consecutive to the first one, gives the output shown in Fig. 2.20. Its outstanding advantage is excellent freedom from base-line shift at high input rates. The use of single and double delay lines is discussed fully in Chapter 6. Delay shaping is also used in ultra-fast nanosecond pulse formers for systems with very short dead times (between 10 and 100 ns). The delay is then largely determined by a length of coaxial cable. Such applications are mentioned in Chapters 8 and 10. In this section we consider the performance of a delay line, concentrating on lumped parameter delay lines.

The characteristic impedance of a distributed parameter lossless transmission line having inductance L and capacitance C equals $(L/C)^{1/2}$ and the wave is propagated with a velocity $(LC)^{-1/2}$. The delay T_d per unit distance equals $(LC)^{1/2}$ ($1/V$ where V is the above-mentioned propagation velocity) and amounts to $3\cdot3$ ns/m. Standard miniature coaxial cable is suitable for delays up to about 30 ns. The space taken up by the coaxial cable becomes excessive for larger delays. Conventional coaxial cables have characteristic impedances between 50 and 125 Ω, the former being the preferred value for nanosecond working. In a modified transmission line, a helical high-impedance cable, the straight conductor of the ordinary line is replaced by a continuous coil of wire in the form of a helix. L is now increased relative to C so that both Z_0 and T_d, the characteristic impedance and the delay, are much larger. In typical commercial helical delay lines Z_0 ranges from 100 to 1000 Ω and T_d from 180 to 1000 ns/m. The characteristic impedance is often made equal to a standard resistance value: the cable most frequently used here has Z_0 equal to 560 Ω and gives a delay of 90 ns/m. One serious disadvantage of helical delay lines is that they are difficult to handle, great care being needed when soldering the helical inner connection. Because of this, and the relatively low bandwidth of such cables, their use is on the decline. Furthermore, helical lines produce phase distortion due to two effects, the variation of inductance with frequency and the self capacitance between adjacent turns.

An alternative method of obtaining delays consists of using lumped parameter networks (sometimes called lines). These are inductor–capacitor combinations. The networks occupy much less space than transmission or helical delay lines, but tend to give more distortion. The general network in Fig. 2.21 has a characteristic impedance given by

$$Z_0 = \{Z_1 Z_2 (1 + Z_1/4Z_2)\}^{1/2}. \tag{2.29}$$

The attenuation is zero up to a frequency f_c given by

$$f_c = (1/\pi)(LC)^{-1/2}. \tag{2.30}$$

If all the significant Fourier spectrum frequencies of the input transient are much less than f_c, the time delay t_d per section of filter in Fig. 2.21 approximates to

$$t_d \sim (1/\pi f_c) = (LC)^{1/2} \tag{2.31}$$

27

and

$$Z_0 \sim (L/C)^{1/2}. \qquad (2.32)$$

The rise line t_r for a step-function input is related to t_d. Standard theory can be used to establish the precise relationship for a single section, but the calculations become extremely complicated for an actual multi-section network. In the following expressions it is assumed that the lumped parameter delay line is matched at both ends as in Fig. 2.22. Such matching

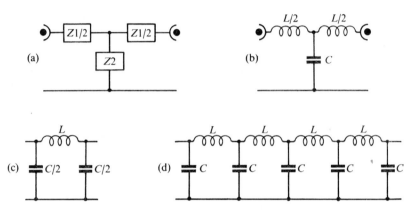

FIG. 2.21. Lumped parameter networks. (a) Basic network. (b) Prototype T section. (c) Prototype π section. (d) Actual filter.

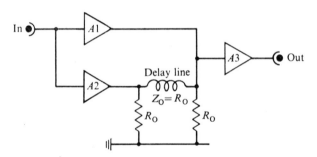

FIG. 2.22. Delay-line shaping—matched line. $A1$, non-inverting high-output impedance. $A2$, inverting, high-output impedance. $A3$, low input impedance.

minimizes the amplitude of multiple reflections due to a mismatch between R_0 and Z_0. The majority of delay-line arrangements take the form of Fig. 2.22. It can be shown (Millman and Taub 1956) that

$$t_d = 1 \cdot 07 (LC)^{1/2} \qquad (2.33)$$

(corresponding closely with (2.31)) and that

$$t_r = 1 \cdot 13 (LC)^{1/2}. \qquad (2.34)$$

The delay T_d for a filter with n sections is $n \times t_d$. It is found empirically that the overall rise-time T_r is given by (Millman and Taub 1956)

$$T_r = t_r \times (n)^{1/3}. \qquad (2.35)$$

Hence

$$n = \{(T_d \times t_r)/(T_r \times t_d)\}^{3/2}. \qquad (2.36)$$

Effecting substitutions from (2.33) and (2.34),

$$n = 1 \cdot 1 (T_d/T_r)^{3/2}, \qquad (2.37)$$

$$C = T_d/(1 \cdot 07 n R_0), \qquad (2.38)$$

$$L = T_d R_0/(1 \cdot 07 n). \qquad (2.39)$$

Note the general nature of (2.37), which expresses the number of sections for a specified ratio of delay time to rise time. The designer first decides on a value of n and then calculates C and L. The value of n can be quite large. The following example brings out typical magnitudes involved in practice. Consider a matched delay line as in Fig. 2.22 forming part of an amplifier with an overall rise time of about 25 ns. Assume that the amplifier consists of six cascaded sections, including the delay line. Let us make the simplifying assumption that all stages have equal individual rise times. Then, using the relation

$$t_r = \left\{ \sum_{n=1}^{n} (t_n)^2 \right\}^{1/2}$$

for n cascaded networks of rise times t_1 to t_n (Wells and Lewis 1959), T_r equals 12·6 ns. According to (2.37) a delay of 500 ns (T_d) would thus require eighty-seven sections (n). Such a delay filter would be too expensive and bulky in many commercial applications. Delay lines for high-grade nuclear pulse amplifiers sometimes contain up to fifty sections. They usually are mounted on plug-in cards that can be quite compact when made up with modern miniature components.

2.5. Shaping for optimum timing

In timing measurements it is important to extract the most accurate timing information from the signal. One important factor involved is brought out in Fig. 2.23. Pulses I and II are two events of peak amplitude $\hat{V}$ and $M\hat{V}$ detected by the counter at different times during an experiment. Due to the difference in amplitude there will be a displacement AB, known as 'time walk' or 'time jitter', for a fixed discriminator threshold. The dynamic range of inputs is often quite large and AB approaches T_R for large M with V_{TH}, the threshold voltage, close to $\hat{V}$, the minimum amplitude triggering the discriminator. The discriminator is assumed to introduce no

4

time walk of its own: the time interval between input and output is independent of amplitude. The signal inevitably undergoes discrimination whether in a discriminator module or in the threshold element of a co-incidence unit and the point under discussion is quite general.

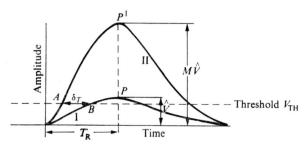

FIG. 2.23. Discriminator time walk.

The initial part of the input signal is statistically more constant in time than any other section. For that reason, timing measurements should be made using the lowest practicable amplitude of the leading edge. The limits here are noise, the sensitivity obtainable with a given design and, allied to the latter, threshold level stability. The timing accuracy is greatly improved under certain conditions by the 'zero-crossing' or 'cross-over' principle illustrated in Fig. 2.24. The input in (a) is shaped into the bipolar pulse

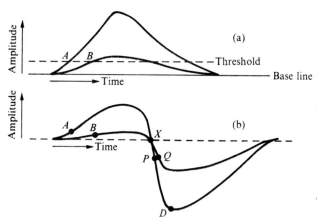

FIG. 2.24. Zero-crossing timing. (a) Inputs. (b) Bipolar shaped pulses.

in (b) by the delay-line process, described in the previous section and now applied for a different purpose. An alternative but rarely used method, tuned LC zero crossing, is described by Bjerke *et al.* (1962). It will be shown presently that the time of zero-crossing relative to commencement of the signal, AX in Fig. 2.24, is independent of signal amplitude. The threshold

V_{TH} of timing discriminator, which then operates on the negative edge XD of the bipolar pulse, is close to zero. An outstanding advantage of the technique is the facility to perform accurate amplitude discrimination at high level on the leading edge and arrange the logic so that the 'zero-crossing' detection at P and Q takes place only if the preceding amplitude threshold has been exceeded. The likelihood of false timing when operating with the leading edge discrimination (Fig. 2.24(a)) is thereby greatly reduced as is the time walk due to amplitude variations. The threshold level is very close to zero and the effective rise time XD of the bipolar pulse is considerably less than the leading-edge rise time AB. This will be proved presently.

The bipolar formation is obtained with the circuit of Fig. 2.19 and will now be analysed for triangular-shaped pulses. Such an analysis is much simpler than that for pulses with exponential rise and fall times and gives results accurate enough for most purposes (Bjerke *et al.* 1962). Generally

$$V_s = V_i + V_r, \tag{2.40}$$

where V_i and V_r are the incident and reflected waves and V_s is the resultant bipolar output voltage applied to the discriminator. If the reflected pulse

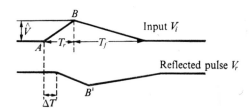

FIG. 2.25. Bipolar formation. $\Delta T < T_r$.

returns after a delay ΔT (twice the electrical length of the cable in Fig. 2.19) the following equations apply from Figs. 2.19 and 2.25:

$$V_i = \hat{V}(t/T_r), \qquad 0 < t < T_r; \tag{2.41}$$

$$V_i = \hat{V}\{1 - (t - T_r)/T_f\}, \qquad T_r < t < T_f; \tag{2.42}$$

$$V_r = -\hat{V}(t - \Delta T)/T_r, \qquad \Delta T < t < (\Delta T + T_r); \tag{2.43}$$

$$V_r = -\hat{V}\{1 - (t - \Delta T - T_r)/T_f\}, \qquad (\Delta T + T_r) < t < (\Delta T + T_r + T_f). \tag{2.44}$$

Also, because the amplifier output divides equally between R_0 and the delay line, $\hat{V}$ equals $(\hat{V}_{in}/2)$, $\hat{V}_{in}$ being defined in Fig. 2.19. The loss of input amplitude may be avoided by the pulse formation in Fig. 2.22. Apart from that the treatment holds equally for both methods. Two cases will now be considered, with the delay ΔT less than and greater than T_r.

31

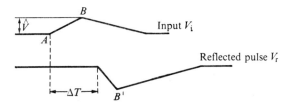

FIG. 2.26. Bipolar formation. $\Delta T > T_r$.

(a) $0 < \Delta T < T_r$.

The equations for V_s are

$$V_s = \hat{V}t/T_r, \qquad 0 < t < \Delta T; \qquad (2.45)$$

$$V_s = \hat{V}\Delta T/T_r, \qquad \Delta T < t < T_r; \qquad (2.46)$$

$$V_s = \hat{V}[\{1 + (T_r/T_f) + (\Delta T/T_r)\} - t\{(1/T_r) + (1/T_f)\}], \qquad T_r < t < (\Delta T + T_r); \qquad (2.47)$$

$$V_s = -\hat{V}\Delta T/T_f, \qquad (\Delta T + T_r) < t < T_f; \qquad (2.48)$$

$$V_s = -\hat{V}\{1 - (t - \Delta T - T_r)/T_f\}, \qquad T_f < t < (T_f + \Delta T). \qquad (2.49)$$

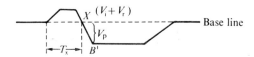

FIG. 2.27. Output for Fig. 2.25.

V_s is plotted in Fig. 2.27 and the time of zero-crossing is obtained from (2.47).

$$T_x = T_r + \Delta T \Big/ \left(1 + \frac{T_r}{T_f}\right). \qquad (2.50)$$

The effective rise time XB' of the negative section in Fig. 2.27 is

$$T_{r(\text{eff})} = \Delta T + T_r - T_x. \qquad (2.51)$$

Combining (2.50) and (2.51),

$$T_{r(\text{eff})} = \Delta T \times T_r/(T_r + T_f). \qquad (2.52)$$

Eqn (2.52) confirms the statement made earlier that the effective rise time of the bipolar section is less than the rise time of the leading edge. In all cases of interest T_r is less than T_f. For the outputs of fast scintillators followed by photomultipliers, $T_f \sim 2T_r$. The peak of the negative section, $\hat{V}\Delta T/T_f$, is considerably below $\hat{V}$.

32

(b) $T_r < \Delta T < T_f$.

This is the more common formation and the equations are

$$V_s = \hat{V}t/T_r, \qquad 0 < t < T_r; \tag{2.53}$$

$$V_s = \hat{V}\{1 - (t - T_r)/T_f\}, \qquad T_r < t < \Delta T; \tag{2.54}$$

$$V_s = \hat{V}[\{(1 + (T_r/T_f) + (\Delta T/T_r)\} - t\{(1/T_r) + (1/T_f)\}],$$
$$\Delta T < t < (\Delta T + T_r); \tag{2.55}$$

$$V_s = -\hat{V}\Delta T/T_f, \qquad (\Delta T + T_r) < t < (T_r + T_f); \tag{2.56}$$

$$V_s = -\hat{V}\{1 - (t - \Delta T - T_r)/T_f\}, \qquad (T_r + T_f) < t < (T_r + T_f + \Delta T). \tag{2.57}$$

Eqns (2.47) and (2.55) are identical and eqns (2.50)–(2.52) apply for both (a) and (b). The preferred arrangement (b), whose bipolar pulse is shown

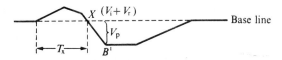

FIG. 2.28. Output for Fig. 2.26.

in Fig. 2.28, is more suited for operation with large statistical time variations than (a). The time walk δT in Fig. 2.23 is given by

$$\delta T = \{(M - 1)/M\}T_r. \tag{2.58}$$

δT approaches T_r for large M. The reduction in δT due to bipolar formation can be appreciable. For $T_f = 2T_r$ and $\Delta T = T_r/2$, $T_{r(eff)}$ of eqn. (2.52) comes to $T_r/6$, resulting in a proportional time-walk reduction of bipolar over leading-edge timing for a given dynamic amplitude range.

References

BJERKE, A. E., QUENTIN, A., KERNS, Q. A., and NUNAMAKER, T. E. (1962) *Nucl. Instrum. Meth.* **15**, 249.

CUTLER, P. (1960) *Electronic circuit analysis*, Vol. 1, *Passive networks* p. 175. McGraw-Hill, New York.

MILLMAN, J. and TAUB, H. (1956) *Pulse and digital circuits*, p. 291. McGraw-Hill, New York.

WELLS, F. H. and LEWIS, I. (1959) *Millimicrosecond pulse techniques*, 2nd edn, p. 7. Pergamon Press, London.

33

3 Bipolar and Field-effect Transistors

By M. O. DEIGHTON

3.1. Introduction to transistors

A BIPOLAR transistor consists essentially of two semiconductor diodes manufactured very close together. This definition applies equally to the original point-contact transistor invented by Shockley (now obsolete), to the later alloy-junction and diffused-base types, and to the modern planar transistor. These differ mainly in method of fabrication; improved characteristics, reliability, and reproducibility are features of the more recent types. Field-effect transistors, considered later, are essentially different in operating principle and more like valves.

The useful properties of transistors depend on the interaction between the two diodes; these are made so that either the anodes (npn transistor) or the cathodes (pnp transistor) form a very narrow region common to both diodes. This is called the 'base' and one lead of the device is connected to it; the other terminals of the two diodes are called 'emitter' and 'collector'. In normal operation the emitter-base diode is forward-conducting and the collector-base diode has an applied reverse bias of several volts. This situation is shown for a pnp transistor in Fig. 3.1. The essence of transistor

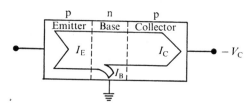

FIG. 3.1. Normal current flow in a pnp transistor.

action is that, instead of flowing out via the base lead, nearly all the emitter current flows right across the base region (whose thickness is exaggerated in Fig. 3.1) and out via the collector lead. The ratio I_C/I_E is called the grounded-base (or common-base) current gain and denoted by α. Typical values of α are between 0·95 and 1·0. If a moderately large load resistor is inserted between collector and the supply, V_C, there will be appreciable voltage gain from emitter to collector and hence also power gain, since the signal currents are approximately equal. Provided the collector is never

34

driven positive, collector voltage changes have very little effect on I_C and almost none on the emitter potential. It is this property, in combination with an α near unity, that makes the transistor a useful device capable of power amplification.

Fig. 3.2 shows the circuit symbols used to represent (a) a pnp transistor and (b) an npn transistor. Note that the *emitter* lead is identified by an arrow, the direction of which indicates that of emitter current flow and hence the type of transistor. The diagram also shows the normal polarity of collector voltage relative to base and the directions and magnitudes of all electrode currents in terms of I_E and α.

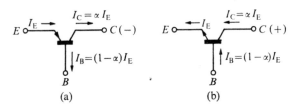

FIG. 3.2. Symbols for pnp and npn transistors. (a) pnp. (b) npn.

Notice that $I_B = I_E - I_C$, since there can be no net current into or out of a transistor, and therefore I_B is much smaller than either of the other currents since these are nearly equal. This fact, combined with the relatively high collector impedance noted above, allows one to obtain a current gain much greater than unity by earthing the emitter and using the base as input electrode. A transistor used thus is said to be in the grounded-emitter (or common-emitter) configuration. The current gain in this mode is I_C/I_B and is denoted by β. Clearly this can become very large if α approaches unity. Fig. 3.2 shows that $\beta = \alpha/(1 - \alpha)$ and values in the range 20–200 are typical. Notice that, for both pnp and npn transistors, the common-emitter connection always gives phase inversion between input and output voltages, whereas the common-base connection does not.

In manufacturers' data sheets and other publications, the symbols h_{fb} and h_{fe} are often used for α and β respectively. Here h_f signifies a forward gain parameter and the second subscript letter indicates the electrode that is common to input and output circuits.

3.2. Structure and properties of semiconductors

Silicon and germanium are the principal materials from which transistors are made. Both are tetravalent elements that form crystals in which each atom has four close neighbours placed at the corners of a tetrahedron. For convenience, this can be represented in one plane by the rectangular

35

structure of Fig. 3.3. Here the four outer valence electrons are shown separately from the nucleus and inner electron shells, which together have a net charge $+4q$, where q denotes the magnitude of the electron charge. By associating together as shown, each atom has a part share in a total of eight electrons, this being the number required to complete its outermost shell. The arrangement is thus stable and each pair of electrons belonging to adjacent atoms constitutes a 'covalent bond', indicated by dotted lines. At room temperature, thermal agitation causes the electrons to vibrate slightly about their mean positions and, very rarely, to jump right out of their proper place (as shown for one electron at the right of Fig. 3.3). When

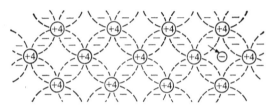

FIG. 3.3. Structure of semiconductor crystal.

this happens, not only does the electron become free to move about within the crystal as a *conduction* electron, but it leaves behind a vacancy, which can easily be filled by a nearby *valence* electron, whose position in turn is filled by another, and so on. Thus, by removing an electron from a valence bond, we have produced a mobile electron together with a mobile electron vacancy. The latter behaves exactly like a positively charged particle and is called a 'hole'. The process just described is known as 'thermal generation of an electron–hole pair'. If a mobile electron and a hole happen to meet anywhere in the crystal, the free electron falls into the vacancy and they both cease to exist, for practical purposes. This is known as 'recombination'.

Considering that the energy required to produce a hole–electron pair is, for germanium, some thirty times the average energy of thermal vibration at room temperature (for silicon it is over forty times the average energy), it is clear that such events will be rare and a single crystal of pure semiconductor material will contain comparatively few conduction electrons and holes. Such a crystal is called 'intrinsic'. In practice one can employ artificial means for greatly increasing the average population of mobile holes or electrons, since useful semiconductor electronic devices consist of crystals having two or more regions in which these two types of conductivity have been induced to various degrees. The method adopted is the introduction of small controlled quantities of impurities into the crystal lattice,

36

usually by diffusion from a gaseous atmosphere maintained at a high temperature.

If a pentavalent substance such as phosphorus is suitably introduced, then some of the crystal lattice positions are occupied by phosphorus atoms instead of by germanium. The result is depicted in Fig. 3.4(a), where four of the five valence electrons of phosphorus fit naturally into the germanium bond structure; the fifth one, finding nowhere to stay, wanders freely in the crystal. Each atom of phosphorus introduced results in the presence of one conduction electron within the crystal. By thus 'doping' the crystal, it has been converted into n-type semiconductor (n for negative). This refers,

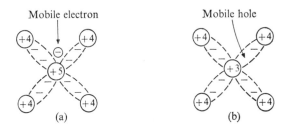

Fig. 3.4. Effect of pentavalent and trivalent impurity atoms in crystal. (a) Pentavalent. (b) Trivalent.

it should be carefully noted, only to the polarity of the *mobile* carriers— the crystal as a whole is still electrically neutral, each mobile negative charge being balanced by an extra fixed positive charge. In like manner, the crystal can be doped with a trivalent impurity such as indium, in which case the situation at the sites of indium atoms is as shown in Fig. 3.4(b). Here the three valence electrons complete three of the bonds, leaving a vacancy in the fourth one, which can move freely in the crystal in the same way as described earlier and which behaves as a mobile positive charge. In such a p-type semiconductor there will be one hole for each impurity atom introduced and the 'positive' charge of each hole is counterbalanced by an extra fixed negative charge,† equivalent to one electron, at the site of the indium atom. Hence overall electrical neutrality of the crystal is again preserved.

Table 3.1 gives some idea of typical densities (per cm³) of holes and free electrons in specimens of p-type and n-type germanium or silicon of stated resistivity, compared with intrinsic material, the degree of doping being

† The nature of the fixed charge, in both cases, is best appreciated by imagining the mobile carrier to have wandered some distance away. The bond structure round the indium atom, for example, is then complete and has a net associated charge $+3q - 4q = -q$.

usually specified by the resistivity of the resulting material. It will be noticed that, besides the large number of holes (equal to impurity density) present in p-type material, there is also a much smaller number of free electrons present and, conversely, there are a few holes present in the n-type material. This is because hole–electron pairs are being continuously generated by thermal agitation throughout the crystal, so that any semiconductor must contain some of both. The average density n, p of each is determined by the fact that, when dynamic equilibrium is reached, the rate of recombination, which varies as the product np, is just equal to the rate of generation. Now a germanium crystal contains about 4.4×10^{22} atoms of germanium per cm^3, so the doped materials in Table 3.1 contain fewer than ten impurity

TABLE 3.1

Typical carrier densities in doped and intrinsic germanium and silicon at 300°K

Semiconductor type	Hole density (p)	Electron density (n)	pn
p-type (0.01 Ω-cm) Ge	3.7×10^{17} cm^{-3}	1.7×10^{9} cm^{-3}	6.25×10^{26} cm^{-6}
intrinsic (47 Ω-cm) Ge	2.5×10^{13} cm^{-3}	2.5×10^{13} cm^{-3}	6.25×10^{26} cm^{-6}
n-type (0.01 Ω-cm) Ge	3.6×10^{9} cm^{-3}	1.75×10^{17} cm^{-3}	6.25×10^{26} cm^{-6}
p-type (1.0 Ω-cm) Si	2.5×10^{16} cm^{-3}	1.85×10^{5} cm^{-3}	4.6×10^{21} cm^{-6}
intrinsic (63 600 Ω-cm) Si	6.8×10^{10} cm^{-3}	6.8×10^{10} cm^{-3}	4.6×10^{21} cm^{-6}
n-type (1.0 Ω-cm) Si	8.9×10^{5} cm^{-3}	5.2×10^{15} cm^{-3}	4.6×10^{21} cm^{-6}

atoms for every million atoms of germanium. (Note that the electrical resistivity is evidently a very sensitive test of purity.) The thermal generation of carriers is therefore virtually unaffected by impurities present and mainly determined by the temperature and parent material. At a given temperature, the product np is therefore a constant of the material, being about 6.25×10^{26} for germanium and 4.6×10^{21} for silicon, both at 300°K (see Table 3.1). Strictly, it is the *difference* between n and p that is dictated by the impurity density, from considerations of overall charge neutrality. However, in most cases, one is so much greater than the other that no appreciable error arises if one assumes the larger or 'majority carrier' density is equal to the impurity density; the 'minority carrier' density then follows from the known value of np. It also follows that the majority carrier density is practically independent of temperature, whereas the minority carrier density increases markedly as the temperature rises.

The temperature dependence of np is of the form

$$np = A \exp(-E_0/kT), \tag{3.1}$$

where A is a constant, k is Boltzmann's constant, and E_0 is the energy required to remove an electron from within the bond;† this is about 0·72 eV for germanium and 1·1 eV for silicon. The temperature dependence of the exponent on the right-hand side of eqn (3.1) can be obtained by differentiation. Thus $\partial(-E_0/kT)/\partial T = E_0/(kT^2)$, assuming E_0 is constant. For germanium near room temperature this is about

$$0·72/(0·025 \times 300) \simeq 0·1/°\text{K}.$$

We conclude that a rise of 10°C causes the product np and therefore, in general, the minority carrier density to increase by a factor of about e in germanium and somewhat more in silicon owing to the larger E_0. However, because of the larger E_0, the minority carriers are initially much fewer in silicon.

3.3. The junction diode

Fig. 3.5 shows a crystal of germanium, the two halves of which are of opposite conductivity types. The fixed charges are indicated by circled

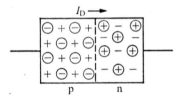

FIG. 3.5. pn junction diode.

symbols and mobile holes and electrons by simple plus or minus signs. For clarity, the minority carriers have been omitted. It is apparent that positive holes can easily cross the junction from left to right and electrons from right to left. Both contribute to the total forward current of the diode, denoted by I_D, which flows when the left-hand lead is made positive with respect to the other. This is the direction of easy current flow, in which the diode is a *low* impedance.

If one reverses the polarity of applied voltage it is found that a small reverse current flows which can be attributed to minority carriers on either side which can now cross the junction easily. The magnitude of this current is not very dependent (within limits) on voltage, but it changes with temperature in the same way as the minority carrier densities, that is by

† Known as the 'energy gap' of the semiconductor.

a factor of about 2·7 for every 10°C rise. An approximate formula for the diode current I_D at any voltage V is

$$I_D = I_S \exp(qV/kT) - I_S, \tag{3.2}$$

where V is positive in the forward direction and negative in the reverse direction. The first term on the right is associated with majority carriers and the second term I_S, known as 'reverse saturation leakage current', with minority carriers. The shape of this characteristic is sketched in Fig. 3.6. Note that it always passes through the origin; if it did not, the diode could be used as a continuous source of power (by connecting a resistance between the terminals), which is contrary to the laws of thermodynamics. The magnitude of I_S has been exaggerated in Fig. 3.6 and the exponential term

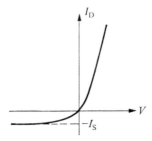

FIG. 3.6. Ideal junction diode characteristic.

in eqn (3.2) changes by a factor e for a change in V of 25 mV, at room temperature. The observed characteristics of germanium and silicon diodes, therefore, look more like the curves in Fig. 3.7, Ge diodes having an

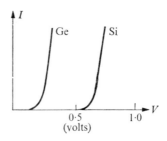

FIG. 3.7. Typical Ge and Si diode characteristics.

apparent threshold of about 200 mV and Si about 600 mV. The high Si threshold is, of course, due to the extremely small effective value of I_S— several orders of magnitude less than in germanium. It should be mentioned here that only part of the measured reverse current of a silicon diode is the component I_S due to minority carriers diffusing from either side of the junction. Much of the leakage of silicon diodes is due to a different process, viz. thermal generation and collection of hole–electron pairs very near the junction in a region called the 'depletion layer'. In other cases, surface leakage due to contamination, etc. may predominate. Therefore the measured reverse leakage of silicon diodes is generally much larger than the proper value of I_S for insertion in eqn (3.2) and usually varies appreciably with voltage. Hence it is better to deduce I_S from the forward characteristic.

An important characteristic of a p–n junction is the incremental or slope resistance in the neighbourhood of a given operating point at which the forward current is I_D. This is easily obtained by differentiating eqn (3.2) and we have

$$1/r_{eff} = \partial I_D/\partial V = (q/kT)\, I_S \exp(qV/kT),$$

whence

$$r_{eff} = kT/\{q(I_D + I_S)\}. \tag{3.3}$$

In practice one is interested in currents much greater than I_S and it is sufficient to use $r_{eff} = kT/qI_D$; the slope resistance thus varies inversely with current, being about 25 Ω at 1 mA, 2·5 Ω at 10 mA, and so on. This wide variation of r_{eff} is merely a reflection of the non-linearity of the I–V curve in Fig. 3.6. At high currents the law is not followed closely, due to the presence of series resistance in the bulk of the semiconductor material.

The temperature dependence of voltage drop across a p–n junction, at a defined forward current I_D, is another important characteristic, since it affects the stability of many circuits. This can also be derived from eqn (3.2), where it must be remembered that the temperature affects both I_S and the exponent. With an appreciable forward current the second term on the right can be neglected and we have $I_D = I_S \exp(qV/kT)$, where I_S, being proportional to the minority carrier densities, can be replaced by $B \exp(-E_0/kT)$, by virtue of eqn (3.1), and B is a constant. One thus obtains $I_D = B \exp\{(qV - E_0)/kT\}$. If I_D is maintained constant, it follows that $(qV - E_0)/kT$ is also constant; hence, differentiating with respect to temperature and assuming E_0 does not change,

$$kTq\left(\frac{\partial V}{\partial T}\right)_{I_D} - k(qV - E_0) = 0,$$

giving

$$\left(\frac{\partial V}{\partial T}\right)_{I_D} = \frac{qV - E_0}{qT} = -\frac{(E_0/q) - V}{T}. \tag{3.4}$$

Eqn (3.4) shows that the temperature coefficient of voltage at a fixed current is generally negative, provided V is less than E_0/q. In a germanium diode the typical temperature coefficient is $0·75/300 = 2·5$ mV/°C, while the corresponding figure for silicon is about 2 mV/°C, in both cases at an assumed room temperature of 300°K. Notice however that, due to the presence of V on the right-hand side of eqn (3.4), the temperature coefficient will *decrease* in magnitude at higher standing currents.

3.4. The depletion layer of a reverse-biased junction

When a reverse voltage is applied across a p–n junction the majority carriers tend to be pushed away from the junction on either side. As a result, the fixed charges associated with impurity atoms near the junction are left uncompensated and these 'uncovered charges' form an electrostatic double

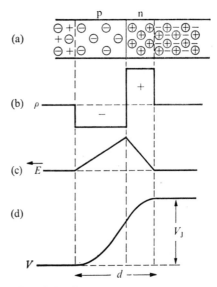

FIG. 3.8. Diagram of depletion layer. (a) Disposition of charges. (b) Net charge density. (c) Field distribution. (d) Potential variation across junction.

layer, negative on the p-side, positive on the n-side. The process of recession from the junction continues rapidly until the electrostatic field associated with the charge layers is just sufficient to absorb the applied potential difference. The latter then appears almost entirely across this narrow region known as the 'depletion layer', since it is entirely depleted† of mobile current carriers. The situation is depicted in Fig. 3.8, with the same symbols as before and assuming uniform impurity densities on each side of the junction.

† It can be shown, from elementary electrostatics, that any region of a semi-conductor must either be electrically neutral or fully depleted of carriers; inter-mediate states are inconsistent with equilibrium. Thermal generation of hole-electron pairs continues within the layer, as elsewhere, but these are swept away by the field as soon as they appear.

Note that in (a) the mobile carriers have been pushed back from the junction, leading to the net charge distribution shown in Fig. 3.8(b). Application of the laws of electrostatics leads to the field and potential distributions shown in (c) and (d), respectively. The field is strongest at the junction and the potential function consists of two parabolic segments. Because of this, the total depletion layer width d varies as $V_J^{1/2}$, where V_J is the actual voltage drop across this region.

Now, if the reverse voltage is changed, the movements of majority carriers, which occur as the depletion region widens or narrows, give rise to transient currents in the external circuit, the effect being similar to that of a capacitance across the junction. The incremental junction capacitance C_J can be shown to be that between two parallel conducting plates spaced apart a distance d. It therefore *varies with voltage* in accordance with

$$C_J \propto V_J^{-1/2}. \tag{3.5}$$

Nowadays the majority of semiconductor junctions are made by a diffusion process, resulting in an approximately linear as opposed to an abrupt transition from p-type material to n-type as one crosses the p–n junction. Fig. 3.8 is then modified somewhat. The most important change is that the depletion layer width is now proportional to $V_J^{1/3}$, so that

$$C_J \propto V_J^{-1/3}. \tag{3.6}$$

Thus the voltage dependence of capacitance is less marked than in an abrupt junction diode.

An important point to be noted is that V_J differs from the externally applied potential difference, because of small contact potential differences at the points where the two leads are attached to the semiconductor material on each side of the junction and also due to small electric fields in the undepleted semiconductor.† These potentials have a resultant u, which is a built-in constant of the diode ($\simeq 0.4$ V for Ge; $\simeq 0.6$ V for Si). As a result, even when a moderate external forward voltage V is applied, there is still a *reverse* voltage V_J present across the junction, given by

$$V_J = u - V \tag{3.7}$$

and the depletion region persists, although somewhat narrower. It is the potential barrier thus presented to majority carriers that limits the voltage-dependent component of forward current to the value given by the first term in eqn (3.2); it also explains the presence of the reverse-leakage term

† It can be shown that such fields must exist in any region where the doping is non-uniform (see section 3.8). They must therefore be included in analysis of a diffused-junction diode.

I_S, even during forward conduction, since diffusing minority carriers still tend to run downhill.

3.5. Transistor action

Consider the three-layer pnp transistor shown diagrammatically in Fig. 3.9. The collector is biased negatively relative to the base, the depletion

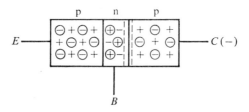

FIG. 3.9. pnp transistor.

layer at the junction being indicated by dotted lines. Now suppose the emitter E is raised in potential relative to base B. This causes continuous injection of holes into the base region from the emitter and also injection of electrons into the emitter from the base. The latter make no contribution whatever to final collector current and represent a loss, which is minimized by doping the emitter region much more heavily than the base. The proportion of total emitter current which is injected as holes into the base is known as 'emitter efficiency'. What happens to these holes after entering the base is the essential feature of transistor action; we therefore concentrate attention on the base region. Here the minority carriers are positive holes (not shown in Fig. 3.9), initially few in number. When a large number of extra holes arrive from across the emitter junction a similar number of electrons initially flow in via the base lead in order to maintain charge neutrality. These constitute a transient base current, which subsequently decays rapidly to a very small value. It is a remarkable property of semiconductors that such abnormal numbers of holes and electrons can exist together for quite a long time (10–100 μs), recombination being relatively slow. Eventually, if left alone, their numbers would decay to thermal equilibrium values like those in Table 3.1. However, conditions in the base of an operating transistor are far removed from *thermal* equilibrium. When the currents have reached their final values (i.e. electrical equilibrium), the hole (p) and electron (n) densities vary through the base as in Fig. 3.10. For comparison the thermal equilibrium values p_0, n_0 are shown dotted. Electrons cannot cross the collector junction, hence the electron population is essentially static, but holes do so readily. The field E (Fig. 3.8) sweeps them rapidly across as soon as they reach the edge of the depletion layer.

This acts as a perfect sink for holes, hence the zero value of p here. While within the base, the holes are in a virtually field-free region, because of the large number of conduction electrons present, and their motion is largely random, but with a general movement towards the collector, caused by *diffusion* resulting from the density *gradient*. Thus, unlike the static electron distribution, that of holes represents a population in transit, with nearly as many leaving at the collector side as enter from the emitter. If the base is thin enough, most of the holes diffuse across without recombining and the resulting diffusion current is virtually uniform across the base. This is the reason for the more or less uniform density gradient. Any holes that do recombine in transit consume an electron that has arrived via the base

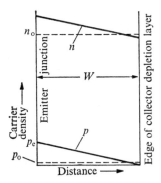

FIG. 3.10. Carrier distribution in base of pnp transistor.

lead. They therefore contribute to base current and are lost to the collector. The fraction of holes crossing the emitter junction which reach the collector is called the 'transport factor'. The current gain α of the transistor is thus given by

$$\alpha = (\text{emitter efficiency}) \times (\text{transport factor}).$$

A good transistor is designed so that both factors are as near unity as possible, consistent with other requirements.

An npn transistor operates essentially as outlined above, except that 'electron' should be substituted for 'hole' and vice versa, wherever these words appear. Also all voltages and currents have the opposite polarity, as in Fig. 3.2(b); plus and minus signs must be interchanged everywhere in Fig. 3.9 and p and n must be interchanged in Fig. 3.10 and elsewhere, as appropriate.

3.6. The collector characteristics. Saturation

In the absence of emitter current the relation between collector current and collector-base voltage is that of a reverse-biased diode (see eqn 3.2).

5

Therefore the collector current in normal operation is obtained by adding a term αI_E to this leakage and, using the sign conventions of Fig. 3.2, one obtains the common-base collector characteristic

$$I_C = \alpha I_E + I_{co}\{1 - \exp(-qV_{CB}/kT)\}, \tag{3.8}$$

where I_{co} is the magnitude of saturation reverse collector leakage, with emitter open, and V_{CB} is the collector-base voltage, reckoned positive if in the normal operating polarity.

A typical set of characteristics is shown in Fig. 3.11, with I_E as running

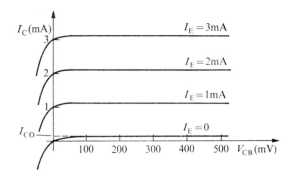

FIG. 3.11. Common-base collector characteristics.

parameter. Notice that the collector current falls off sharply as soon as V_{CB} becomes appreciably negative, that is as soon as the collector junction becomes *forward* biased, but that collector current is practically independent of collector voltage over quite a wide range of the normal *reverse* bias.

Such curves are not very useful, however, since all transistors look much alike when plotted thus. Clearly one has to examine accurate curves very closely to find how far α falls short of unity, and thus the value of β. It is better, therefore, to plot collector current versus collector-emitter voltage V_{CE} for various values of *base* current. The resulting common-emitter collector characteristics are sketched for a typical low-power transistor in Fig. 3.12 and this should be compared with Fig. 3.11.

Evidently the β of this transistor is about 50, since there is a change of about 1 mA in collector current for every 20 μA of base current. Two other features are also noticeable: the marked upward slope of the 'flat' portion of each curve and also the much larger leakage I'_{CO} at zero base current. Both effects are undesirable in practical applications and must be minimized. The leakage effect can be explained as follows. In normal operation eqn (3.8) simplifies to $I_C = \alpha I_E + I_{CO}$. Now $I_E = I_C$, when $I_B = 0$, therefore, at this condition, $I_C = I_{CO}/(1 - \alpha) \simeq \beta I_{CO}$. The physical reason for the much larger leakage I'_{CO} is that I_{CO} flows between base and collector and therefore constitutes an input current at the base which is magnified by the grounded-emitter transistor. (The two leakage currents I_{CO} and I'_{CO} are often designated I_{CBO} and I_{CEO}, respectively, in manufacturers' literature and must be carefully distinguished.) It is thus clear that open-circuiting the base lead is an unsatisfactory way of cutting off a transistor. Any voltage-dependent leakage, for example due to resistive contamination bridging the collector junction, will be similarly magnified by the transistor in the common-emitter connection, giving rise to a finite slope of the characteristics as in Fig. 3.12. This would also give rise to a corresponding but much smaller slope in the curves of Fig. 3.11. Effects other than contamination can, however, produce a similar result.

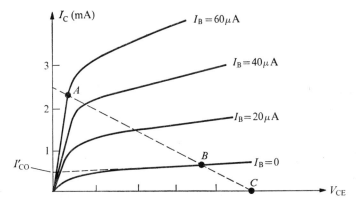

FIG. 3.12. Common-emitter collector characteristics.

One such important effect was first described by Early (1952). Referring to Fig. 3.9, the width of the depletion layer at the collector junction and its penetration into the base region must both increase with increasing collector-base voltage (section 3.4), hence the width of the active region of the base (illustrated in Fig. 3.10) must decrease correspondingly. One consequence of this is that holes injected from the emitter have a shorter distance to diffuse in order to reach the collector, and therefore less chance of loss by recombination. This results in a slight increase in value of the transport factor and so in α, which in turn causes a significant increase in β with V_{CE}. This mechanism accounts satisfactorily for the commonly observed increase in slope of the collector characteristics as I_B takes higher values (see Fig. 3.12); it is to be noted that ohmic leakage shunting the collector junction would not have this result unless β were strongly dependent on collector current.

In Fig. 3.12, the collector current falls off rapidly towards zero when the collector-emitter voltage approaches within about 100 mV of zero. This condition is known as 'saturation' or 'bottoming'. If the transistor is connected to a suitable collector supply voltage through a load resistor, then it will operate at some point on a load line such as that shown in Fig. 3.12. With 60 μA base current it will be at point A, and with zero I_B at point B. At A, the emitter-to-collector voltage drop can be as little as a few tens of millivolts and the impedance of the transistor is very low. However, at B the current may be appreciable and, moreover, rather ill-defined and temperature-dependent. If it is desired to operate the transistor as a switch, that is as a very low impedance in the 'on' condition and as a high impedance when 'off', the poorly-defined condition at B is clearly undesirable and one should ensure virtually complete cut-off of collector current by reverse biasing the emitter-base junction by a small amount. The operating point then moves very close to the V_{CE}-axis (point C in Fig. 3.12) and a small reverse base current will flow. Both at A and C the power dissipation within the transistor is very small, so that it closely resembles an ideal switch. For this reason, transistors are widely employed as switching elements in digital computer logic systems and as binary counting devices in nucleonic instrumentation.

It is sometimes required to estimate the electrode voltages or currents from known external circuit conditions other than those given in the manufacturer's data, or where these data are not sufficiently accurate. This need arises mainly when the transistor is cut-off, in saturation, or operating in the inverse mode (see below), hence it is important in relation to the pulse and digital circuits with which this book is chiefly concerned. Ebers and Moll (1954) derived general non-linear equations relating voltages to currents, which are applicable over a wide range of conditions with certain limitations;† a convenient form of these equations is as follows:

$$I_C + \alpha I_E = I_{CO}\{\exp(qV_{CB}/kT) - 1\}, \tag{3.9}$$

and

$$I_E + \alpha_i I_C = I_{EO}\{\exp(qV_{EB}/kT) - 1\}, \tag{3.10}$$

where α, α_i, I_{CO} and I_{EO} are positive constants of the transistor, two of which have already been encountered. It is important to note that the sign conventions employed in these equations are different from those used elsewhere in this chapter; collector and emitter currents, I_C and I_E, are here reckoned positive if crossing their junctions in the forward direction (from p to n), while the junction voltages, V_{CB}, V_{EB}, are likewise reckoned positive

† Chiefly that the three semiconductor regions are uniform, with abrupt junctions between them, that ohmic voltage drops are negligible, and that low-level injection is assumed, i.e. minority carrier densities are everywhere small compared with majority carrier densities.

if they constitute forward bias. Thus, for normal operation of a pnp or npn transistor, I_C and V_{CB} in eqns (3.9) and (3.10) are both negative quantities, while I_E and V_{EB} are positive. Allowing for the different sign conventions used, it can be seen that eqns (3.8) and (3.9) are identical. Eqn (3.10) expresses the essential symmetry of any pnp (or npn) structure, that is the fact that a transistor can, in principle, be operated in the inverse mode with the collector acting as emitter and the emitter as collector. (α_i represents the current gain that would be obtained thus and I_{EO} is the 'collector' leakage, that is the reverse saturation leakage current of the normal emitter junction when the collector is open circuit.) The majority of practical transistors, however, are not designed to be so used and have considerably degraded inverse performance, typical values of α_i being in the range 0·3–0·8 (β_i from 0·5 to 5). The low emitter junction breakdown voltage (<5 V) often encountered would, moreover, restrict any output signal obtainable to less than this amount. Nevertheless, some transistors, known as 'symmetrical' or 'bidirectional' types, are intended for use in either the normal or inverse mode and have roughly equal performance in both. The importance of the subject stems from the fact that saturation represents the transition region between normal and inverse operation, where both occur to some degree, so that the parameters of both are involved.

We illustrate the preceding discussion by using the Ebers–Moll equations to derive the equation for the collector saturation characteristic and that for the base input characteristic in normal grounded-emitter operation, these being important characteristics from the viewpoint of the circuit designer. For this purpose we need two additional equations:

$$I_B = I_E + I_C, \tag{3.11}$$

where I_B is base current, reckoned positive in its normal direction, and

$$\alpha I_{EO} = \alpha_i I_{CO}, \tag{3.12}$$

a fundamental relation first derived by Shockley, Sparks, and Teal (1951).

The collector-emitter voltage drop is obtained from eqns (3.9) and (3.10) as

$$\exp\{q(V_{EB} - V_{CB})/kT\} = \exp(qV_{EB}/kT)/\exp(qV_{CB}/kT)$$

$$= (I_{CO}/I_{EO}).(I_E + \alpha_i I_C + I_{EO})/(I_C + \alpha I_E + I_{CO})$$

$$= (\alpha/\alpha_i).\{I_B - (1 - \alpha_i) I_C\}/\{\alpha I_B + (1 - \alpha) I_C\},$$

using eqns (3.11) and (3.12) and neglecting I_{EO}, I_{CO} in comparison with I_E or I_C. This reduces to

$$V_{CE} = \pm(kT/q)\ln\{(\beta/\beta_i).(\beta_i + 1 - I_C/I_B)/(\beta + I_C/I_B)\}, \tag{3.13}$$

49

which gives actual collector voltage with respect to emitter, taking the positive sign for an npn transistor and the negative sign for a pnp transistor. As an example, if $\beta = 50$, $\beta_i = 5$ and the transistor is operated with collector current limited to $10I_B$, then, with the present sign convention, $I_C/I_B = -10$ and the collector-emitter voltage drop with this degree of saturation is $(kT/q)\ln 4 \simeq 34.6$ mV, assuming $kT/q = 25$ mV. Note that if $I_C = 0$ with a finite I_B, V_{CE} is not zero but becomes $(kT/q)\ln(1 + 1/\beta_i)$, which is about 5 mV in the above example and known as the 'offset' voltage.† For a more detailed discussion of eqn (3.13) the reader is referred to a paper by the present author (Deighton 1964).

For analysis of the base input characteristic, normal operation is assumed, with V_{CB} negative and large compared with kT/q. Eqn (3.9) then simplifies to $I_C = -(\alpha I_E + I_{CO})$, and substituting this result in (3.11) gives

$$I_B = (1 - \alpha) I_E - I_{CO}. \tag{3.14}$$

Substitution in (3.10) leads to

$$I_E(1 - \alpha\alpha_i) = I_{EO}\{\exp(qV_{EB}/kT) + \alpha - 1\}, \tag{3.15}$$

using eqn (3.12).

Combining (3.14) and (3.15),

$$I_B/I_{EO} = (1 - \alpha)/(1 - \alpha\alpha_i) . \exp(qV_{EB}/kT) - (1 - \alpha)^2/(1 - \alpha\alpha_i) - \alpha/\alpha_i.$$

This result can be simplified by expressing it in terms of the betas. Writing $\alpha = \beta/(\beta + 1)$ and $\alpha_i = \beta_i/(\beta_i + 1)$, one obtains

$$I_B/I_{EO} = (\beta_i + 1)/(\beta + \beta_i + 1) . \{\exp(qV_{EB}/kT) - \beta/\beta_i - 1\}. \tag{3.16}$$

Eqn (3.16) shows that the general shape of the base input characteristic is like that of a junction diode—see Fig. 3.6. The reverse saturation base leakage current is about equal to I_{EO}, provided β and β_i are both appreciably greater than unity, while the current still increases exponentially (if $I_B \gg I_{EO}$) by a factor e for every 25 mV; the main difference is a horizontal displacement of the characteristic along the voltage axis, the zero-current point being not at the origin but at $V_{EB} = (kT/q)\ln(1 + \beta/\beta_i)$. In the numerical example above, this displacement amounts to about +60 mV and constitutes a not very important addition to the apparent thresholds illustrated in Fig. 3.7. Thus, broadly speaking, a germanium transistor requires forward bias of about 250 mV and a silicon transistor about 600 mV across the base-emitter junction before it conducts appreciably.

Although the theory above is strictly valid only when the three regions

† This quantity is important in certain switching applications, for example d.c. chopper-amplifiers.

of the transistor are uniform, agreement with measured characteristics is often quite good even in modern diffused transistors where the regions are essentially non-uniform. Such discrepancies as occur are usually due to additional ohmic voltage drops arising from bulk resistance of the various regions, an effect that tends to be more important in silicon devices. A further factor one has to allow for is that I_{EO} and I_{CO} in the Ebers–Moll theory refer to bulk reverse leakage of the junctions, whereas the measured leakages may include appreciable surface components, as was seen in section 3.3. However, the essential results, eqn (3.13) for instance, depend chiefly on the ratio I_{EO}/I_{CO}, which one can derive from eqn (3.12) as the ratio of two easily measured alphas, instead of resorting to leakage measurements of dubious accuracy.

3.7. High-frequency performance

It was seen in section 3.5 that the flow of current through a transistor is associated with a certain distribution of excess charges, of both kinds, in the base region. In the pnp transistor the electron component is accounted for by charges that have entered via the base lead and the hole component by emitted holes that have not reached the collector. It follows that during the charge build-up collector current is appreciably less than emitter current and the base current is much greater than its final value. This is illustrated in Fig. 3.13, where a unit step of emitter current is assumed and leakage currents are neglected. Notice that $I_E = I_B + I_C$ at all times, using the sign conventions of Fig. 3.2. Evidently the shaded area in the upper part of Fig. 3.13 represents the final excess charge required by the distribution in the base, provided recombination is neglected. This charge can be related to final collector current by reference to Fig. 3.10. If p_e is the value of p at the emitter edge of the base, then the density gradient is p_e/W and the rate of diffusion flow is Dp_e/W, where D is the diffusion constant for holes and W the base width. Let A be the area of the base and $+q$ the charge of each hole (numerically equal to the electron charge). Then the collector current is $I_C = AqDp_e/W$. Also the total hole charge in the base is $Q_B = AW \cdot \frac{1}{2}p_e q$, since the average hole density is $\frac{1}{2}p_e$, owing to the triangular distribution. From the above results,

$$Q_B = I_C \cdot \frac{W^2}{2D}. \tag{3.17}$$

Base charge is thus proportional to I_C, the constant of proportionality being $W^2/2D$. This is a very important transistor parameter, having the dimensions of time and called 'mean carrier transit time'. It is denoted by T_0 and, in high-frequency alloy-junction transistors, has values of the order 10^{-8} s.

In modern silicon planar transistors transit times one or two orders of magnitude smaller than this are obtained (see section 3.8).

Returning now to Fig. 3.13, the time variation of collector current can be represented approximately by the dashed exponential curve, which has a time constant T_0 to preserve the original area $\alpha_0 I_E T_0$ and therefore base charge. The increase of I_C with time is then given approximately by

$$I_C = \alpha_0 I_E \{1 - \exp(-t/T_0)\}. \tag{3.18}$$

Hence the mean carrier transit time is a measure of the response time of collector current to a step of emitter current.

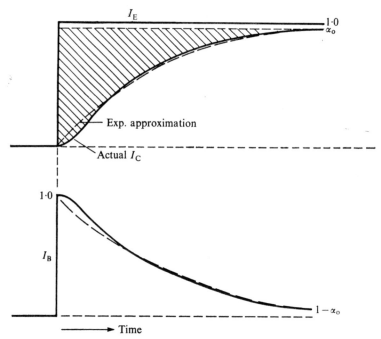

FIG. 3.13. Transient response of I_C and I_B to unit step of I_E.

Eqn (3.18) implies that α is dependent on frequency, having both a reduced magnitude and a phase lag at higher frequencies. By analogy with a passive CR integrating circuit (Chapter 2, section 2.3), one can write

$$\alpha(\omega) = \alpha_0/(1 + j\omega/\omega_\alpha), \tag{3.19}$$

where α_0 is the low-frequency value of α and ω_α is the radian frequency at which α is attenuated 3 dB; $\omega_\alpha/2\pi$ or f_α is known as the 'alpha cut-off frequency' and, with the exponential approximation above, equals $1/(2\pi T_0)$. Allowing for the true shape of the curves in Fig. 3.13, the actual value of f_α may vary from $1.22/(2\pi T_0)$ for an alloy-junction transistor to about $2/(2\pi T_0)$ for a 'diffused' transistor (see section 3.8).

Let us consider the frequency dependence of β that results from eqn (3.19). We have

$$\beta(\omega) = \alpha(\omega)/\{1 - \alpha(\omega)\} = \{\alpha_0/(1 - \alpha_0)\}[1/\{1 + j\omega/(1 - \alpha_0)\omega_\alpha\}]. \quad (3.20)$$

This can also be written

$$\beta(\omega) = \beta_0/(1 + j\omega/\omega_\beta), \quad (3.21)$$

where β_0 is the d.c. beta and $\omega_\beta \equiv (1 - \alpha_0)\omega_\alpha$. The frequency $\omega_\beta/2\pi$ is called the β-cut-off frequency (f_β) and is evidently approximately equal to f_α/β_0. In fact, $\alpha_0\omega_\alpha = \beta_0\omega_\beta$ and the current gain-bandwidth product in this case is the same whether base or emitter is used as input electrode, a result that is not surprising.

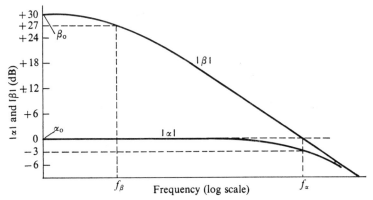

FIG. 3.14. Typical variation of $|\alpha|$ and $|\beta|$ with frequency.

Fig. 3.14 shows the variation of $|\alpha|$ and $|\beta|$, in dB, with frequency, for an ideal transistor with $\beta_0 = 31\cdot6$. Frequency is plotted on a logarithmic scale and the diagram illustrates the inverse dependence of $|\beta|$ on frequency over quite a wide range extending above and below f_α. The frequency at which $|\beta| = 1$ is denoted by f_1 or f_T and is equal to the product $f.|\beta(f)|$ over the frequency range for which $|\beta| \propto 1/f$. For ideal transistors, eqn (3.20) shows that $f_T \sim f_\alpha$ and the phase lag of β is approximately 90° throughout the frequency range above. Practical transistors, however, exhibit more phase shift relative to attenuation in the α/frequency characteristic and, as a result, the frequency f_T is appreciably less than f_α, while f_β is still approximately f_T/β_0. The reason for this will be apparent from the phasor diagram, Fig. 3.15, which shows the curves traced by the end-point of the α-phasor OB in the two cases, as the frequency is raised from zero to high values. Clearly the f_α-point on either characteristic is at its intersection with a circle of radius $\alpha_0/\sqrt{2}$, whereas the f_T-point is at the intersection with the perpendicular bisector of OA, the unit real phasor:

53

this follows from the fact that, if $|\beta| = 1$, then $|\alpha| = |1 - \alpha|$, hence $OB = AB$ at $f = f_T$. The manufacturer normally specifies f_T, either directly or by quoting a value for $|\beta|$ at some standard high frequency, usually 100 MHz.

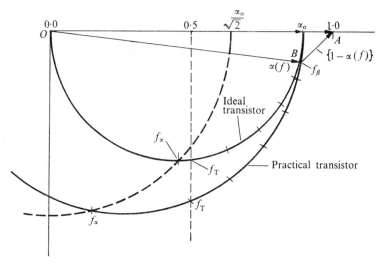

FIG. 3.15. Phasor diagram illustrating frequency dependence of $\alpha(f)$ and relation between f_α, f_T, and f_β.

3.8. Diffused transistors

The realization of high-frequency amplifiers and of fast switching circuits clearly depends on reducing the mean transit time T_0 to the smallest possible value. With alloy transistors the thinnest (uniform) base region that can be attained with reasonable control is about 10^{-3} cm thick. Substituting this value of W in eqn (3.17) and assuming a Ge pnp transistor, we obtain $T_0 = 1\cdot15 \times 10^{-8}$ s, corresponding to a cut-off frequency (f_T) around 15 MHz. Such a transistor is therefore very slow by present-day standards. It has been found possible to manufacture much faster transistors by diffusing the required impurities, usually in succession and under carefully controlled conditions, into one side of the slice so as to form first the base and then the emitter regions. Such a device is known as a 'diffused transistor'. If the starting material is n-type silicon, the result is an npn transistor.

By this means the transit time can be reduced to around 10^{-10} s, yielding cut-off frequencies in excess of 1000 MHz. This improvement of some two orders of magnitude arises chiefly in two ways:

(i) Since all diffusions are from the same surface, better control of base width and hence a substantial reduction thereof is possible.

(ii) The nature of the impurity diffusion process results in non-uniform doping of the base region, this being generally heavier near to the emitter and relatively light near the collector junction. This results in a 'built-in' electric field throughout the base region, the direction of which aids the transit of minority carriers injected by the emitter. These are now subject to a 'drift field' in addition to normal diffusion and, in consequence, cross the base region and reach the collector much more quickly than they would with a uniform base of the same width. The mean transit time is thus reduced by about another order of magnitude.

A third factor, which contributes to high-frequency performance, is the afore-mentioned low doping density near the collector side of the base. This results in a relatively wide depletion layer at the collector junction and hence the junction capacitance is quite low.

Virtually all modern discrete transistors and integrated circuits are manufactured by variants of the diffusion process outlined above. It should be appreciated, however, that the marked improvement in speed of operation is not achieved without some sacrifice of other characteristics. Most notable of these are the (usually) high capacitance and low reverse breakdown voltage of the emitter junction, due to rather high doping levels on both sides, the somewhat reduced emitter efficiency arising from the same cause, which yields lower current gain, and some degree of dependence on collector-emitter voltage to maintain the built-in field. As a consequence of the last, some diffused transistors saturate at relatively high values of V_{CE} (about 0·5 V) and, in this vicinity, f_T is strongly dependent on V_{CE}.

3.9. High-speed switching

In high-speed switching circuits the fundamental factors limiting transistor switching speed are the same as those which limit its high-frequency response to sinusoidal signals, namely the build-up or decay of stored charge carriers in the base region (and sometimes in the collector as well) and the presence of parasitic shunt capacitances across the junctions or series inductances in the leads of the device.

We have already considered the collector-current response to an emitter-current step (Fig. 3.13) and have seen that in modern transistors it is very fast, with typical rise times of only a fraction of a nanosecond. Most switching circuits, however, employ the common-emitter configuration, with driving signals applied to the base, hence we shall concentrate attention on arrangements of this type.

Fig. 3.16 shows the idealized collector current pulse waveform due to

55

a rectangular pulse of base current, in several cases. Curve (a) is the response for a small amplitude base current pulse and the broken curve (b) is for a larger current; both are characterized by an exponential rise and fall, whose time constant is the reciprocal of ω_β, that is $\beta_0 T_0$ or $\beta_0/(2\pi f_T)$, and it can be shown that this remains substantially true even when the α-characteristic has considerable excess phase-shift. If a collector load-resistance R_L is used such that the peak voltage $\beta_0 I_B R_L$ developed across it would exceed the collector-supply voltage V_{CC}, then the transistor saturates at some point on the front edge of the waveform and the resulting output current (or voltage) pulse has the flattened form shown in curve (c). By increasing the amount of base drive and, in effect, only using a small initial part of the collector

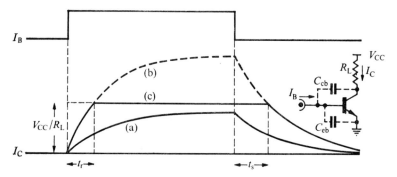

Fig. 3.16. Idealized collector current response to rectangular base current pulse. (a) Small I_B, (b) Large I_B, no collector saturation. (c) Large I_B, collector saturation.

transient, one can thus obtain a shorter rise time (t_r in Fig. 3.16); however, this results in an appreciable delay (t_s) in the start of the trailing edge and its subsequent decay is no faster than in curve (a). Moreover, it can be seen that a reduction in t_r achieved by increasing I_B must always be accompanied by an increase in t_s. The use of heavy saturation in the transistor therefore only improves the turn-on transient at the expense of the turn-off time.

The turn-off delay t_s is actually due to accumulation of very large numbers of minority carriers in the base region (and some in the collector) when the transistor is saturated, since both emitter and collector junctions are then forward-biased and inject carriers into the base. These decay principally by recombination, any initial drift field being largely neutralized by their presence, so that a relatively long time elapses after cessation of the base current before their numbers fall to the levels normally associated with a collector current V_{CC}/R_L and thus allow the collector to come out of saturation. For these reasons the observed value of 'storage time' t_s is often considerably greater than one would estimate from the simple picture in

Fig. 3.16. Heavy saturation should clearly be avoided in high-speed switching circuits.

The capacitances C_{cb} and C_{eb}, which shunt the collector and emitter junctions and include lead strays as well as junction capacitances, impose further limitations on the switching speed. Normally a cut-off transistor will have zero voltage or possibly a small reverse bias applied to its emitter junction, so the capacitance C_{eb} must first be charged through at least 0·6 V (in Si) before the transistor commences to conduct. This causes a delay t_d, after the start of the I_B pulse, before I_C begins to rise. t_d can be estimated from

$$I_B t_d = C_{eb} . \Delta V_{EB}, \tag{3.22}$$

where ΔV_{EB} is the initial voltage change required at the base. Strictly, C_{cb} is charged in parallel, but C_{cb} is usually small compared with C_{eb}. Once conduction commences in the transistor, V_{EB} changes very little thereafter and charging of C_{cb} predominates; this capacitive current diverts part of I_B away from its task of filling the base region with excess carriers, so the rise time t_r is also lengthened somewhat. Assuming fairly heavy saturation, so that the initial part of the transient is substantially linear, then a negligible part of the charge supplied to the base by I_B during t_r is consumed by carrier recombination; almost all of the charge is used either to charge C_{cb} or to build up the necessary charge distribution in the base and we have (approximately)

$$I_B t_r = I_{C, max} T_0 + C_{cb} V_{CC} = I_{C, max}(T_0 + C_{cb} R_L), \tag{3.23}$$

where R_L denotes the collector load resistance. Thus the rate of rise of collector current is approximately $I_B/(T_0 + C_{cb} R_L)$, instead of I_B/T_0 which one obtains when $C_{cb} = 0$. More generally, where C_{cb} has to be taken into account and the transistor is not driven to saturation, it can be shown that the transient current (or voltage) at the collector remains substantially exponential, but the time constant is modified to $\beta_0(T_0 + C_{cb} R_L)$.

The switching speed of transistors is normally specified by the manufacturer in terms of typical values of delay time (t_d), rise time (t_r), storage time (t_s), and fall time (t_f), for given drive conditions and a given collector-load impedance. As might be inferred from some of the preceding discussion, it is often impossible to extrapolate these data (particularly t_s) from one set of conditions to another and one must then rely on experimental measurement.

3.10. Field-effect transistors

Field-effect transistors (FETs) are assuming considerable importance in nuclear pulse-processing equipment. The primary advantage of these devices is their very high input impedance, typical input currents under operating conditions being as low as 10^{-10} A. In this respect they resemble or surpass thermionic valves and are much superior to any ordinary pnp or npn transistor in applications such as the monitoring of voltages stored on small capacitances. They are semiconductor devices and, like ordinary 'bipolar' transistors, are available in two opposite-polarity types known as p-channel and n-channel FETs. Their mode of operation, however, is quite different and depends on modulation of a conducting channel by voltage-induced changes in width of an adjacent depletion layer. Like valves, they are therefore voltage-controlled devices and certain types are now finding wide application as input devices in low-noise nuclear pulse amplifiers (see Chapter 7).

Fig. 3.17(a) shows, in diagrammatic form, the structure of a p-channel

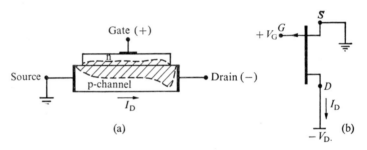

FIG. 3.17. Junction field-effect transistor (p-channel). (a) Structure. (b) Symbol.

FET. Electrical conduction is between two electrodes, called 'source' and 'drain', which are attached by ohmic contacts to the two ends of a piece of p-type semiconductor. A p–n junction is formed on one (or both) sides of the specimen and the control electrode or 'gate' is connected to the n-type material. In practice, these devices are made by diffusing source and drain regions as parallel strips of strongly p-type material, with a shallow strip of n-type material interposed as gate between them, the main bulk of the channel being first formed by diffusion of p-type impurity into the surface layers of a nearly intrinsic piece of material. The manufacturing techniques are thus very similar to those used for making diffused transistors of conventional type.

In the device illustrated in Fig. 3.17(a), a positive voltage is applied to the gate, in order to reverse bias the p–n junction and so establish a depletion layer, shown shaded in the figure. Since this part of the channel is depleted of mobile holes, it cannot contribute to electrical conduction between source and drain, which is therefore confined to the unshaded part of the channel. Increase of the positive gate potential widens the depletion layer further

58

and therefore constricts the current channel still more, while reduction of the gate potential has the opposite effect. The effective resistance between source and drain is therefore modulated by the gate potential, while the current demanded by the gate itself is only the reverse leakage of a p–n junction and usually negligible.

Fig. 3.17(b) shows the circuit symbol used to represent such a p-channel field-effect transistor. The arrow in the gate lead denotes the sense of the p–n junction (i.e. the direction from p to n) so that an n-channel device would be indicated by reversing the direction of the arrow. The diagram also shows the normal polarities of the gate and drain potentials (with respect to the source) and the direction of the drain current I_D. For an n-channel FET these are all reversed.

One of the important parameters of an FET is the 'pinch-off' voltage; this is the reverse voltage between gate and channel regions required to extend the depletion region fully across the conducting channel. It is denoted by V_p and is usually a few volts. If the gate-source bias is equal to or greater than V_p, the channel is pinched off at the source and no drain current can flow. With zero gate-source voltage, however, the depletion layer is initially very narrow and drain current increases rapidly at first, when an increasing negative potential is applied to the drain of the p-channel device. With higher drain potentials, however, the depletion layer widens appreciably at the drain end causing a constriction in the conducting channel near the drain, as illustrated in Fig. 3.17(a). The apparent resistance of the channel thus increases and the I_D/V_D curve flattens out (see Fig. 3.18), eventually saturating at a certain current I_{DO} (sometimes denoted by I_{DSS}) when the drain voltage reaches V_p. At voltages greater than V_p no significant increase in I_D occurs and the device is then said to be in the 'pinch-off region'. Evidently the drain current must then be confined within a very narrow conducting filament at the drain end of the channel. At intermediate values of gate bias the drain characteristic has a lower initial slope and saturates at lower values of drain current and voltage, as indicated in Fig. 3.18. In all cases, the point at which saturation occurs is that where the gate-drain voltage is V_p, that is where $V_G + V_D = V_p$. The pinch-off region is thus bounded by the dashed curve in Fig. 3.18.

Shockley (1952) first derived a theoretical expression for I_D as a function of V_G and V_D, assuming uniform p- and n-regions with an abrupt junction and using the approximation of essentially one-dimensional geometry. The expression is complicated and becomes more so if one allows for the non-uniform diffused structure of modern FETs and especially if account is taken of the fact that channel depth may be a significant fraction of the length (from source to drain). It is found, however, that most practical devices conform quite closely to a simple square-law relationship between current and voltage. In the region below pinch-off (to the left of the dashed curve in Fig. 3.18), this can be written

$$I_D/I_{DO} = (2V_D/V_p)(1 - V_G/V_p) - (V_D/V_p)^2. \qquad (3.24)$$

As V_D increases from zero, I_D increases at first, reaching a peak at $V_D = V_p - V_G$. At this point I_D is given by

$$I_D = I_{DO}(1 - V_G/V_p)^2, \qquad (3.25)$$

an equation that applies also at higher drain voltages, because of the saturation of drain current in the pinched-off region. Here, therefore, the

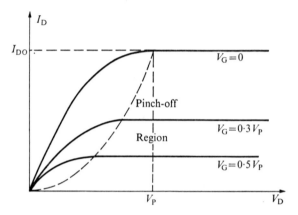

FIG. 3.18. Drain characteristics of FET.

device has a square-law characteristic, given by eqn (3.25) and sketched in Fig. 3.19. The mutual conductance in the pinch-off region is obtained by differentiating eqn (3.25) and is

$$g_m = (2I_{DO}/V_p)(1 - V_G/V_p). \qquad (3.26)$$

It is thus proportional to the square root of the standing current I_D.

There are three principal temperature effects on FET characteristics. In the first place the gate current, being the leakage of a p–n junction, is subject to large increases as the temperature rises (see section 3.3.) Secondly, the majority carrier mobility in the channel, which determines the conductivity of the material, decreases with rising temperature, hence the values of I_D and g_m, at given electrode potentials, both decrease. Thirdly, there are contact potentials present where the leads join the semiconductor, so that the p–n junction voltage is not quite the same as the externally applied bias. These contact potentials have a temperature coefficient of about 2 mV/°C, in a sense that opposes the mobility/temperature effect. As a result, it is usually possible to choose a particular gate bias for a device, at which I_D

is practically independent of temperature. This condition is seldom useful, because it rarely coincides with operating conditions dictated by other factors.

Operation of field-effect transistors at high frequencies is normally limited only by the distributed capacitance across the gate-channel depletion layer. For approximate circuit calculations, it is possible to consider this as equivalent to two lumped capacitances, one (C_{gs}) between gate and source and the other (C_{gd}) between gate and drain. The former acts merely as a capacitive load on the voltage source driving the FET; the latter, although only about one quarter of C_{gs}, may be more important in voltage amplifiers, because of the Miller effect.

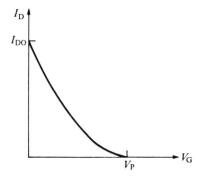

FIG. 3.19. $V_G - I_D$ characteristic of FET.

This chapter would not be complete without a brief mention of the MOST (metal-oxide-silicon transistor). This is a variant of the FET in which a thin insulating layer of silicon dioxide is interposed between the gate electrode (usually evaporated aluminium) and the semiconductor. Conduction in an 'inversion' layer at the surface of the semiconductor is then influenced electrostatically by the gate voltage. The primary advantage of such a device is the high degree of insulation of the gate, resulting in gate currents as low as 10^{-15}A. However, the physics of surface conduction is not yet well understood. Although these devices offer promise of eventually replacing electrometer valves, their characteristics are at present too unstable and noisy for that purpose. They are, however, finding application as switching elements, particularly in integrated arrays.

References

DEIGHTON, M. O. (1964) *Solid-st. Electron.*, **7**, 531.
EARLY, J. M. (1952) *Proc. Inst. Radio Engrs.* **40**, 1401.
EBERS, J. J. and MOLL, J. L. (1954) *Proc. Inst. Radio Engrs.* **42**, 1761.
SHOCKLEY, W., SPARKS, M., and TEAL, G. K. (1951) *Phys. Rev.* **83**, 151.
SHOCKLEY, W. (1952) *Proc. Inst. Radio Engrs.* **40**, 1365.

6

4 *The Transistor as a Circuit Element*

By L. J. HERBST

4.1. Introduction

THE principles of transistor operation were outlined in the previous chapter. The treatment consisted of deriving the static i/v characteristics and was followed by a section on high-frequency performance. In contrast the next chapter contains a brief account of some basic circuit techniques. The purpose of this chapter is to bridge the gap between fundamental device action and the outline of circuit techniques. The emphasis is on equivalent circuits and their manipulation for calculating amplifier performance.

4.2. Bipolar transistors

4.2.1. *Low-frequency equivalent circuits*

The equivalent circuits shown in this chapter all apply for small signal conditions. Small signal operation implies that input and output signal swings are small compared with the quiescent d.c. operating voltages and currents. It then follows that the latter remain virtually constant throughout the signal cycle. Under these conditions amplifier performance will be largely independent of signal amplitude and the various parameters of the equivalent circuit may be assumed to be constant. In practice small signal equivalent circuits are often used to estimate amplifier behaviour under large signal conditions. The calculated amplifier gain might then be accurate to within, say, 20 per cent. Such an approach is usually employed to obtain some idea of amplifier performance before proceeding to more accurate calculations.

Low-frequency equivalent circuits for the basic amplifier configurations are shown in Figs. 4.1–4.3. Alternative low-frequency equivalent circuits employ different parameters. The advantage of the circuits in Figs. 4.1–4.3 is that they physically resemble the device. The parameter r_e was defined in eqn (3.3); numerical substitutions at 300° K give

$$r_e = \{25\cdot6/I_e(\mathrm{mA})\}\,\Omega, \qquad (4.1)$$

where I_e is the operating emitter current. Since $I_e \sim I_c$, I_e in (4.1) can be replaced by I_c. The value of the base resistance r_b is usually specified by

62

the manufacturer. For general purpose low-level silicon transistors r_b lies between 50 and 100 Ω. For small signal v.h.f./u.h.f. amplifier transistors employed in nanosecond amplifiers and switching applications r_b is typically between 5 and 30 Ω. The output resistance r_c is generally in the range 0·5 to 2 MΩ. Figs. 4.1–4.3 contain the h parameter h_{fe}, the common-emitter

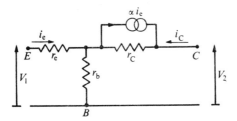

FIG. 4.1. Common-base low-frequency T equivalent circuit.

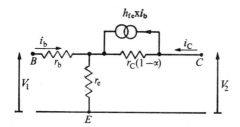

FIG. 4.2. Common-emitter low-frequency T equivalent circuit.

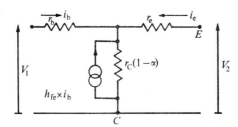

FIG. 4.3. Emitter-follower low-frequency T equivalent circuit.

current gain into zero collector load. By definition h_{fe} is synonymous with β in the previous chapter. The h parameters are mentioned again in section 4.2.3. The reason for preferring h_{fe} to β is that the former has now largely replaced the latter in general usage. The term 'low frequency' implies that all parameters shown in Fig. 4.1–4.3 are independent of frequency.

The manipulation of an equivalent circuit is illustrated with the aid of Fig. 4.4, which is evolved from Fig. 4.2 by including generator and output

63

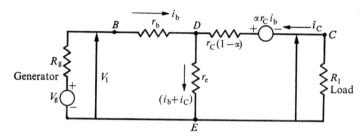

FIG. 4.4. Common-emitter equivalent circuit including generator and load.

resistors and by reducing the current generator $h_{fe} \times i_b$ across $r_c(1 - \alpha)$ in Fig. 4.2 to a voltage generator $h_{fe} \times i_b \times r_c(1 - \alpha)$. The latter reduces further to $\alpha r_c i_b$, using the identity $h_{fe} \equiv \alpha/(1 - \alpha)$. This type of reduction is strongly recommended for all equivalent circuits containing current generators. The equations for loops BDE and DCE are

$$V_1 = (r_b + r_e)i_b + r_e i_c, \tag{4.2}$$

$$0 = r_e(i_b + i_c) + i_c R_1 + i_c r_c(1 - \alpha) - \alpha r_c i_b$$
$$= (r_e - \alpha r_c)i_b + \{r_c(1 - \alpha) + r_e + R_1\}i_c. \tag{4.3}$$

Further, V_g, the open-circuit generator voltage of R_g, is related to V_1 by

$$V_1 = V_g - i_b R_g. \tag{4.4}$$

The above equations yield voltage, current, and power gain as well as input and output resistance. The current gain is obtained from (4.3), which gives

$$A_i = i_c/i_b = -[(\alpha r_c - r_e)/\{r_c(1 - \alpha) + r_e + R_1\}]. \tag{4.5}$$

Eqn (4.5) can be simplified since r_e is very much smaller than $\alpha r_c, r_c(1 - \alpha)$, or—in most cases—$R_1$. Ignoring r_e in (4.5) gives

$$A_i \sim h_{fe} \bigg/ \left\{1 + \frac{R_1}{r_c}(1 + h_{fe})\right\}. \tag{4.6}$$

The identity $1/(1 - \alpha) \equiv 1 + h_{fe}$ has been used to obtain (4.6). Similarly the voltage gain is

$$A_v = v_2/v_1 = -\frac{R_1(\alpha r_c - r_e)}{r_b\{r_c(1 - \alpha) + r_e + R_1\} + r_e(R_1 + r_c)}. \tag{4.7}$$

Once again approximations are possible. First r_e may be ignored in the factors within the brackets of (4.7), which then becomes

$$A_v \sim -\frac{R_1 \times h_{fe}}{r_b\left\{1 + \frac{R_1}{r_c(1 - \alpha)}\right\} + r_e\left\{(1 + h_{fe}) + \frac{R_1}{r_c(1 - \alpha)}\right\}}. \tag{4.8}$$

If, as is often the case, $R_1 \ll r_c(1 - \alpha)$, eqn (4.8) reduces to

$$A_v \sim -\frac{R_1 \times h_{fe}}{r_e(1 + h_{fe}) + r_b},\qquad(4.9)$$

an expression which may also be derived from a simple equivalent circuit (see next chapter). Eqn (4.4) is used to obtain the voltage gain V_2/V_g when required. To calculate input and output resistance a voltage is injected at the terminal in question and the resistance is the ratio $V_{injected}/$current flowing into terminal. For example, the output resistance of the common-emitter amplifier is calculated with the aid of Fig. 4.4 modified by removing R_1 and making V_g equal to zero. The loop equations are

$$0 = (R_g + r_b + r_e)i_b + r_e i_c,\qquad(4.10)$$

$$V_2 = i_c\{r_c(1 - \alpha) + r_e\} + i_b(r_e - \alpha r_c),\qquad(4.11)$$

giving

$$r_0 = V_2/i_c = r_c(1 - \alpha) + r_e \cdot \frac{R_g + r_b + \alpha r_c}{R_g + r_b + r_e}\qquad(4.12)$$

$$\sim r_c(1 - \alpha) + r_e \cdot \frac{R_g + \alpha r_c}{R_g + r_b + r_e}\qquad(4.12a)$$

since $r_b \ll \alpha r_c$.

4.2.2. *High-frequency equivalent circuits*

The main reactive effects that influence high-frequency models of the transistor are emitter-base and collector-base depletion layer capacitances and the diffusion of minority carriers through the base. The last mentioned phenomenon may be presented in the equivalent circuit in two ways. It can either be indicated by a capacitance or by inserting a current generator whose phase and amplitude vary with frequency.

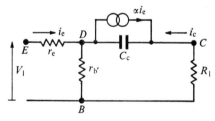

FIG. 4.5. High-frequency common-base T equivalent circuit.

A common-base high-frequency equivalent circuit derived from Fig. 4.1 is shown in Fig. 4.5. The variation of α with frequency is as in eqn (3.19), namely

$$\alpha = \alpha_0/\{1 + jf/f_\alpha\}.\qquad(4.13)$$

It was pointed out in section 3.7 that the phase change of α with frequency is greater than indicated in (4.13). A more accurate expression is

$$\alpha = \{\alpha_0/(1 + jf/f_\alpha)\} \exp(-\phi f/f_\alpha), \tag{4.14}$$

where ϕ ranges from 0·2 to 1·0 approaching 0·2 for alloy junction and 0·5 for silicon planar diffused-junction transistors. To illustrate the use of Fig. 4.5 we obtain the current gain from the equation for loop DBC:

$$(i_e + i_c)r_b' + i_c/j\omega C_c + \alpha i_e/j\omega C_c + i_c R_1 = 0. \tag{4.15}$$

Eqn (4.15) is obtained by reducing the current generator αi_e to a voltage generator $\alpha i_e \times (1/j\omega C_c)$, a method employed previously in changing Fig. 4.2 to Fig. 4.4. Re-arranging (4.15),

$$A_i = i_c/i_e = -(\alpha + j\omega C_c r_b')/\{1 + j\omega C_c(r_b' + R_1)\}. \tag{4.16}$$

To complete the calculation of current gain, the expression of (4.13) or (4.14) has to be used for α in (4.16). Eqn (4.16) contains an important high-frequency parameter, $C_c \times r_b'$. To a first approximation r_b' is identical with r_b in Figs. 4.1–4.3. The precise value of r_b', and indeed the representation of the ohmic base resistance, are controversial matters (Pritchard 1967).

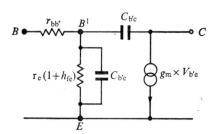

FIG. 4.6. Common-emitter hybrid π equivalent circuit.

Common-emitter amplification is generally analysed using the hybrid equivalent circuit of Fig. 4.6. The various parameters are related to the low- and high-frequency T equivalent circuit and other parameters as follows: $r_{bb'} \sim r_b$, $C_{b'c} \sim C_c$, $C_{b'e} \sim (1/2\pi f_1 r_e)$ and $g_m = (\alpha_0/r_e)V_{B'C}$. The parameter h_{fe} is understood to be the low-frequency current gain $\alpha_0/(1 - \alpha_0)$. A more accurate representation includes $r_{b'c}$ and r_{ce} across $C_{b'c}$ and $g_m \times V_{B'E}$ respectively, but these resistances are relatively large. For that reason they may be omitted at even only moderately high frequencies. The frequency dependence of current gain is expressed by $C_{b'e}$, which introduces a reactive term in $V_{B'E}$.

The emitter-follower has an inherently large bandwidth due to heavy negative current feedback. The emitter-follower is prone to parasitic oscillations because it presents an input impedance containing a negative resistance under certain conditions when the output load consists of a resistor R in parallel with a capacitor C. An analysis has been made using Fig. 4.3 and the following expression for h_{fe}:

$$h_{fe} = h_{feo} \Big/ \Big\{ 1 + j\frac{f}{f_c} \Big\}, \qquad (4.17)$$

where $h_{feo} = \alpha_0/(1 - \alpha_0)$ and $f_c = f_\alpha/(1 + h_{feo})$, α_0 and h_{feo} being low-frequency current gains. The condition for stability is that (Lawrence 1967)

$$CR < 1/2\pi f_\alpha. \qquad (4.18)$$

Eqn (4.18) shows that the danger of instability increases with faster transistors (higher f_α). Eqn (4.18) is quoted again in the next chapter (eqn (5.19), which contains f_t in place of f_α).

4.2.3. Four-pole parameters

An analytical treatment of amplifiers, complementary to the equivalent circuit approach, consists of relating input and output voltages and currents for the general network of Fig. 4.7 by linear equations containing four-pole

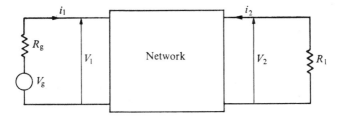

FIG. 4.7. General network.

parameters. A number of combinations are possible; the sets most used are given by the following matrices:

$$\begin{bmatrix} v_1 \\ i_2 \end{bmatrix} = \begin{bmatrix} h_{11} & h_{12} \\ h_{21} & h_{22} \end{bmatrix} \begin{bmatrix} i_1 \\ v_2 \end{bmatrix} \qquad (4.19)$$

$$\begin{bmatrix} v_1 \\ v_2 \end{bmatrix} = \begin{bmatrix} z_{11} & z_{12} \\ z_{21} & z_{22} \end{bmatrix} \begin{bmatrix} i_1 \\ i_2 \end{bmatrix} \qquad (4.20)$$

$$\begin{bmatrix} i_1 \\ i_2 \end{bmatrix} = \begin{bmatrix} y_{11} & y_{12} \\ y_{21} & y_{22} \end{bmatrix} \begin{bmatrix} v_1 \\ v_2 \end{bmatrix}. \qquad (4.21)$$

Of the above, the h parameters are known by the name hybrid parameters because they contain an impedance (h_{11}), and admittance (h_{22}), and voltage (h_{12}) and current (h_{21}) ratios. Expanding (4.19),

$$v_1 = h_{11} i_1 + h_{12} v_2, \qquad (4.22)$$

$$i_2 = h_{21} i_1 + h_{22} v_2. \qquad (4.23)$$

The number notation for the h parameter suffices is frequently replaced by double letters i(input), o(output), f(forward) and r(reverse) taking the place of 11, 22, 21, and 12. An additional suffix denotes the common electrode of the configuration. The common-emitter parameters are thus expressed by

$$v_{be} = h_{ie} \times i_b + h_{re} \times v_{ce}, \tag{4.24}$$

$$i_c = h_{fe} \times i_b + h_{oe} \times v_{ce}. \tag{4.25}$$

Eqn (4.25) gives the derivation of h_{fe}. Equivalent circuit representations of the y and h parameters are shown in Figs. 4.8 and 4.9. Amplifier per-

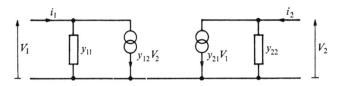

FIG. 4.8. Y parameters equivalent circuit.

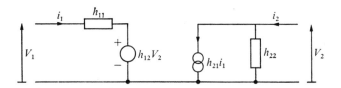

FIG. 4.9. h parameters equivalent circuit.

formance can be calculated using the h, y, or z parameters for Fig. 4.7 together with auxiliary equations like (4.4) if required. The reason for using h parameters is that they are easily measured: i_1 and v_2 can readily be changed independent of i_2 and v_1. The choice of parameters is usually determined by the manufacturer's data for the particular transistor being used. Some years back the h parameters were documented more than any other set, with the exception of the y parameters. The latter were and still are preferred for tuned r.f. applications. Now the tendency is to revert to the z parameters of Figs. 4.1–4.3. Generally the parameters are assumed to be independent of signal amplitude. The analysis can, however, be extended to cases where the parameters are functions of i_1, i_2, v_1, and v_2. Sometimes, for example in calculation of distortion, this is done. The value of using the parameters at all is then questionable. The justification for their retention is that the analytical technique employed for the initial approximate calculations can be extended to obtain more accurate results.

4.2.4. *Saturated switching*

The process of high-speed switching has already been outlined in section 3.9. Here we supplemented that treatment by quoting a few formulae for rise, fall, and storage time. The three states of a transistor in switching are:

(i) OFF—emitter-base and collector-base junctions reverse biased.

(ii) ON—emitter-base and collector-base junctions forward biased. The transistor is then in saturation.

(iii) ACTIVE—emitter-base forward and collector-base junction reverse biased. This is the state for small signal amplification and for non-saturated current mode switching described in the next section.

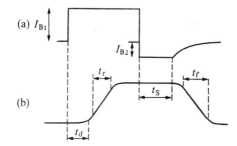

FIG. 4.10. Waveforms—saturated common-emitter switch. (a) Base current. (b) Collector current.

The following expressions are largely based on the classical paper of Moll (1954). They apply to a common-emitter inverter with an ON current of I_c, a collector load R_1, an ON drive of I_{B1}, and an OFF drive of I_{B2}. The various switching times are illustrated in Fig. 4.10.

$$t_r = h_{fe(o)} \times (1/2\pi f_1 + C' R_1) \ln \{h_{fe(o)} \times I_{B1}/(h_{fe(o)} \times I_{B1} - 0{\cdot}9I_c)\}, \quad (4.26)$$

$$t_f = h_{fe(o)} \times (1/2\pi f_1 + C' R_1) \ln \{(I_c + h_{fe(o)} \times I_{B2})/(0{\cdot}1I_c + h_{fe(o)} \times I_{B2})\}, \quad (4.27)$$

$h_{fe(o)}$ being the common-emitter current gain at the edge of saturation. C' is the sum of $C_{b'c}$ (Fig. 4.6) and circuit stray capacitance across R_1. Further approximations based on (4.26) and (4.27) are

$$t_r \sim 0{\cdot}9I_c(1/2\pi f_1 + C' R_1)/\{I_{B1} - (I_c/2h_{fe(o)})\}, \quad (4.28)$$

$$t_f \sim 0{\cdot}9I_c(1/2\pi f_1 + C' R_1)/\{I_{B2} + (I_c/2h_{fe(o)})\}. \quad (4.29)$$

T_d is given by

$$t_d = Q_{OB}/I_{B1}, \quad (4.30)$$

where Q_{OB}, the change in the charge of input and output capacitances, may be expressed by

$$Q_{OB} = \int C_i \, dV_{BE} + \int C_0 \, dV_{CB}. \tag{4.31}$$

C_i, the input capacitance, may be equated to the emitter-base depletion capacitance C_{TE} and $C_{ob} \sim C_{b'c}$. The limits of integration in (4.31) extend over the voltage levels of V_{BE} and V_{CB} in the OFF and ON conditions. Finally, t_s is given by

$$t_s = \tau_x \ln \{(I_{B1} + I_{B2})/(I_c/h_{fe(o)} + I_{B2})\}, \tag{4.32}$$

$$\tau_x = Q_{BS}/I_{BS}. \tag{4.33}$$

Q_{BS} is the excess base charge, that is the charge in the base in addition to the charge present when the transistor operates in the ACTIVE region with the same collector current. I_{BS} in (4.33) is given by

$$I_{BS} = I_{B1} - I_c/h_{fe(o)}. \tag{4.34}$$

The set of equations in this section is a hybrid system combining the classical approach and the charge-control theory (Beaufoy and Sparkes 1957).

The effect of minority carrier storage in the collector region in the ON condition has been ignored. This effect has been greatly reduced by doping the collector of npn transistors with gold, thereby greatly reducing minority carrier lifetime. Collector charge storage is also reduced by the epitaxial structure often employed for the collector layer.

4.2.5. *Current mode switching*

In current mode switching the transistors operate in the OFF or in the ACTIVE region. In the basic switch of Fig. 4.11 the condition for J_1 to remain in the ACTIVE region is

$$(V_s - IR_1) > V_B + E_{ON}. \tag{4.35}$$

The rise time of the collector current pulse is given closely by (Roehr 1963)

$$t_r = \frac{0.8I(1/2\pi f_1 + C' R_1)}{E_{ON}/(R_s + 2r_b')}, \tag{4.36}$$

where R_s is the resistance of the source driving J_1 and C', the sum of C_c (Fig. 4.5) plus circuit capacitance. A similar expression holds for fall time. If, as is standard practice in digital integrated circuits, E_{ON} and E_{OFF} in Fig. 4.11 are equal, rise and fall times will be equal also. The storage time t_s (in the previous section) has disappeared. Current mode logic is faster than saturated logic but consumes more power.

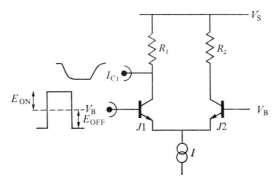

FIG. 4.11. Current mode switch.

4.3. Field-effect transistors

Most FET amplifiers operate in the common-source mode. A common-source equivalent circuit is illustrated in Fig. 4.12. At low frequencies C_{DG}

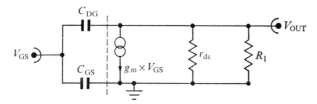

FIG. 4.12. Common-source equivalent circuit.

and C_{GS} may be ignored, which amounts to omitting the section to the left of the dotted line. The low-frequency voltage gain is thus

$$A_v = (g_m \times R_1 \times r_{ds})/(R_1 + r_{ds}). \tag{4.37}$$

Usually $r_{ds} \gg R_1$, leading to

$$A_v \sim g_m \times R_1. \tag{4.38}$$

At high frequencies the input capacitance C_{is} is obtained from Fig. 4.12:

$$C_{is} = C_{GS} + C_{DG}(1 + A_v), \tag{4.39}$$

the contribution $C_{DG}(1 + A_v)$ being due to the Miller effect. The 3 dB bandwidth B of an amplifier having a gain independent of frequency and an output load consisting of R(resistor) in parallel with C(capacitor) is

$$B = 1/2\pi CR. \tag{4.40}$$

Combining (4.38) and (4.40),

$$A \times B = g_m/2\pi C. \tag{4.41}$$

71

$A \times B$ is known as the figure of merit. The performance of a single stage may be calculated by taking C to be the sum of transistor output capacitance plus the input capacitance for a similar stage.

References

BEAUFOY, R. and SPARKES, J. J. (1957) *A.T.E. Jl.* **13**, 310.

LAWRENCE, E. E. (1967) Integrated circuits. *Instn elect. Engrs. Conf. Publ. No.* 30, p. 267.

MOLL, J. L. (1954) *Proc. Inst. Radio Engrs.* **42**, 1773.

PRITCHARD, R. L. (1967) *Electrical characteristics of transistors*, p. 410. McGraw-Hill, New York.

ROEHR, W. D. (editor) (1963) *Switching transistor handbook*, p. 91. Motorola Inc., Arizona.

5 *Basic Circuit Techniques*

By L. J. HERBST

5.1. Small signal low-frequency amplifiers

THE three basic amplifier configurations are shown in Figs. 5.1–5.3, where

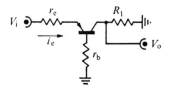

FIG. 5.1. Common-base amplifier.

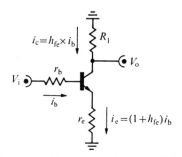

FIG. 5.2. Common-emitter amplifier.

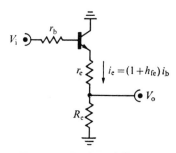

FIG. 5.3. Emitter follower.

r_b and r_e are T equivalent circuit small signal parameters. Input impedance, voltage, and current gain can be derived from Figs. 5.1–5.3. For example, the common-emitter input resistance is obtained from

$$v_i = i_b \times r_b + (1 + h_{fe}) i_b \times r_e,\qquad(5.1)$$

73

giving

$$r_{in} = r_b + (1 + h_{fe})r_e. \tag{5.2}$$

Another example is that of emitter-follower voltage gain. From Fig. 5.3

$$v_o/v_i = (1 + h_{fe})R_e/\{(1 + h_{fe})(R_e + r_e) + r_b\}. \tag{5.3}$$

The performance of the three amplifier configurations is summarized in Table 5.1.

TABLE 5.1

Amplifier performance

	Common-base	Common-emitter	Emitter-follower
Current gain	$h_{fe}/(1 + h_{fe})$	h_{fe}	$1 + h_{fe}$
Voltage gain	$\dfrac{h_{fe} \times R_1}{r_e(1 + h_{fe}) + r_b}$	$\dfrac{h_{fe} \times R_1}{r_e(1 + h_{fe}) + r_b}$	$\dfrac{R_e}{R_e + r_e}$
Input resistance	$r_e + \dfrac{r_b}{1 + h_{fe}}$	$r_e(1 + h_{fe}) + r_b$	$(r_e + R_e)(1 + h_{fe}) + r_b$
Output resistance	r_c	$r_c/(1 + h_{fe})$	$r_e + \left(\dfrac{r_b + R_g}{1 + h_{fe}}\right)$

R_g is the resistance of the generator driving the emitter-follower.

In nuclear instrumentation the common-base stage is mainly used as a buffer at the input of a module. In high-speed systems, a resistor, R_e in Fig. 5.4, matches the amplifier input to the characteristic impedance of the

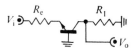

FIG. 5.4. Common-base stage with matching resistor R_e.

transmission system. The emitter-follower provides a high-input impedance. It is also used for transmitting a signal over a long path via a coaxial cable to another unit; the low-output impedance makes it suitable for this purpose. Low-frequency amplification is generally carried out with the common-emitter amplifier. The output resistance of the common-emitter stage equals $r_c/(1 + h_{fe})$ and this can shunt R_1 to a significant extent. From the equivalent circuit of Fig. 5.5, A_i', the effective current gain, is given by

$$A_i' = h_{fe} \times r_c(1 - \alpha)/\{r_c(1 - \alpha) + R_1\} \tag{5.4a}$$

$$= h_{fe}/\{1 + (R_1/r_c)(1 + h_{fe})\}. \tag{5.4b}$$

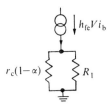

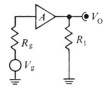

FIG. 5.5. Common-emitter equivalent output circuit.

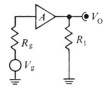

FIG. 5.6. Amplifier connected to source.

The identity

$$(1 - \alpha) \equiv 1/(1 + h_{fe}) \tag{5.5}$$

has been used to transform (5.4a) into (5.4b). The voltage gain V_0/V_g of the amplifier in Fig. 5.6 is calculated from

$$i_b = v_g/(R_g + r_{in}) \tag{5.6}$$

and

$$v_0 = A'_i \times R_1 \times i_b. \tag{5.7}$$

Using (5.4b) and the expression for r_{in} given in Table 5.1,

$$v_0/v_g = [h_{fe}/\{1 + (R_1(1 + h_{fe})/r_c)\}] \times R_1/\{r_b + r_e(1 + h_{fe}) + R_g\}. \tag{5.8}$$

5.1.1. *D.c. biasing*

In the standard circuit of the common-emitter amplifier in Fig. 5.7, bias

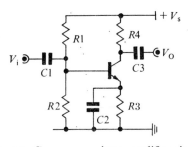

FIG. 5.7. Common-emitter amplifier circuit.

stability is ensured by the emitter resistor $R3$ bypassed by $C2$. $R1$ and $R2$ are chosen to draw a current much greater than the base current, thereby

75

making the base voltage largely independent of transistor parameters. R_3 is then chosen to give the required emitter current. The larger the voltage across R_3 the smaller will be the change in I_c due to the spread in V_{BE} and in h_{FE}.

A common-base bias arrangement with temperature compensation is shown in Fig. 5.8. For constant emitter current, $\partial V_{BE}/\partial T$ is about $-1\cdot 8$

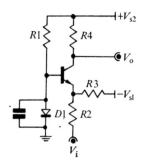

FIG. 5.8. Common-base amplifier with d.c. bias compensation.

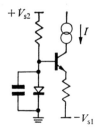

FIG. 5.9. Constant current source.

mV/°C. Temperature compensation is effected by balancing this voltage variation called for with a diode in the base of J_1. It is assumed that $V_{D1} \ll V_{S2}$. For perfect compensation the I/V characteristics of D_1 and J_1 (emitter-base) should be identical. Also D_1 and J_1 should draw equal currents, i.e. $(V_{S2} - 0\cdot 7)/R_1$ should equal V_{S1}/R_3. If V_{D1} and V_{EB} (J_1) are equal, the emitter input will be at ground potential. A potentiometer is sometimes included in the base circuit to ensure zero d.c. voltage at the emitter. Fig. 5.9, a circuit similar to Fig. 5.8, is a constant current source used in differential amplifiers and elsewhere.

5.2. High-frequency amplification

High-frequency amplification is analysed with the aid of the equivalent circuits in Figs. 5.10–5.12. H_{fe} in this section is the complex high-frequency

76

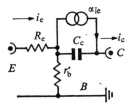

FIG. 5.10. Common-base equivalent circuit.

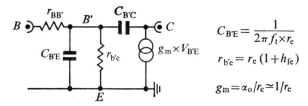

$$C_{B'E} = \frac{1}{2\pi f_t \times r_e}$$

$$r_{b'e} = r_e (1 + h_{fe})$$

$$g_m = \alpha_0 / r_e \simeq 1/r_e$$

FIG. 5.11. Common-emitter hybrid π equivalent circuit.

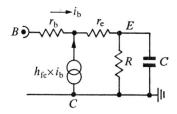

FIG. 5.12. Emitter-follower equivalent circuit. R, C constitute external load.

common-emitter current gain into short circuit collector load given by

$$h_{fe} = \alpha/(1 - \alpha), \tag{5.9}$$

where

$$\alpha = \alpha_0/(1 + jf/f_\alpha). \tag{5.10}$$

5.2.1. Common-base amplification

The equation for the amplifier represented by Fig. 5.10 and operating into a collector load R_1 is

$$(i_c - i_e)r_b' + i_c/j\omega C_c - i_e \times \alpha/j\omega C_c + i_c R_1 = 0, \tag{5.11}$$

giving

$$i_c/i_e = (\alpha + j\omega C_c + r_b')/\{1 + j\omega C_c(R_1 + r_b')\}. \tag{5.12}$$

Eqn (5.12) shows up the importance of the $r_b' \times C_c$ product, a parameter omitted in the simplified treatment following later. For high gain $r_b' \times C_c$ should be as small as possible. Good high-frequency performance requires that

$$r_b' \times C_c \ll (1/2\pi f_\alpha). \tag{5.13}$$

For current v.h.f. transistors used in nanosecond amplification and switching, $f_\alpha \sim (500\text{--}1000)$ MHz, giving $(1/2\pi f_\alpha)$ equal to 150–300 ps, so that $r_b' \times C_c$ should be less than 50 ps.

The current gain into a collector load R_1 can be estimated from eqn (5.10) and Fig. 5.13:

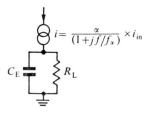

FIG. 5.13. Common-base simplified equivalent circuit.

$$i_c/i_{in} \simeq \frac{\alpha_0}{(1+jf/f_\alpha)} \times \frac{1}{(1+j\omega C_c \times R_1)} \qquad (5.14)$$

$$\simeq \frac{1}{1+jf(1/f_\alpha + 2\pi C_c \times R_1)}. \qquad (5.15)$$

For $f < f_\alpha B_{CB}$, the common-base 3 dB bandwidth, is

$$B_{CB} = 1/\{(1/f_\alpha) + 2\pi C_c R_1\}. \qquad (5.16)$$

5.2.2. Common-emitter amplification

For common-emitter analysis, Fig. 5.11 can be transformed into Fig. 5.14. The term $C_{b'c} \times (1 + g_m R_1)$ is due to the well-known Miller effect by which

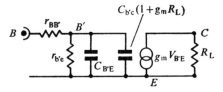

FIG. 5.14. Common-emitter simplified equivalent circuit.

the feedback capacitance between amplifier input and output terminals is multiplied by $(1 + A)$ where A is the voltage gain. An analysis of Fig. 5.14 leads to

$$A_i \sim \frac{h_{feo}}{1 + jh_{feo} \times f\{1/f_t + 2\pi C_{b'c} \times R_1\}} \qquad (5.17)$$

which closely resembles (5.15). At high frequencies, where $f > (f_t/h_{feo})$,

$$A_i \sim \frac{1}{f\{1/f_t + 2\pi C_{b'c} \times R_1\}}. \qquad (5.18)$$

Thus the high-frequency 'gain-bandwidth' product $A_i \times f$ equals the 3 dB common-base bandwidth of eqn (5.16). Common-base and common-emitter wide-band amplifiers have much the same bandwidth. The latter is more flexible because gain can be traded for bandwidth by local negative feedback; the common-base amplifier is restricted to unity current gain.

5.2.3. *Emitter-follower*

The emitter-follower is inherently a broad-band circuit with a bandwidth greater than that of the common-base or common-emitter amplifier. Frequently the emitter-follower feeds the output of one module to another, which might be up to several hundred feet away, via a coaxial cable. If the cable termination at the far end is a resistance equal to the characteristic impedance of the cable, the emitter-follower looks into a purely resistive load. Quite often this is not the case and the effective emitter-follower load is equivalent to a resistor R shunted by a capacitor C as in Fig. 5.12. The real part of the input impedance may then become negative over a certain range of frequencies resulting in parasitic oscillations. The emitter-follower will be stable if (Lawrence 1967)

$$CR < (1/2\pi f_t). \tag{5.19}$$

For a given CR, instability is more likely to occur in fast transistors having high f_t. Where necessary, stability can be ensured by inserting an unbypassed base 'stopper' resistor in the base lead. The emitter-follower bandwidth will thereby be decreased by an amount that can usually be tolerated. Stability is also improved by placing a ferrite bead in the emitter and/or base lead.

5.3. FET amplifiers

Two features make the junction field effect transistor suitable for certain nuclear pulse amplifiers. First it has a very high input resistance, of the order $10^9 \ \Omega$, and secondly it has a very low noise figure.

The high input impedance is needed when operating with gas-filled detectors. The time constant of the parallel resistance–capacitance network shunting the detector should be as high as possible: the FET is the ideal element for this purpose.

The low noise figure helps to give a low overall noise factor for amplifiers with a FET input stage. A typical noise spectrum for a junction FET is shown in Fig. 5.15. The corner frequencies at B and C are about 500 Hz and 50 MHz respectively. Good low-noise bipolar transistors have a noise factor of 2 dB over the region BC with corner frequencies around 10 kHz (B) and 30 MHz (C).

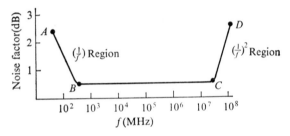

FIG. 5.15. Noise spectrum for junction FET.

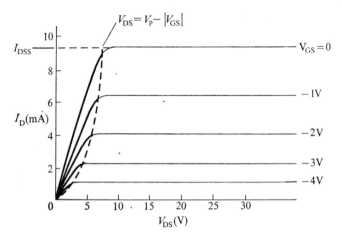

FIG. 5.16. Static characteristics N channel FET.

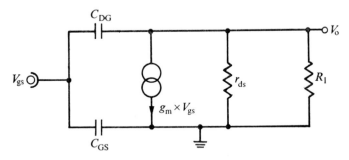

FIG. 5.17. Junction FET—common-source equivalent circuit.

The FET is a voltage-controlled device. The static characteristics of Fig. 5.16 and the equivalent circuit of Fig. 5.17 could apply equally well to a pentode. In most cases $R_1 \ll r_{ds}$ where r_{ds} is the common-source output resistance, and

$$A_V \sim g_m \times R_1. \tag{5.20}$$

80

At high frequencies the input impedance will be a capacitance in shunt with the input resistance and given by

$$C_{is} = C_{gs} + C_{dg}(1 + A_V). \tag{5.21}$$

The bandwidth Δf of an amplifier with a load consisting of R and C in parallel is

$$\Delta f = 1/\{2\pi R_1 \times C_c\}. \tag{5.22}$$

Substituting for R_1 from (5.20),

$$A_V \times \Delta f = g_m/2\pi C_c. \tag{5.23}$$

For fast junction FETs ($g_m \simeq 3.5$ mA/V, $C = C_{iss} + C_{oss} \simeq 10$ pF), $A_V \times \Delta f$ comes to about 60 MHz and is thus at least an order of magnitude below the gain-bandwidth product of fast bipolar transistors. In many applications, however, the bandwidth of the FET is adequate.

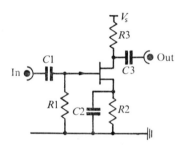

FIG. 5.18. Common-source FET amplifier.

The MOST (metal oxide silicon transistor) is used where the highest input resistance possible is needed. It has an input resistance of about $10^{14}\,\Omega$. Its disadvantage, compared with the junction FET, is a higher noise figure which is more unstable with life and temperature than the noise figure for the junction FET.

A diagram of a common-source FET amplifier is shown in Fig. 5.18. Due to the large production spread in g_m and I_{DSS} for a given transistor, d.c. conditions will extend over a considerable range in production.

Constant current bias is obtained by inserting a bipolar transistor in the source path of J_1 as shown in Fig. 5.19. A similar arrangement in Fig. 5.20 gives a source follower with an extremely high input resistance. The input resistance is

$$r_{in} = R_g(1 + g_m \times R_1), \tag{5.24}$$

where R_1 is now the output impedance of the common-emitter bias transistor J_2 with negative feedback via R_f. R_1 will be $10^6\,\Omega$ or more, so that $(1 + g_m R_1)$ can be of the order 10^3.

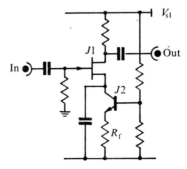

FIG. 5.19. FET amplifier with constant current bias.

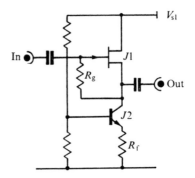

FIG. 5.20. Source follower with constant current bias.

5.4. Operational IC amplifiers

The operational amplifier is a feedback arrangement in the form of Fig. 5.21. Necessary conditions for proper performance are

(a) a high input impedance,
(b) phase reversal between input and output, and
(c) a high open-loop voltage gain A.

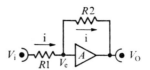

FIG. 5.21. Operational amplifier.

The equations for Fig. 5.21 are

$$v_i = iR_1 + v_e, \qquad (5.25)$$

$$v_e - v_o = iR_2, \qquad (5.26)$$

$$v_o = Av_e, \qquad (5.27)$$

provided the signal current flowing into the amplifier input terminal is negligible compared with i. Eqns (5.25)–(5.27) give

$$v_o/v_i = -R2/\{R1 - (R1 + R2)/A\} \tag{5.28}$$

which, for $|A| \geqslant 1$, reduces to

$$v_o/v_i = (-R2/R1). \tag{5.29}$$

The actual voltage gain is then determined by the passive elements $R1$ and $R2$ and is virtually independent of variations in amplifier operating gain with temperature, supply voltages, etc.

IC elements generally employ a differential input amplifier and the appropriate schematic in Fig. 5.22 is a modification of Fig. 5.21. The

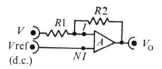

Fig. 5.22. Operational amplifier with differential input. I, inverting input. NI, non-inverting input.

differential amplifier usually has a constant current source like $J3$ in Fig. 5.23. The non-inverting input terminal has to be taken to a d.c. reference potential (frequently ground). The differential input stage has two advantages over the single-ended amplifier. First it is a means of operating with

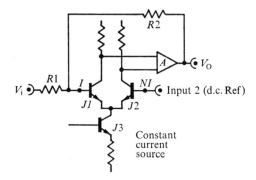

Fig. 5.23. IC operational amplifier with differential input stage.

stable d.c. bias currents unaccompanied by excessive negative signal feedback. Moreover it allows non-inverting as well as inverting amplification. The connections for non-inverting amplification are given in Fig. 5.24.

The major problem with operational amplifiers is how to obtain stable operation with negative feedback. A typical Bode plot of open $|A|$ and closed-loop $|G|$ gain versus frequency is shown in Fig. 5.25. G_1, G_2, and G_3 are three different values of closed-loop gain. The criterion for unconditional stability is that the slope of the response curve at the point of intersection with G should be less than 12 dB per octave. The response in Fig. 5.25 has been approximated by three linear sections AB, BC, and CD with slopes of 6 dB, 12 dB, and 18 dB per octave. In Fig. 5.25 only closed-loop gains intersecting the response curve above B will be unconditionally stable.

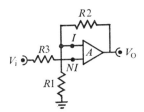

FIG. 5.24. IC operational amplifier (non-inverting operation).

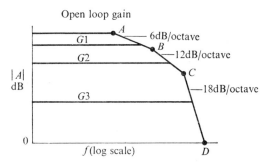

FIG. 5.25. Bode plot feedback amplifier.

Most commercial operational IC amplifiers contain various forms of compensation giving a slope of less than 12 dB per octave over a wide section of AD in Fig. 5.25. The manufacturer usually also recommends external feedback arrangements containing capacitors as well as resistors to achieve stable operation.

Two widely used operational IC amplifiers are the Fairchild μA 702 and 709. Their performance is summarized in Table 5.2. These amplifiers fulfil many of the functions performed previously by discrete circuitry. The major restrictions are limited bandwidth and maximum signal output voltage swings. An increasing number of differential IC amplifiers have a much smaller open-loop gain, 100 or less, obtained by heavy negative local feedback on individual stages of the IC chip. These amplifiers will give stable closed-loop voltage gains of 5 or less with bandwidths around 20 MHz.

TABLE 5.2

Typical performance of μA 702 and 709 amplifiers

	μA 702	μA 709
Open loop voltage gain	3600	45000
Input resistance	40 kΩ	400 kΩ
Output resistance	200 Ω	150 Ω
Common mode rejection	95 dB	90 dB
Max. output voltage swing	±5·3 V	±15 V
Input offset voltage temp. coeff.	2·5 μV/°C	3 μV/°C

5.5. Monolithic bipolar digital ICs

Monolithic IC gates operate in the saturated mode or in a non-saturated current steering mode. The saturated group consists of:

(a) RTL—resistor transistor logic.
(b) MDTL—modified diode transistor logic, also called DTL diode transistor logic. The latter designation should strictly speaking be reserved for an earlier system.
(c) T²L—transistor-transistor logic.

Two systems are available in current logic:

(d) ECL—emitter coupled logic, also known as CML, current mode logic.
(e) E²CL—emitter–emitter coupled logic.

The basic gates of the various systems will now be described.
The RTL two-input NOR gate in Fig. 5.26 follows discrete component

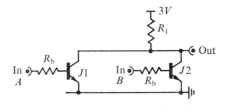

FIG. 5.26. RTL NOR gate.

practice and consists of two saturated inverters with a common collector load R_1. R_1 and R_b equal 640 Ω and 450 Ω respectively for ordinary RTL. For mW RTL (milliwatt RTL), a slower version consuming less power, R_1 and R_b equal 3·6 kΩ and 1·5 kΩ respectively. The output will go low to about 0·1 V when either A or B has a high input. With A and B both at the low signal level, about 0·1 V, the output will be high.

MDTL was, like RTL, evolved from discrete circuitry and a three-input NAND gate is shown in Fig. 5.27. Gate operation is best understood by

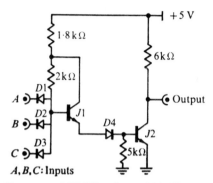

FIG. 5.27. MDTL 3-input NAND gate.

imagining the gate inputs to be fed from the output and the gate output driving the input of a similar gate. When all three inputs are high diodes D_1–D_3 are cut off and the chain J_1–D_4–J_2 will be turned on, resulting in an output of 0·1–0·2 V. Otherwise the output will be high. D_4 increases the noise immunity by adding 0·7 V to the voltage level necessary for turning on J_1 and J_2.

T^2L was evolved from MDTL and was specifically designed for integrated circuitry. In the three-input T^2L NAND gate of Fig. 5.28, J_1 is

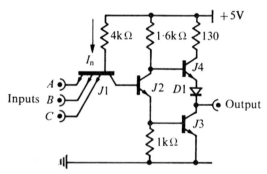

FIG. 5.28. T^2L 3-input NAND gate.

a multiple emitter-transistor whose three emitters take the place of diodes D_1–D_3 in the MDTL gate of Fig. 5.27. The crux of the action is the switching of the input transistor J_1. The base node current I_n is diverted into one or more of the input terminals, or into J_2 base. If I_n remains the same

86

throughout, the excess base saturation charge will remain constant and switching will be extremely fast because the base charge does not change. In practice I_n changes by about 20 per cent when switching from one state to the other; this still gives very fast switching for J_1. When J_2 and J_3 are ON, the collector voltage of J_2 will be about 0·9 V ($V_{BE(sat)}$, 0·8 V, for J_3 plus $V_{CE(sat)}$, 0·1 V for J_2). The output voltage will be around 0·1 V and the voltage drop of 0·8 V across the base-emitter path of J_4 in series with D_1 will be insufficient for conduction along that path. The reader can now see the need for D_1. Without that diode, J_4 and J_3 could conduct simultaneously. The output impedance, consisting of the saturation resistance of J_3, will be low. When J_2 is OFF due to one or more of the input gates

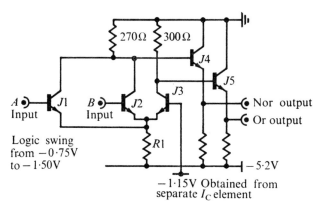

FIG. 5.29. ECL gate.

A–C being low, J_4 and D_1 will conduct. The gate output impedance, now consisting of the output impedance of emitter-follower J_4 in series with the impedance of D_1, will again be low. In this respect $T^2 L$ has an important advantage over RTL and MDTL which have a high output impedance in the high-voltage state.

In the ECL gate of Fig. 5.29 the current through R_1 is switched from J_3 into J_1 and/or J_2 when any of the inputs go high. Being non-saturated the ECL gate is faster than any of the previous systems. Logic capability is enhanced by having two outputs with opposite logic levels. Note that both logic levels are negative.

In $E^2 CL$, which is sketched in Fig. 5.30 and operates similarly to ECL, the outputs are taken from the collectors of J_5 and J_6. The arrangement is less flexible than Fig. 5.29 because the collector loads of J_5 and J_6 must be equal to the characteristic impedance of the transmission system, whereas the emitter-follower outputs in Fig. 5.29 are not tied to a particular impedance. The logic levels are 0 and 0·7 V.

87

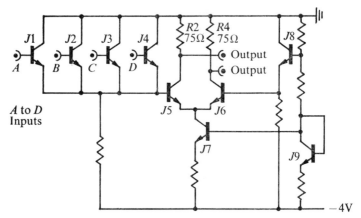

FIG. 5.30. E²CL gate.

5.5.1. *Binary storage elements*

Data processing is carried out with binary elements that store a logic '1' or '0' and nearly always its complement. These complementary outputs are designated Q and $\bar{Q}$; if Q is at '1', $\bar{Q}$ is at '0' and vice versa. The distinction between Q and $\bar{Q}$ is that a logic '1' on R resets Q to '0' and $\bar{Q}$ to '1'. IC binary elements are adaptations of the clocked R–S flip-flop in Fig. 5.31

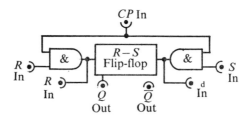

FIG. 5.31. Clocked R–S flip-flop.

where the R and S terminals operate on the flip-flop on arrival of the clock pulse.

The truth table for the clocked R–S flip-flop of Fig. 5.31 is given in Table 5.3.

In addition to the R and S inputs, most IC flip-flops have d.c. coupled direct access R_D and S_D inputs which override the AND gate logic in Fig. 5.31. The R_D and S_D inputs will not be mentioned again because they are not involved in the flip-flop action to be described.

The binary storage element should be capable of clocked R–S action and also of simple toggling, that is changing state on receipt of a clock pulse. Toggling is needed for binary counting. The two binary elements capable

TABLE 5.3

Truth table for clocked R–S flip-flop, Fig. 5.31

R_n	S_n	Q_{n+1}	$\overline{Q}_{n+1}$
1	0	0	1
0	1	1	0
0	0	No change	
1	1	Undefined	

The suffices n and $(n + 1)$ refer to the states before and after receipt of the clock pulse.

of performing both the above functions are the *J–K* and *D* flip-flops. In the *J–K* flip-flop the outputs are fed back to the inputs as shown in Fig. 5.32. The *R* and *S* terminals are designated *J* and *K*; the reason for this

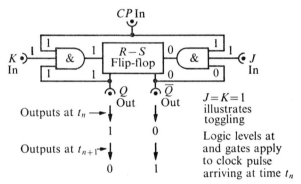

FIG. 5.32. Basic *J–K* flip-flop.

choice is not known. The operation for all input combinations other than $R = S = 1$ is that of the flip-flop in Fig. 5.31. When $R = S = 1$ the bistable will toggle as illustrated in Fig. 5.32. The truth table for the *J–K* flip-flop is given in Table 5.4.

TABLE 5.4

Truth table for J–K flip-flop

J_n	K_n	Q_{n+1}	$\overline{Q}_{n+1}$
0	1	0	1
1	0	1	0
0	0	Q_n	$\overline{Q}_n$
1	1	$\overline{Q}_n$	Q_n

In the D flip-flop (D for data) of Fig. 5.33(a) there is only one data input (Herbst 1969). The internal inverter feeds the input from the AND gate on the S to the AND gate on the R side. On receipt of a clock pulse, Q_{n+1}

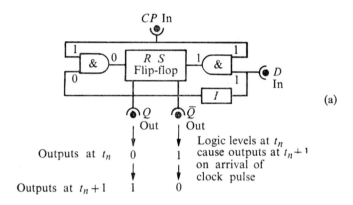

(a)

Outputs at t_n	0	1	Logic levels at t_n cause outputs at t_n+1 on arrival of clock pulse
Outputs at t_n+1	1	0	

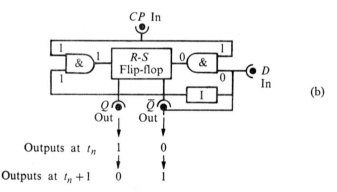

(b)

Outputs at t_n	1	0
Outputs at t_n+1	0	1

FIG. 5.33. D flip-flop. (a) Basic flip-flop. 1, input; I inverter. Logic levels at t_n cause outputs at t_{n+1} on arrival of clock pulse. (b) Flip-flop connected for toggling. $\bar{Q}$ connected to D for toggling. Logic levels at t_n cause toggling on arrival of clock pulse.

takes the state of the D input prior to the clock pulse irrespective of the previous state Q_n. The characteristic equation is

$$Q_{n+1} = D_n. \qquad (5.30)$$

Connecting the D terminal to the $\bar{Q}$ output gives the toggling action illustrated in Fig. 5.33(b). Thus the D flip-flop accomplishes data transfer and toggling. In many applications the D flip-flop can replace the J–K flip-flop, being simpler and cheaper. The two J–K input terminals are advantageous for more sophisticated logic.

In the above explanations of the J–K and D elements, steady-state inputs

90

were assumed to exist at the J, K, and D terminals. If we consider the toggle action, the flip-flop will keep on toggling if the clock pulse is still present after the initial toggling action is over. Clearly there must be some restriction on the clock pulse duration. Furthermore, changes in $\bar{Q}$ and Q must not take place immediately on arrival of the clock pulse because this would cause a race between them and indeterminate output. A variety of methods are used to satisfy the above conditions.

In the edge-triggered J–K and D bistables, the outputs change state after a fixed propagation delay from the leading edge of the clock pulse. The timing waveforms are shown in Fig. 5.34. The inputs must be present for a

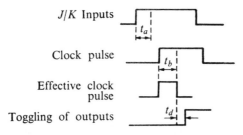

FIG. 5.34. Timing waveforms for edge triggered J–K flip-flop. t_a, minimum 'set-up' time; t_b, minimum 'hold' time. Logic arrangement restricts effective clock pulse duration to t_b; t_d, delay between trailing edge of effective clock pulse and toggling.

minimum 'set-up' time prior to the arrival of the clock pulse and must remain for a minimum 'hold-up' time afterwards.

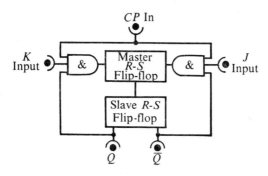

FIG. 5.35. Master-slave J–K flip-flop.

An elegant method of making data transfer virtually independent of rise time, fall time, and duration of the clock pulse is the master-slave (sometimes called dual rank) system of Fig. 5.35. Two separate flip-flops are used and the outputs change state on the trailing edge of the clock pulse.

91

Information enters the master bistable, a temporary memory store, on the leading edge and is transferred to the slave bistable on the trailing edge of the clock pulse. The logic sequence of the Texas Master Slave J–K type SN7473N is given in Fig. 5.36.

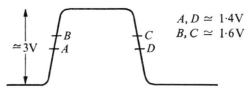

$$A, D \simeq 1.4\text{V}$$
$$B, C \simeq 1.6\text{V}$$

$\simeq 3\text{V}$

FIG. 5.36. Sequence of action Texas SN 5473/7473. Master-slave J–K flip-flop. A, slave isolated from master; B, information from AND gate inputs entered into master; C, AND gate inputs disabled; D, information transferred from master to slave.

Edge-triggered and master-slave flip-flops are d.c. coupled throughout. In the a.c. coupled J–K the memory consists of a capacitor that charges on arrival of the leading edge of the clock pulse. The discharge on the trailing edge of the clock pulse initiates the toggling action. Like the master-slave, the a.c. coupled J–K changes state on the trailing edge of the clock pulse.

5.5.2. *Comparison of various systems*

The various systems will now be compared with reference to Table 5.5. T^2L is the outstanding general purpose system, being very economical in terms of speed versus input power. The levels are fully compatible with MDTL. RTL was the first system available commercially on a large scale but is now over-shadowed by MDTL and T^2L. Poor noise immunity and limited fan-out capability are its chief limitations.

The fastest performance is naturally obtained with emitter-coupled logic where E^2CL has been introduced comparatively recently as an alternative to ECL. The latter system is well established and contains J–K and D flip-flops with clock rates in excess of 100 MHz. Faster T^2L elements are also becoming available and, while ECL is faster, it is worth noting that current T^2L J–K flip-flops have clock rates of 50 MHz (min) and 80 MHz (typ).

Low-power versions for several of the saturated systems are intended for applications where the high packing-density and/or low-power consumption are needed. They are inevitably much slower than the standard elements. Low-power elements are available in RTL, MDTL, and T^2L. However, MOST ICs described in the next section will probably be preferred in low-power applications.

TABLE 5.5

Comparison of IC systems

Logic	Typical output swing (V)	Power supply (V)	Gate performance		Binary element (J–K or D) performance	
			Typical propagation delay (ns)	Gate power (mW)	Typical counting rate (MHz)	Power consumption (mW)
MDTL	3	5	25	9	8	50
RTL	1·2	3	12	12	12	60
mW RTL	1·2	3	45	2·5	1·3	12
ECL (MECL)	0·7	−5·2	8	30	20	80
					30	110
ECL (MECL II)	0·7	−5·2	5	40	70†	80
					100†	—
E²CL (Mullard)	0·7	−4	2	40	90	240
T²L (standard)	3	5	13	15	35	65
T²L (fast)	3	5	7	22	50	80
T²L (low power)	3	5	33	1·5	3	4

† Guaranteed minimum rate.
Data for ECL (MECL II) and E²CL is tentative.

5.6. MOST digital ICs

MOST ICs are slower than bipolar elements because the field effect transistor is a slower device. Nevertheless they are competing with bipolar transistor ICs now that the technology of MOST manufacture is sufficiently advanced. The feature that makes the MOST IC so attractive is the very low-power consumption, which gives a much higher function density per chip area than is possible with bipolar ICs. MOST ICs are especially suitable for medium to slow speed large scale integration.

The two standard gate elements are the P enhancement inverter in Fig. 5.37 and the complementary inverter, containing one P and one N enhancement MOST, in Fig. 5.38. In Fig. 5.37, J_2 is the drain load for J_1. When J_1 is ON (negative V_{GS}) J_2 has a forward gate-source bias and offers a resistance of 100 kΩ or more. The ON drain-to-source resistance of J_1 is typically between 1 and 5 kΩ so that the ON output voltage will be between −0·15 and −0·5 V when V_{SS} equals −15 V. In the OFF state, J_1 (V_{GS} equal to zero) is a very high resistance (10^8 Ω or more) connecting J_2 to ground.

$J2$ is biased very close to the threshold voltage V_{TH}, about 5 V, and the gate output will be $-(V_{GG} - V_{TH})$ V.

Separate voltage supplies are therefore sometimes used to give an output equal to V_{SS}, though many ICs operate with a common supply voltage for V_{GG} and V_{SS}. Turning $J1$ ON is very much faster than switching it OFF, mainly because the load impedance offered by $J2$ exceeds the $J1$ ON resistance by a factor of 10 or more.

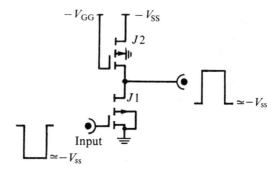

FIG. 5.37. *P* enhancement MOST inverter.

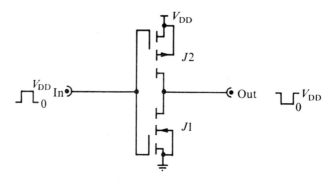

FIG. 5.38. Complementary MOST inverter.

Complementary MOST logic has only very recently become feasible. The difficulty is the fabrication of the N enhancement MOST. A tremendous amount of research was required to master the necessary technology. The action of the complementary inverter in Fig. 5.38 is quite simple. The gate output is zero (to within a millivolt or so) or equal to V_{DD}. With the input at zero, $J1$ is OFF and $J2$ ON. Similarly with the input at V_{DD}, $J1$ is ON and $J2$ OFF. It is best to visualize Fig. 5.38 connected to similar gates at input and output. As for the $T^2 L$ gate of Fig. 5.28 one or other of the transistors in Fig. 5.38 is ON so that the gate has a low output impedance in both states. Rise and fall times are equal and the fan-

94

out capability is superior to the P enhancement gate. Complementary MOST ICs include gates with propagation delays of 70 ns and J–K and D flip-flops with a minimum clock rate of 4 MHz.

References

HERBST, L. J. (1969) *Discrete and integrated semiconductor circuitry*, p. 124. Chapman and Hall, London.

LAWRENCE, E. E. (1967) Integrated circuit convention. *Instn elect. Engrs. Publ. No.* 30, p. 267.

6 *Nuclear Pulse Amplifiers*

By A. B. GILLESPIE

6.1. Basic principles

ANALOGUE pulse signals from nuclear radiation detectors are very small ($\approx \mu$V to mV) and require amplification and impedance transformation before subsequent analogue and/or digital processing by later instruments in the pulse-counting and analysing chain.

The exception is the scintillation counter, where most of the gain is obtained from the photomultiplier tube. However, even here a small amount of external gain is frequently required and this is provided by an electronic amplifier that conforms in most other respects to the design criteria described in this chapter.

Nuclear radiation detectors, in which the detection of a high-speed charged particle or photon produces liberated charge, can be represented by the simple equivalent electrical circuit shown in Fig. 6.1. C is the

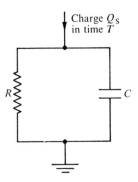

FIG. 6.1. Equivalent circuit of radiation detector.

capacitance of the detector plus associated components (photomultiplier anode capacitance in the case of the scintillation counter) and R is the resistance through which the necessary polarizing potential is applied to the detector (anode load of the photomultiplier for the scintillation counter). For the purpose of pulse signal formation these two components can be considered to be connected in parallel.

If the liberated charge per detection is Q_s, this is swept to the plates of the ionization type of detector by the polarizing field in a time T known as the collection time of the detector. (In the case of the scintillation counter, the photomultiplier charge arrives at the final anode during a time determined by the decay time of the particular phosphor in use.)

In both cases this produces a pulse signal across C rising to a magnitude Q_s/C in a time T and thereafter subsequently decaying slowly to zero as the charge leaks away through R. This is shown in Fig. 6.2.

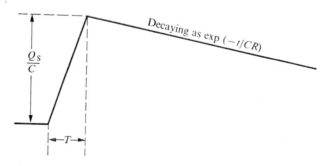

FIG. 6.2. Typical detector voltage pulse.

Nuclear events are distributed at random in time according to the Poisson relationship

$$p_m = \{(\bar{m}\,\delta t)^m \exp(-\bar{m}\,\delta t)\}/m! \tag{6.1}$$

where p_m is the probability of exactly m events occurring in a time interval δt if the mean rate of arrival is $\bar{m}$ per second. Such a random succession of detections in a nuclear counter produces two voltages across the equivalent circuit in Fig. 6.1, a mean value with a superimposed fluctuation. The mean value has a magnitude $(Q_s/C)\bar{m}CR$, while the r.m.s. value of the fluctuation has a magnitude $(Q_s/C)\{(\bar{m}CR)/2\}^{1/2}$.

If now the detector time constant CR is very large compared with $\dfrac{1}{\bar{m}}$, the mean interval between events, then the mean voltage across R is relatively large compared with the fluctuating component. This is shown in Fig. 6.3. This mean voltage, which is proportional to the radiation strength or activity, can be amplified by a d.c. amplifier and subsequently measured, and this is an example of a nuclear detector functioning as a direct ratemeter. This is not representative of the ultimate conditions existing in a pulse-counting system and so is not discussed in detail in this chapter. It is included, however, because such conditions can tend to arise in the early stages of the pulse amplifier and the above relationships are then important.

If, on the other hand, the detector time constant CR is small compared

97

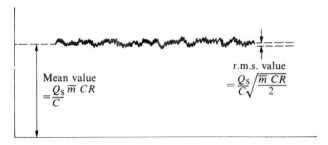

FIG. 6.3. Equivalent circuit voltages for $CR \gg 1/\overline{m}$.

with $\dfrac{1}{\overline{m}}$ then the mean voltage across R is also small, while the fluctuating component is relatively large. In fact the voltage waveform across CR simply consists of a random succession of impulses similar to that shown in Fig. 6.2. This is shown in Fig. 6.4.

FIG. 6.4. Equivalent circuit voltages for $CR \ll 1/\overline{m}$.

These small pulses require amplification by a high-gain pulse amplifier, which must be a.c. coupled to prevent drift, and a subsequent measurement of their mean rate of arrival is an alternative way of measuring the radiation activity. This is an example of the nuclear detector functioning as a pulse counter, and the present chapter is primarily concerned with systems of this type. The pulse-counting system is clearly characterized by the presence of a short time constant, which aims to reduce the pulse amplitude quickly to zero as soon as the charge collection is complete and so minimize pulse overlap. Note, however, that because of the random time distribution, some pulses will always overlap, but this probability can be reduced within limits the smaller CR is made compared with $\dfrac{1}{\overline{m}}$.

The advantages of the pulse-counting system over the ratemeter d.c. amplifier system for activity measurements are improved sensitivity together with the possibility of removing unwanted background radiation or noise by the use of pulse-amplitude selection. Additionally, if the detector generates pulses whose amplitudes are accurately proportional to the energies of the incident-charged particles or photons, then by performing an amplitude distribution measurement on the pulses at the amplifier output this will give an energy spectrum of the incident radiation. This latter advantage, which is highly desirable to the experimentalist, must of course be paid for in terms of accurately linear and stable amplification together with the additional complex instrumentation needed to derive the pulse-amplitude distribution. Such instruments are discussed in detail in later chapters.

6.2. Elementary considerations of bandwidth and counting speed

It was shown in the previous section that the pulse-counting system is characterized by the presence of a detector time constant that is kept small compared with $\frac{I}{m}$ to prevent overlapping of successive pulses. Clearly the higher the counting rates involved so the shorter this time constant must be made, but in the limit the maximum counting speed is set by the collection time T of the detector, since theoretically no pulses can have a shorter duration.

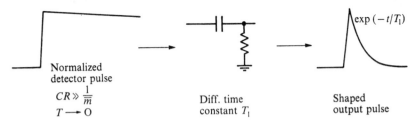

Normalized
detector pulse

$CR \gg \frac{1}{m}$

$T \longrightarrow 0$

Diff. time
constant T_1

$\exp(-t/T_1)$

Shaped
output pulse

FIG. 6.5. Pulse shaping using a differentiating time constant.

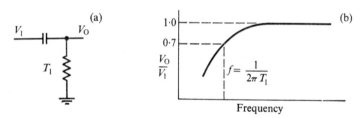

FIG. 6.6. Gain frequency response of differentiating time constant. (a) Arrangement. (b) Response.

It is not essential that the short time constant should be that of the detector itself. It is possible to leave the detector time constant relatively large compared with $\frac{I}{m}$ and instead to include the pulse-shaping time constant at a later stage in the amplifier. Provided no overloading or serious non-linearity occurs in the amplifier stages prior to the pulse-shaping time constant, the two systems are, to a first order, directly comparable. This time constant, which is mainly responsible for determining the pulse width, is usually referred to as the differentiating time constant T_1. The pulse-shaping function of the differentiating time constant is shown in Fig. 6.5 and the gain frequency response of such a circuit is shown in Fig. 6.6. The differentiating time constant, which is now replacing one of the longer a.c. interstage coupling circuits, attenuates the lower frequencies, and is thus instrumental in determining the lower limit of the amplifier pass band. This time-constant arrangement in a high-gain amplifier where electrical noise is important is a preferred one, since the thermal noise generated in R and passed by the amplifier is that fraction that appears across C, and this varies as $\frac{\text{I}}{\sqrt{R}}$. Thus by keeping R large and using a short differentiating time constant in the amplifier the signal-to-thermal noise ratio is improved. Similarly, any other noises from the detector and early stages of the amplifier, having a preponderance of low-frequency components, will be limited by the differentiating time constant in the same way. For these reasons, together with the need to vary the differentiating time constant to suit the

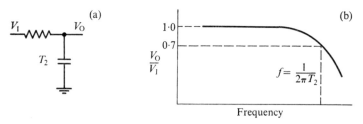

FIG. 6.7. Gain frequency response of integrating time constant. (a) Arrangement. (b) Response.

conditions of particular experiments, it has become almost universal practice in high-gain nuclear pulse amplifier technology to locate the differentiating time constant at an early stage within the amplifier and to leave the counter time constant relatively large.

The amplifier will also possess an upper limit to its pass band, and as it is likewise desirable to be able to vary this to suit experimental conditions and also to reduce high-frequency noise, a second time-constant circuit is included to perform this function. This is referred to as the integrating time constant T_2 and is shown in Fig. 6.7 together with its gain frequency characteristic.

Thus between them the differentiating and integrating time constants completely determine the pass band of the nuclear pulse amplifier. This is the simplest and possibly the most versatile pulse-shaping system in use and is generally regarded as the standard against which all the more exotic variations are compared. Some of these variations are discussed in later sections of this and the next chapter.

As a basis for computation and comparison it is convenient to have an equivalent circuit of the nuclear pulse amplifier and this is shown in Fig.

6.8. Here $\sum C$ represents the total cold input capacitance due to the detector and amplifier and R is the effective shunt resistance. T_1 and T_2 are the differentiating and integrating time-constant circuits, the dotted connections indicating that these circuits are isolated and do not load each other. The gain G represents the active circuit amplification and is of course constant at all frequencies, as the gain variation with frequency is controlled by T_1 and T_2.

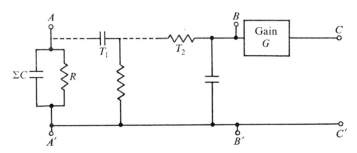

FIG. 6.8. Equivalent circuit of nuclear pulse amplifier.

6.3. Pulse measurements after amplification

It is the function of the nuclear pulse amplifier to raise the small detector pulses to a maximum level usually in the range 5–50 V, the former figure being representative of modern transistor amplifiers, the latter figure representative of the older amplifiers that used thermionic valves. For gas ionization chambers and semiconductor radiation detectors, maximum gains in the range 10^5–10^6 are not uncommon and the amplifier circuits have to be designed to provide these gain requirements with a high degree of linearity and stability. In addition, because such high gains have to be used, the electrical noise generated in the detector and the first stage of the amplifier cannot be ignored. In fact, one of the major problems in nuclear pulse-amplifier design is that of achieving a good signal-to-noise ratio at the amplifier output, and this problem is discussed in some detail in Chapter 7.

If measurements of radiation activity only are required it is customary to plot an experimental curve showing the number of pulses arriving in a given time and exceeding a given amplitude as a function of that amplitude. This is done using a pulse-amplitude discriminator (see Chapter 8) and is known as the integral bias curve. A typical curve is shown in Fig. 6.9. At low discriminator bias levels the curve turns up rapidly due to the counting of noise crests. At higher levels it is hoped to achieve a flatish region or plateau where the measured pulse rate is reasonably insensitive to bias level variation. Above this the pulse rate falls rapidly towards zero, corresponding to the discriminator bias exceeding the maximum pulse amplitude. The flatness of the plateau depends on many factors, chiefly the counter geometry, source preparation, and the type of radiation being measured. The optimum operating point is that corresponding to the middle of the plateau as here

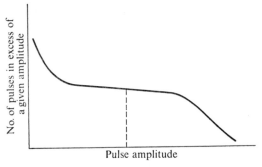

FIG. 6.9. Integral bias curve.

the measured count rate is least dependent on instrumental variations. For a reasonable plateau slope, the amplifier long-term gain stability need not be exceptional, and generally a figure of some 2–4 per cent is adequate.

On the other hand, where measurements of both activity and energy are required, the experimental curve that is drawn is the differential bias curve, which shows the number of pulses with amplitudes falling within an incremental range or channel, occurring in a given time, and plotted as a function of the pulse amplitude. This measurement requires a complex

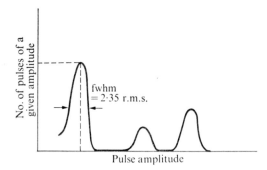

FIG. 6.10. Differential bias curve.

pulse-processing instrument known as a multichannel analyser (see Chapter 11) and a typical amplitude spectrum curve is shown in Fig. 6.10. The position of the peaks on the pulse amplitude scale is a measure of the relative energies of the radiations present, while the height of the peaks on the ordinate scale is a measure of their relative activities. The peak width or

102

dispersion is due chiefly to electrical noise (but see Chapter 7, section 7.1) and this clearly sets a limit to the ability of the analysing system to distinguish between source energies that are close together. The full peak width is usually measured at half height and this is known as the full width at half maximum (fwhm). For Gaussian noise this is 2·35 times the r.m.s. value. For measurements of this type, particularly where semiconductor detectors are used, the requirements on the stability of gain and the linearity of the output–input characteristic of the amplifier, are usually quite severe. A long-term gain stability of 0·01 per cent is frequently required with a similar specification on the differential linearity. (The latter may be defined as the difference between the slope of the output–input characteristic for two input levels, one corresponding to a very small output amplitude and the other corresponding to an output amplitude near to the maximum.)

6.4. Amplifier block schematic

A block schematic lay-out of a typical general-purpose nuclear pulse amplifier is shown in Fig. 6.11. The head amplifier is a small compact unit,

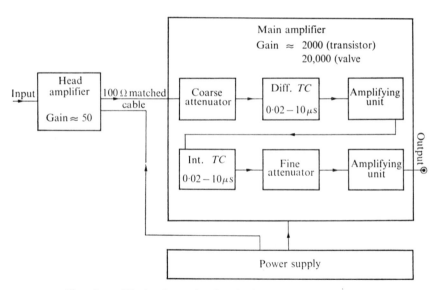

FIG. 6.11. Block schematic of typical nuclear pulse amplifier.

separated from the main assembly and intended to be mounted close to the radiation detector. This keeps the total input capacitance $\sum C$ as small as possible and in consequence the input pulse signal and the signal-to-noise ratio as large as possible. The head amplifier is designed to feed its output signal to the main amplifier along a matched cable. Matching is desirable, as long cable lengths are frequently necessary if the detector and head amplifier are to be positioned in areas where the local radiation level is very high. For the same reasons, it is not customary to include in the head amplifier any controls that may need altering during an experimental run. The head amplifier gain must be high enough to raise the signal and its associated noise well above any additional noise that may be introduced in the long cable or by the main amplifier. On the other hand, it must not be too large, since at high counting rates considerable signal overlap and pile-up, with the attendant danger of overloading, will be occurring in all stages prior to the differentiating time constant.

All controls are contained in the main amplifier unit. These include differentiating and integrating time-constants, variable or switched, usually over the range 0·02–10 μs, this being suitable for use with radiation detectors having collection times lying within the same range. Fine and coarse attenuators are also included to permit setting of the amplifier gain. Those parts of the block schematic marked 'amplifying unit' contain the active valve or transistor circuits that provide the gain, and these are described in some detail in later sections of this chapter.

The power unit and its d.c. supply lines must, of course, be appropriate to the transistor or valve circuits used in the amplifier. In both cases the supply lines are invariably stabilized, as this minimizes the possibility of overall feedback through the power supply impedance, and also materially assists in the problem of achieving very high linearity and gain stability in the active amplifier circuits.

6.5. Some fundamental aspects of pulse shaping

6.5.1. *Unipolar pulse shaping*

The nuclear pulse amplifier just described, with single differentiating and integrating time constants for bandwidth limitation and pulse shaping, has been used in this form for upwards of 30 years. It has been shown that to obtain an optimum signal-to-noise ratio consistent with a required counting speed the differentiating and integrating time-constants should be made equal (Gillespie 1953). Most nuclear pulse amplifiers are operated with equal time constants since this effectively removes one variable from the system. The resulting equal time-constant amplifier is regarded as a standard, and all variations, principally those involving other methods of pulse shaping, are now compared with this. Different methods of pulse shaping have been investigated, chiefly from the point of view of signal-to-noise ratio and the introduction of signal amplitude errors at high counting rates. The two are related, but it is proposed to consider primarily

104

the latter effect in this section and to discuss the signal-to-noise aspects in Chapter 7.

In what follows it helps in the understanding if the pulse shapes produced by the various networks are considered initially as originating from a unit-amplitude step function as input signal. This effectively assumes that the counter collection time is short compared with the amplifier-shaping time constants. The pulse produced by the equal time-constant network is defined by $t/T_1 \exp(-t/T_1)$ and is shown in Fig. 6.12. This has an amplitude

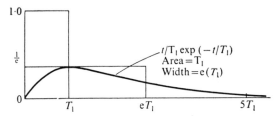

FIG. 6.12. Pulse shape with equal time constants.

of $1/e$, occurring at a time T_1. Its area is also T_1 and its effective duration, defined as $\dfrac{\text{area}}{\text{height}}$ is eT_s. These properties are shown in the diagram, together with equal-area rectangular pulses, one having unit amplitude and one having the same effective duration.

6.5.1.1. *Pulse undershoot.* Unfortunately in a practical amplifier this assumption of a step-function input is not completely valid, and the input pulse will have a decaying top due to the detector time constant or a combination of this with the interstage low-frequency coupling circuits. It is reasonable to assume that only one time constant is important, referred

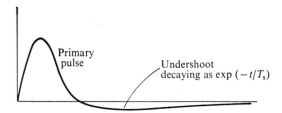

FIG. 6.13. Undershoot due to secondary differentiating time constant.

to as the secondary differentiating time-constant T_S, and that all others can be made many times larger. The effect of this secondary time-constant is to produce a small amplitude undershoot on the primary pulse, as shown in Fig. 6.13. For a small undershoot, the secondary time constant must be large, compared with the differentiating time constant, but the undershoot

105

then lasts for a time comparable with the secondary time constant. It is this long undershoot, in association with possible pulse overlap at high count rates, that causes errors in the measured pulse amplitude.

The other assumption of the detector collection time being short, compared with the shaping time constants, if not valid, can also lead to errors in pulse amplitude. These errors are caused by random variations in the detector collection time, but this effect is not a function of pulse rate and can usually be reduced to negligible proportions by making the amplifier-shaping time constants some three to ten times the detector collection time (Gillespie 1953). This is not usually a severe requirement.

Returning then to the pulse undershoot problem, this has assumed much greater significance with the advent of the semiconductor radiation detector, since these detectors have made possible a marked reduction in the width of individual peaks in an energy spectrum plot. This stems fundamentally from a figure of around 3 eV to produce a hole–electron pair in a semiconductor detector compared with 30 eV for an ion pair in a gas ionization detector. For particle and gamma-ray spectrometry in the energy range several MeV down to hundreds of keV the semiconductor detector is thus fundamentally capable of giving line widths of o·1–o·3 per cent, compared with the more usual o·5–1·5 per cent for a gas ionization detector and 2·5–7·5 per cent for a scintillation counter. (See Chapter 7, section 7.1.) For the semiconductor and gas-ionization detectors, such energy resolution figures are not always realizable, and depend critically on the detector capacitance and the electrical noise level of the detector and the amplifier. However, for energies above 1 MeV the fundamental resolution figures for the semiconductor detector may be approached very closely, and thus any additional broadening of the spectrum peaks, due to undershoot and other effects in the amplifier, should be kept to a minimum and preferably should not exceed o·02 per cent r.m.s.

Pulse undershoot can also be serious in amplifiers used for the measurement of source activity. A feature of such measurements is the very large pulse-amplitude range encountered, due to simplified detector design and source preparation. This leads to serious overloading of the amplifier by many of the pulses, with consequent amplification of the undershoot, resulting in pulses that may be hundreds of times longer than the primary pulse. These very wide pulses will then limit the maximum counting speed that is possible with the equipment. This problem is discussed again later in this section.

Suppose, in a practical amplifier, the equal pulse-shaping time constants are each 1 μs and the secondary time constant is 1 ms. From the unit-amplitude rectangular pulse in Fig. 6.12 it is obvious that the pulse droop after 1 μs, caused by this 1 ms time constant, is 0·001 and that this results in a pulse undershoot of the same magnitude, decaying to zero with a time constant of 1 ms. Similarly, for the other rectangular pulse of amplitude 1/e and duration e μs the droop and consequently the undershoot amplitude is again 0·001. It is the primary pulse area which to a first order determines the undershoot amplitude, and so for the equal time-constant pulse the undershoot will also be 0·001 ($=T_1/T_s$) or 0·27 per cent of the primary pulse amplitude. In this latter case, the time of the undershoot maximum is some 20 μs behind the start of the pulse, but as it subsequently decays to the base line with a time constant of 1 ms, the following simplified calculations will not be significantly affected by considering the undershoot as a true exponential defined by $0·001 \exp(-t/T_s)$.

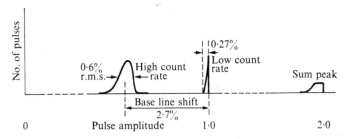

FIG. 6.14. Spectral line distortions at low and high pulse rates.

Consider now a low counting rate of less than 100 per second. A small percentage of pulses occur during the undershoot of previous pulses, and they have their amplitudes effectively reduced by varying amounts, but the majority of pulses will be unaffected. Practically no pulses have their amplitudes increased, since the probability of primary pulse overlap is negligibly small, so the effect on a noise-free line spectrum is as shown in Fig. 6.14. Broadening and smearing takes place over an energy range of 0·27 per cent below the true value. If now the counting rate is raised to 10 000 per second then nearly all the pulses occur during the undershoots of previous pulses, with also a small but significant number of coincidences occurring during the primary pulse itself. This very significant overlap and piling up of the long duration undershoots, will, as explained in section 6.1, produce a d.c. component of magnitude $V_u \bar{m} T_s$ with a superimposed fluctuation having an r.m.s. value of $V_u \sqrt{\{(\bar{m}T_s)/2\}}$, where V_u is the amplitude of the undershoot. The d.c. component behaves as a shift or depression of the base line from which the pulses rise, and this will introduce an error in any instrument that measures pulse amplitude from a reference zero level. Similarly, the fluctuation puts a spread on the individual pulse

107

amplitudes leading to spectral line broadening. These effects are also shown in Fig. 6.14. The centroid of the distribution has been displaced downwards by an amount equal to the d.c. shift, the distribution itself is tending to become more symmetrical about the peak value, but the peak value will still always occur on the high-energy side of the centroid. The primary pulse overlap will additionally produce a small number of greater amplitude pulses, most of these occurring near twice the primary pulse amplitude due to the flatish region at the top of the primary pulse. This results in another very small peak in the amplitude spectrum at twice the level of the main peak and is usually referred to as a sum peak. This is also shown in Fig. 6.14. Its area is equal to the probability of primary pulses coincidence.

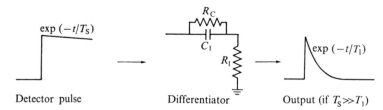

FIG. 6.15. Technique of pole-zero cancellation on time constant shaped pulse.

If now a value of 0·27 per cent of the primary pulse amplitude is substituted for V_u in the above expressions, with $\bar{m} = 10\ 000$ counts/s and $T_S = 1$ ms, this gives a base-line shift of 2·7 per cent and an r.m.s. spread on the pulse amplitude of 0·6 per cent. Both these figures are large compared with the limit of 0·02 per cent recommended earlier, and ways of reducing them must be sought.

6.5.1.2. *Minimizing the effects of pulse undershoot.* From the theoretical expressions for base-line shift and r.m.s. dispersion, it is clear that lengthening the secondary time constant T_S will reduce the latter, as the undershoot amplitude V_u is inversely proportional to T_S. This has, however, no effect on the former, so other methods of reducing base-line shift must be used. These are discussed later in this section. To reduce the r.m.s. dispersion to a value less than 0·02 per cent at a counting rate of 10 000 counts/s will require a secondary time constant of 1 s, with all other coupling time constants in the amplifier at least ten times larger, i.e. >10 s. Practically this is very difficult, since the secondary time constant is usually the detector time constant, and making this many times greater than the 1 ms so far assumed, will result in serious pile-up and certain overloading in the amplifier stages prior to the differentiating time constant. Fortunately, this problem can be resolved by using the technique of pole-zero cancellation (Nowlin, Blankenship, and Blalock 1965). This consists simply of shunting the capacitor C_1 of the differentiating time constant with a compensating resistor R_C and making the time constant $C_1 R_C$ equal to T_S. This is shown in Fig. 6.15. The output pulse has no undershoot and decays exponentially with a time constant that is theoretically $C_1\{R_1 R_C/(R_1 + R_C)\}$ but for all practical purposes is still $C_1 R_1 = T_1$ if T_S is large compared with T_1. What this technique does then is to eliminate the undershoot effect of the secondary time constant (if for practical reasons this must remain small) and make the next shortest coupling time constant the new controlling secondary time constant T_S. Pole-zero cancellation may be applied more than once, but with care, should there be other situations where very long coupling time constants are impracticable, otherwise all coupling time constants should be made as large as possible, preferably >10 s.

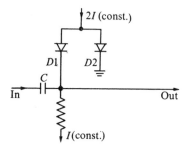

FIG. 6.16. Double diode base-line restorer.

This then results in a nuclear pulse amplifier which adds very little spectral broadening at high count rates but has still several disadvantages. The two most important are d.c. shift of the base line and its inability to work accurately in situations where the counting rate is rapidly varying, a consequence of the settling time transients of the long interstage coupling time constants. However, for many applications with radioactive sources of long half-life, this latter restriction is not very serious.

A simple method of minimizing the effects of base-line shift is the d.c. restoring circuit shown in Fig. 6.16 (Robinson 1961). This circuit has the merit of operating on pulses of either polarity. If the coupling capacitor C is large (say 10 μF upwards) then the restoration of the base line derives from an averaging process over many input pulses, and the following simple calculation is applicable. During a positive pulse the diode D_1 is cut off and the diode current I will then charge the coupling capacitor C. If the pulse duration is δ the charge is $I\delta$. This charge must be removed between pulses. If the mean pulse rate is $\bar{m}$ and i is the average current necessary to remove the charge, then equating the two gives $i = \delta I/\{(1/\bar{m}) - \delta\}$. This current i flows through the two diodes in series and will displace the output

9

level from zero by an amount equal to i multiplied by twice the diode incremental resistance r. The diode resistance r is approximately $1/40I$ so the base-line displacement works out to $\delta/20\{(1/\bar{m}) - \delta\}$. For $\bar{m} = 10\,000$ pulses/s and $\delta = 2\cdot7\,\mu s$ (i.e. $1\,\mu s$ time constants) the base-line displacement is $1\cdot4\,mV$. For positive pulses of $5\,V$ this is less than $0\cdot03$ per cent. The amplifier output impedance Z must also be kept low since, to a first order, the circuit removes a slice of signal of magnitude IZ volts on either side of the base line. For $Z = 10\,\Omega$ this is $1\,mV$. The base-line displacement can be further reduced by connecting the series diodes between the input and output of a wide-band d.c.-coupled operational amplifier (Chase and Poulo 1967). This effectively reduces the diode incremental resistance by the gain of the amplifier without altering the diode current I.

The second mode of operation is with a small coupling capacitor (≈ 1000 pF). Here the charge accumulated on C during the pulse can be removed by the conducting diodes before a further pulse arrives, so the base line is effectively restored to zero after each pulse. This mode of operation may, however, result in a slight worsening of the signal-to-noise ratio by some 15–20 per cent. This occurs if the time constant Cr is smaller than the amplifier pulse-shaping time constants, a condition that almost certainly applies when the restorer is used with an operational amplifier to reduce the diode resistance.

Other more involved methods of minimizing base-line shift have been used. One technique is to measure the magnitude of the base-line shift during a strobe pulse, amplify and smooth this error signal, and feed it back in the right phase to reduce the shift (Goulding and McNaught 1960). Perhaps the most elegant, and certainly the most involved method of all, is that known as spectrum stabilization (Ladd and Kennedy 1961, Dudley and Scarpetti 1964). Here a radioactive source with a well-defined spectrum peak is included with the unknown, and it is arranged using feedback that this peak should always occupy a chosen channel in the spectrum analyser. Any variation above or below this channel produces an error signal that can be used to correct the d.c. bias of the system at some convenient point. Similarly, if a second reference peak is included, then any measured variation in the amplitude gap between the two peaks can be interpreted as a gain change in the system and corrective action can be taken to minimize this.

6.5.1.3. *Other pulse-shaping networks.* The pulse shape produced by equal differentiating and integrating time constants is shown in Fig. 6.12. For computational purposes the duration of this pulse is taken as $\dfrac{\text{area}}{\text{height}}$ and this is equal to eT_1. It is clear, however, that after this time interval, the pulse amplitude is only slightly less than 50 per cent of its maximum, and a substantial part of the pulse has still to occur. This tail part is decaying to the base line very slowly, and takes many time constants to reach an amplitude that is still comparable with the pulse amplitude errors previously considered important, for example 1 per cent after $7 \cdot 5 \ T_1$, $0 \cdot 1$ per cent after $10 \ T_1$ and $0 \cdot 01$ per cent after $12 \cdot 5 \ T_1$. A pulse having the same effective width, but with much faster edges and a rapid decay to the base line, will still cause, through primary pulse overlap at high rates, pulses of larger amplitude, but, of those that do occur, most of them will be in the region of twice the primary pulse amplitude with very few having their amplitudes increased by only $0 \cdot 01$–$1 \cdot 0$ per cent. Pulses of twice the primary pulse amplitude show up in the sum peak and may not be very serious, unless

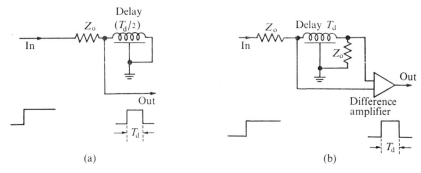

FIG. 6.17. Delay-line pulse shaping. (a) Single-ended matching. (b) Double-ended matching.

coinciding with other peaks in the spectrum, whereas those only slightly raised in amplitude will distort the spectral broadening of the primary peak itself on the high-energy side of the centroid.

Additionally, in amplifiers used for activity measurements, where very large overloads are occurring, the long tail will be amplified and so produce much wider pulses, which will seriously limit the maximum counting speed for a given pulse-rate loss. Thus, in both types of measurement—activity, and energy—the way in which the pulse tail returns to the base line, through the amplitude range of, say, $1 \cdot 0$ per cent down to $0 \cdot 01$ per cent is very important, and ways of speeding this decay are desirable.

One method is to replace the differentiating time constant with a delay line which, of itself, produced a rectangular pulse of duration $T_d = eT_1$. Schematic circuits to do this are shown in Fig. 6.17. Circuit (b) is preferable, since the line is here terminated in the characteristic impedance Z_0 at both ends, and mismatches are of less consequence. If the integrating time constant of the amplifier is unaltered in value ($=T_1$) the output pulse has the shape shown by the full line in Fig. 6.18(a). Calculations show that over the range 1·0 per cent down to 0·01 per cent of pulse amplitude, the pulse duration is not significantly different from the equal time-constant case. If, however, the integrating time constant is reduced in value, say to $T_1/4$, then the output pulse has the shape shown in Fig. 6.18(b), and clearly decays much faster. Here calculations show that over the same amplitude range, the pulse duration is only half of its value with equal time-constant shaping. The price one has to pay for this is, however, increased noise and a worse signal-to-noise ratio (see Chapter 7). For the case quoted here, the signal-to-noise ratio is about 40 per cent worse.

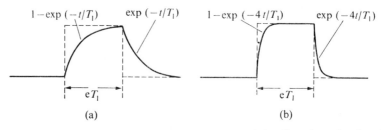

(a)   (b)

FIG. 6.18. Effect of integrating time constant on delay-line shaped pulse. (a) Amplifier integrating time constant T_1. (b) Amplifier integrating time constant $T_1/4$.

The delay-line method of pulse shaping is attractive on account of the flat top and the very steep sides possible. Moreover, the technique of pole zero cancellation to remove undershoot caused by the secondary time constant is still applicable. In this case the resistive loss of the delay line, which produces a positive offset of the base line, is controlled such that the droop caused by the secondary time constant just cancels this step. This is shown in Fig. 6.19.

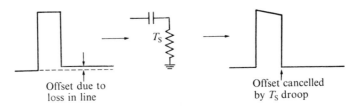

FIG. 6.19. Pole-zero cancellation on delay-line shaped pulse.

The most serious disadvantage in using a delay line for pulse shaping is the practical problem of avoiding small amplitude steps or rings on the trailing edge of the pulse due to mismatches between sections and at the ends of the line. Steps of <1 per cent are significant to the present discussion and yet are representative of an extremely good line. It is this feature of delay-line shaping that no doubt limits its use in practice, but nevertheless there are situations where the advantages may often outweigh the disadvantages.

A simpler, but in the present context, less effective method is that of using a single time-constant differentiator and several time-constant integrators, all having the same value. The pulse shape from such a network having n integrators is defined by $t^n \exp(-t/T_1)/n!T_1^n$ (Fairstein and Hahn 1965). The maximum amplitude of this pulse occurs at a time nT_1 but its area is always T_1 independent of the value of n. Using Stirling's approxi-

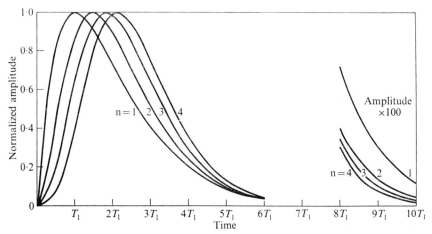

FIG. 6.20. Comparison of pulse shapes using multiple integrating time-constants.

mation, it can be shown that the value of the maximum amplitude is $1[\sqrt{(2\pi n)}\{1 + (1/12n)\}]$ and the pulse duration, defined as before by $\dfrac{\text{area}}{\text{height}}$, is $\sqrt{(2\pi n)}T_1\{1 + (1/12n)\}$. Note, for $n = 1$ this is very closely equal to eT_1. To get a better comparison of the pulse shapes with an increasing number of integrators, it is convenient in each case to use time-constant values of $[T_1\{1 + (1/12)\}]/[\sqrt{(n)}\{1 + (1/12n)\}]$. This simply makes the pulse width defined by $\dfrac{\text{area}}{\text{height}}$ the same in all cases, and so allows a closer examination and comparison of the pulses over the amplitude range 1·0 per cent down to 0·01 per cent. Subject to the above condition, it is also true that the percentage undershoot due to a secondary time constant T_S is, to a first order, not affected by the number of integrators.

This pulse-shape comparison is shown in Fig. 6.20, for $n = 1, 2, 3$, and 4. Each pulse amplitude has been normalized to unity and the curves show clearly the tendency for the pulse shape to become more symmetrical with increasing n, theoretically approaching a Gaussian for $n = \infty$. Otherwise there is no very significant difference in pulse shape until the amplitude drops below about 4 per cent. After a time of approximately $6T_1$ the curves converge and their behaviour thereafter is shown at the right-hand side of the diagram with amplitudes increased 100 times. The significant feature here is that after a time of $8T_1$, when the $n = 1$ pulse has fallen to 0·7 per cent of its maximum amplitude, the corresponding figures are 0·4 per cent for $n = 2$, 0·35 per cent for $n = 3$, and 0·3 per cent for $n = 4$. This amplitude difference is much greater at time $10T_1$, showing that the multiple integrator pulses are approaching the base line much faster than the equal time-constant pulse. Note that over the amplitude range of interest, the improvement is greatest in going from $n = 1$ to $n = 2$ with less to be gained, in relation to the practical complexity, by going above $n = 2$. There is, therefore, a case for having a second integrator in any general purpose amplifier and, because of the practical simplicity, this is to be recommended. This also results in a slight improvement in the signal-to-noise ratio (see Chapter 7).

6.5.2. *Bipolar pulse shaping*

The objective here is to use a pulse-shaping network in the amplifier that generates a positive pulse followed immediately by a large amplitude but short duration undershoot. The ideal arrangement, which can only be approximated in practice, is one where the positive and negative lobes are of equal amplitude and of identical shape. A rectangular pulse with these properties, and having the area of each lobe equal to T_1, is shown in Fig. 6.21. If now this pulse is passed through a secondary differentiating time constant T_S where T_S is 1000 times the duration of the positive lobe, then the long time-constant overshoot produced after the negative lobe has an amplitude of $(0·001)^2 = (T_1/T_S)^2$. This is also shown in Fig. 6.21. The overshoot amplitude varies as $1/T_S^2$ and so lengthening this time constant will, at high counting rates, progressively reduce the d.c. shift of the base line and also the superimposed r.m.s. fluctuation. In fact, since the overshoot amplitude is now the square of its previous undershoot value with unipolar pulse shaping, then with a secondary time constant of 1 ms the d.c. shift and r.m.s. dispersion are already 1000 times less than before and can be neglected in this ideal case.

In a practical situation, where two similar differentiating time constants and one or more integrating time constants may be used to generate a bipolar pulse, the positive and negative lobe symmetry is not easily obtained, but the positive lobe area above the base line is always equal to the negative lobe area below, and the resulting overshoot amplitude due to a secondary time constant T_S still varies as $1/T_S^2$. To allow for the range of switched time constants included in the amplifier, it is customary and adequate to make T_S at least 10 ms and to make all other time constants some ten times greater. Any short time constants which for practical reasons cannot be made as large as 10 ms can be nullified, as before, by the use of pole zero cancellation. This then leads to an amplifier in which both the d.c. shift of the base line and the superimposed r.m.s. fluctuation can be neglected until the count rate becomes sufficiently high for a significant proportion of coincidences to occur during the positive and negative lobes of the primary pulse. When this happens, spectral disturbances reappear in the form of sum and difference peaks, each with areas equal to the probability of coincidences occurring during the respective lobes. With a symmetrical pulse like that in Fig. 6.21 the difference peak will occur at zero amplitude.

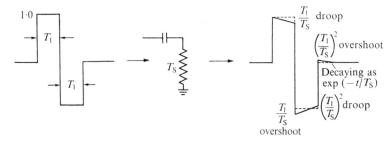

FIG. 6.21. Overshoot of bipolar pulse due to secondary differentiating time-constant.

Delay-line pulse shaping lends itself to the production of a nearly symmetrical bipolar pulse, and it was in this form that bipolar pulse shaping first appeared (Fairstein 1956). This is done by using in cascade two identical delay-line shapers like those shown in Fig. 6.17(a) or (b). The positioning of these delay-line shapers, or for that matter identical time-constant differentiators in the amplifier, can provide some other advantages. One must of course be placed early in the amplifier chain to prevent over-loading in the stages just preceding it, but the other is not limited in this way and can be placed almost at the main amplifier output, where, because of its high-pass frequency characteristic, it will attenuate any low-frequency interference or hum picked up in the main amplifier. A further advantage lies in the constancy of the time, after the nuclear event in the detector, at which the pulse waveform crosses the base line between the positive and negative lobes. This zero-cross-over time is independent of pulse amplitude, and is thus valuable in experiments where accurate timing measurements are required (Fairstein 1956).

NUCLEAR PULSE AMPLIFIERS

Bipolar pulse shaping using delay networks suffers from the practical disadvantage of steps and rings in the waveform due to mismatches. This is more serious than before, as two networks now have to be used. This has led to a study of bipolar pulse generation using two differentiating time constants and n integrating time constants, all having the same time-constant value. The pulse leaving such a shaping network, for a step-function input, is defined by

$$\frac{(n+1)T_1 - t}{(n+1)! T_1^{n+1}} t^n \exp(-t/T_1)$$

(Fairstein and Hahn 1965). This may be rewritten as

$$\frac{t^n}{n! T_1^n} \exp(-t/T_1) - \frac{t^{n+1}}{(n+1)! T_1^{n+1}} \exp(-t/T_1)$$

showing that the output pulse is the difference between a singly differentiated pulse with n integrations and a singly differentiated pulse with $(n+1)$ integrations. The maximum value of the positive lobe occurs at a time $\{n+1 - \sqrt{(n+1)}\}T_1$, the zero cross-over at $(n+1)T_1$ and the maximum amplitude of the negative lobe at $\{(n+1+\sqrt{(n+1)})\}T_1$. The pulse width may be regarded as the width of the positive lobe $\left(=\dfrac{\text{area}}{\text{height}}\right)$ plus the width of the negative lobe, but this is not expressible in a simple analytical form. A definition of pulse width, as twice the geometric mean of the previous two values, is much simpler analytically and is equal to $2T_1\sqrt{\{(n+1)^{n+1}/n^n\}}$. An important parameter is the ratio between the negative and positive lobe amplitudes, since this is indicative of the bipolar symmetry of the pulse. Good symmetry is synonymous with a rapidly decaying pulse tail. This amplitude ratio is defined by

$$\left\{\frac{n+1+\sqrt{(n+1)}}{n+1-\sqrt{(n+1)}}\right\}^n \exp\{-2(n+1)^{1/2}\}$$

and its variation with n is illustrated in Table 6.1. Many integrators are needed for good symmetry and like for the unipolar case a practical compromise is struck with $n = 2$.

Serious overloading (amplitude limiting) of an asymmetric bipolar pulse will result in a marked unbalance of the areas above and below the base line and will give rise to a base-line displacement that will recover with a time constant equal to the subsequent coupling time constant. This is to be avoided, and leads therefore to the second differentiator being placed right at the amplifier output, or sufficiently near to the output to ensure that no further overloading can occur.

116

TABLE 6.1

Bipolar pulse shaping

2 Differentiators No. of integrators (*n*)	*n* Integrators $\dfrac{-\text{ve lobe amplitude}}{+\text{ve lobe amplitude}}$
1	0·343
2	0·438
4	0·54
8	0·64
16	0·725
∞	1·0

Bipolar pulse shaping is thus a way of minimizing many of the undershoot effects associated with unipolar pulses in an a.c. coupled system, and is to be recommended. The chief disadvantage is a poor signal-to-noise ratio, and this is discussed in Chapter 7. In a general purpose nuclear pulse amplifier a good compromise is to have the usual differentiating and integrating time constants supplemented by an optional second differentiator and a second integrator. In this way, for a minimum of practical complexity, the system is then able to operate in both the unipolar and bipolar modes.

6.5.3. *D.c. coupling and pulse processing*

Many of the problems associated with unipolar and bipolar pulse shaping stem from the need to return the pulse to the base line as quickly as possible after the collection of the liberated charge in the detector. Circuit elements to do this usually result in an exponentially decaying pulse tail that reduces the counting rate at which overlap problems become serious. In addition, because a high-gain amplifier is required, a.c. coupling between stages is needed to avoid d.c. drift of the base line. These same a.c. couplings then produce base-line disturbances that vary with counting rate as discussed in the previous two sections. A novel way of operating a detector, amplifier, and pulse-amplitude analyser combination has recently been proposed, and from tests it would appear that potentially this system can remove many of the foregoing problems and minimize others (Kandiah 1966).

If the polarizing resistor effectively shunting the detector capacitance is made very large, ideally infinite, and the subsequent amplifier is d.c. coupled, then the waveform at the amplifier output will have a step function or staircase form as shown in Fig. 6.22. Each step corresponds to the

117

collection of charge in the detector, and the pulse-amplitude analyser is now required to measure the height of each step, irrespective of its mean potential at the amplifier output. Assuming for the moment that circuits can be devised to do this, there are other important requirements associated with this mode of operation. For example, in the absence of a randomly occurring pulse signal from the detector, the mean potential at the amplifier output may drift in either direction depending on the relative magnitudes of, for example, low-energy background radiation, detector leakage, and amplifier drift. This is permitted over a limited range, but movement of the mean potential, above or below one of two boundary levels, operates a discriminator that is arranged to inject a pulse of charge of the appropriate sign into the detector to restore the mean potential of the output approximately to the middle of the permitted range. With the use of very high-speed

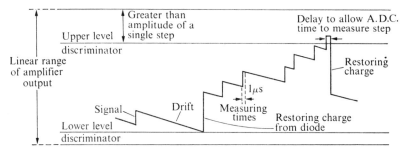

FIG. 6.22. Output of d.c. coupled amplifier with charge restoration.

circuits, this can be done extremely quickly, and in the system described, each discriminator flashes a gallium arsenide lamp which in turn energizes a silicon photodiode connected to the detector. A block schematic of the system is shown in Fig. 6.23.

The pulse shaper after the amplifier is to generate from the staircase waveform a conventional pulse purely for recognition purposes, and the event recognition unit then notifies the pulse-amplitude analyser, via the logic unit, that this is a pulse requiring measurement. The pulse-amplitude analyser used here is an analogue to digital converter preceded by an analogue delay circuit in order that, after receiving a signal from the event recognition unit, it has time to measure the amplitude of the staircase waveform relative to some datum level, just before the step and again just after. The pulse amplitude is then the difference between these two digital readings. Should the amplifier output waveform exceed the upper discriminator level due to the arrival of a genuine pulse, then a short delay is introduced before flashing the gallium arsenide lamp, to allow the analogue-to-digital converter time to measure this pulse. Also, the upper discriminator level must be low enough for the pulse step to remain within the linear range of the amplifier output.

The potential advantages of this mode of operation at high counting rates are thus freedom from spectral distortion and shifts arising from pulse undershoots, instantaneous recovery following very large overloads, and operation at higher counting rates, before primary pulse overlap becomes serious. There may well be other advantages that will appear as the system is further developed. One obvious disadvantage is, of course, the very high

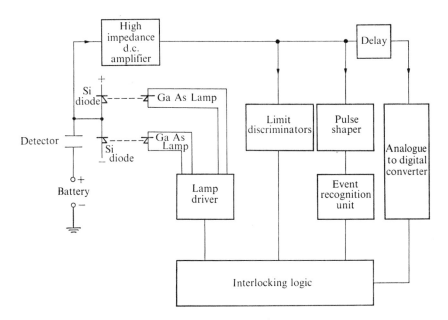

FIG. 6.23. Block schematic of d.c. coupled system.

degree of linearity necessary over the amplifier dynamic range. Two identical steps must be measured as such, although one may occur at the amplifier output near to the lower discriminator level, while the other occurs near to the upper discriminator level. Non-linearity will give rise to an amplitude difference between the steps, and this in turn will produce additional spectral-line broadening. The linearity requirements are here considerably more stringent than are necessary for a conventional type of amplifier.

The effective bandwidth of the system, as far as pulse amplitude measurement is concerned, is limited at the high-frequency end by an integrating time constant placed just before the analogue-to-digital converter, and at the low-frequency end by accurately defining the time between the two measurements. If this latter time is 1 μs, then the signal-to-noise ratio of

119

this system is identical to that of a conventional system using a single 1 μs delay-line shaper with a single time-constant integrator. (See Chapter 7.)

A later development of the d.c. coupled amplifier system uses pulse shaping by switched time constants and gated integration (see Chapter 7, section 7.8) and restores the detector charge after each pulse has been measured. In addition to offering the above advantages at high pulse rates, this system also comes very close to the theoretical optimum in signal-to-noise ratio and resolving time (Kandiah *et al.* 1968).

6.6. Principles of amplifying unit design

6.6.1. *Negative feedback*

The need for good stability of gain and linearity between the input and output signals of a nuclear pulse amplifier has been stressed several times in the preceding sections. Both of these performance criteria are critically

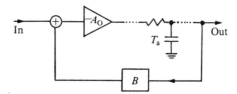

FIG. 6.24. Single-lag feedback amplifier.

dependent on the operational constancy of the parameters of the active elements being used. For example, the mutual conductance g_m of a thermionic valve can vary significantly with the current through the valve and also between valves of the same type. Similarly, the base-to-collector current gain β of a transistor varies with current, appreciably with ambient temperature, and very markedly between transistors of the same type. Stabilization of the power supplies to the active elements and of their steady operating currents is a first step to counter some of these effects, and this is usually done, but to minimize other parameter variations caused by temperature changes or by replacing the active elements themselves, requires the application of negative feedback. In this section it is proposed to outline, without extensive proof, some of the advantages and characteristics of negative feedback as applied to high-performance pulse amplifiers.

6.6.1.1. *Single-lag circuit.* Consider the simple feedback circuit shown schematically in Fig. 6.24. The low-frequency voltage gain of the amplifier is $-A_0$, the negative sign implying a phase reversal between input and output. At higher frequencies, attenuation and phase shift in the amplifier are controlled by a single-lag circuit of time constant T_a, so the bandwidth of the amplifier without feedback for a 3 dB drop in gain is $f = 1/(2\pi T_a)$. A fraction B of the output voltage is then returned to the input circuit in such a way as to oppose the normal input voltage—this is known as negative feedback. The fraction B fed back is usually defined by a passive circuit comprising very stable components. B may also be frequency dependent, its low-frequency value being B_0. If G is the gain of the amplifier with feedback then

$$G = \frac{\dfrac{A_0}{1 + j\omega T_a}}{1 + \dfrac{A_0 B}{1 + j\omega T_a}} \qquad (6.2)$$

and the low-frequency gain G_0 is

$$G_0 = \frac{A_0}{1 + A_0 B_0}. \qquad (6.3)$$

If $A_0 B_0 \gg 1$ then G_0 is very closely equal to $1/B_0$. Thus the gain is primarily dependent on the fraction B_0 defined by the feedback network and much less dependent on the amplifier gain A_0. The exact dependence of G_0 on A_0 is obtained by differentiating eqn (6.3). This gives

$$\frac{dG_0/G_0}{dA_0/A_0} = \frac{1}{1 + A_0 B_0}. \qquad (6.4)$$

The gain stabilization factor is $1 + A_0 B_0$ and this is known as the feedback factor F. The non-linearity of the circuit is also improved by the same amount. From eqn (6.3) $F = A_0/G_0$, so the feedback is also instrumental in reducing the low-frequency gain of the amplifier F times.

From the complex gain expression given in eqn (6.2) it can be shown, if $B = B_0$ (i.e. the feedback network is not frequency dependent), that

$$G = \frac{G_0}{1 + j\omega T_a \left(\dfrac{G_0}{A_0}\right)} \qquad (6.5)$$

and so the lag-circuit time constant T_a, controlling the bandwidth of the amplifier without feedback, is also reduced F times, and consequently the new bandwidth will be F times greater.

The single-lag feedback amplifier may, therefore, be summarized as follows:

(a) The low-frequency gain is reduced F times.
(b) The lag-circuit time constant is reduced F times and the bandwidth increased by the same amount.
(c) The imperfections in gain stability and linearity resulting from variations in A_0 are both improved F times.

6.6.1.2. *Two-lag circuit.* From the foregoing simple analysis, high feed-back factors are clearly desirable. The feedback factor F may be increased by increasing A_0 but this usually demands additional active elements in the amplifier. Each element will have associated with it at least one phase lag, and too many phase lags are dangerous because of the possibility of the cumulative phase shift reaching $180°$ and the overall feedback then becoming positive. If this occurs at a frequency where the loop gain AB has a magnitude greater than 1, then the amplifier will become an oscillator. Two phase lags are quite safe, because the loop gain is zero by the time the phase shift reaches $180°$. Circuits containing three lags can, under specified conditions, be prevented from oscillating, but from a practical point of view, the two-lag circuit forms the basis of many amplifying unit designs and its properties are now described.

The schematic circuit of an amplifier with two identical lags is shown in Fig. 6.25. The 3 dB bandwidth of this amplifier before applying feedback

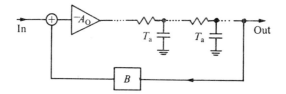

In   Out

Fig. 6.25. Two-lag feedback amplifier.

is $0.64/2\pi T_a$ and with feedback its gain G is given by

$$G = \frac{\dfrac{A_0}{(1+j\omega T_a)^2}}{1 + \dfrac{A_0 B}{(1+j\omega T_a)^2}}. \tag{6.6}$$

At low frequencies this becomes $G_0 = A_0/(1 + A_0 B_0)$ and the feedback factor F, is, as before, $1 + A_0 B_0 = A_0/G_0$. However, for a step-function input this circuit has an output pulse response that contains a superimposed damped oscillation or ring, as shown in Fig. 6.26, and is unsuitable in this form as a pulse amplifier. The circuit requires compensation to modify the pulse response to an optimum, wherein the output rises as rapidly as possible to its steady state without exhibiting rings or overshoots. One way of providing this compensation is to make $B = B_0(1 + j\omega T_F)$ and if a simple resistance potentiometer is used to define B_0, the above relationship is approximately obtained by shunting the upper resistor R_F with a small condenser C_F to give a time-constant value T_F. In practice this also introduces a third phase lag, but for small B_0 this extra phase lag is also small and can usually be neglected in comparison with the two dominant lags. The value of the time constant T_F for exact compensation is

$$T_F = \frac{2T_a}{1 + \sqrt{F}} \tag{6.7}$$

and when the circuit is properly compensated, the complex gain G becomes

$$G = \frac{G_0}{\left\{ 1 + j\omega T_a \sqrt{\left(\frac{G_0}{A_0}\right)} \right\}^2}. \tag{6.8}$$

Eqn (6.8) shows that in a feedback amplifier with two equal lags, when properly compensated the feedback is instrumental in reducing each lag time constant $\sqrt{F}$ times, and so the bandwidth will increase by the same

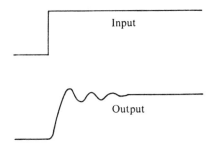

FIG. 6.26. Transient response of two-lag feedback amplifier without compensation.

amount. The low-frequency gain, however, is reduced F times, as was the case for the single-lag circuit.

Another way of realizing proper transient compensation is to stagger the two lag circuit time-constant values. If these are pT_a and T_a/p then the optimum is reached when $p^2 \sim 4F$ and the complex gain is again given by eqn (6.8).

In designing or assessing the gain and bandwidth performance of a feedback amplifier the following rules are helpful.

1. The low-frequency gain is very closely $1/B_0$ and is fixed by the components of the feedback network.

2.(a) If two significant phase lags are present and their time-constant ratio does not exceed $4F$ then some transient response compensation will be necessary on the feedback network. The value of T_F needed will depend on the ratio of the lag circuit time constants and will lie between zero and the value given in eqn (6.7).

(b) The bandwidth is estimated by first working out the geometric mean of the two lag circuit time constants and then dividing this

123

value by $\sqrt{F}$. The final bandwidth will be that of an amplifier having two cascaded lag circuits with time constants equal to this reduced value.

3.(a) Should the two lag circuit time constants have a ratio exceeding $4F$ then no further compensation is needed on the feedback network, i.e. $T_F = 0$.

(b) The bandwidth is then estimated by dividing the larger time constant by F and working out the bandwidth defined by a single-lag circuit with a time constant of this reduced value.

6.6.1.3. Some practical aspects of feedback application. A key assumption in the foregoing sections on feedback theory is that the signal taken from the amplifier output can be returned to the input in such a way as to oppose the normal input signal signal without significantly affecting the parameters of the unfedback amplifier. This is rarely possible and the practical effects

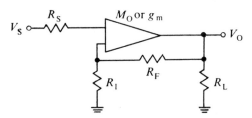

Fig. 6.27. Amplifier schematic—feedback returned to earthy input terminal.

of such complications are now described in some detail for two commonly used feedback configurations.

The first circuit configuration is shown in Fig. 6.27. Here the feedback signal has to be returned to the normally earthy terminal of the input stage (e.g. cathode of a valve or emitter of a transistor) and requires the addition of an appropriate load resistor R_1 in that part of the circuit. R_S is the source resistance (which may be augmented by externally added resistance), R_L is the load on the amplifier output, and R_F is the feedback resistor. The amplifier, before applying feedback (i.e. before adding R_1) either has a voltage gain M_0 with a low output impedance, or has a voltage to current conversion factor g_m with a very high output impedance. It is required to calculate the overall gain with feedback, i.e. V_0/V_S.

In the former case (voltage gain M_0) R_1 in parallel with R_F will apply a small amount of local feedback to the first stage, so in calculating the voltage gain A_0 of the complete amplifier without feedback this factor must be taken into account, i.e. $A_0 = M_0/$local feedback factor. The fraction of the output signal returned to the input is as one might expect $B_0 = R_1/(R_1 + R_F)$.

In the second case (mutual conductance g_m) the value of M_0 is equal to g_m multiplied by R_L in parallel with $R_1 + R_F$, and M_0 is again reduced by the local feedback on the first stage to give the effective value for A_0. Here the local feedback is due to R_1 in parallel with $R_F + R_L$. As before $B_0 = R_1/(R_1 + R_F)$.

In both cases, as R_F and R_L are usually large compared with R_1 it is sufficiently accurate in calculation to take the local feedback on the first stage as due to R_1 alone. Similarly in the second case it is usually adequate to take $M_0 = g_m \times (R_F R_L)/(R_F + R_L)$. These points will be illustrated in later examples.

The second circuit configuration is shown in Fig. 6.28. Here the feedback signal has to be returned to the same input terminal as the normal input signal. R_S, R_F, and R_L are all the same as before and R_1 can here be taken

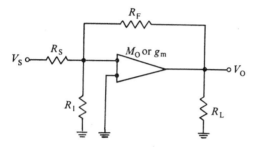

FIG. 6.28. Amplifier schematic—feedback returned to signal terminal.

as representing some physical impedance across the amplifier input which practically may not be avoidable. As before it is required to calculate the overall gain with feedback, i.e. V_0/V_S.

There are at least two ways of looking at this circuit (both of course giving the same answers). The one preferred by the author calculates the unfedback gain A_0 as a passive attenuation factor stemming from a resistance network comprising R_S as the upper resistor and R_F in parallel with R_1 as the lower resistor followed by the active gain M_0, i.e.

$$A_0 = M_0 \times \frac{\dfrac{R_F R_1}{R_F + R_1}}{R_S + \dfrac{R_F R_1}{R_F + R_1}}.$$

The fraction B_0 of the output signal returned to the input then has to be R_S/R_F. This is not obvious and must be accepted in this analysis. It is, however, compatible with the well-known fact that the gain V_0/V_S of this circuit is very closely R_F/R_S ($=1/B_0$).

10

When the amplifier has a high output impedance and its gain is specified by a mutual conductance g_m, then the resistance attenuator now consists of R_S as the upper resistor and R_1 in parallel with $R_F + R_L$ as the lower resistor. Also M_0 is calculated as g_m multiplied by R_F and R_L in parallel. Finally, the fraction of the output signal returned to the input is again given by $B_0 = R_S/R_F$. These points will also be further illustrated in later examples.

6.6.2. *Thermionic valve amplifiers*

Most present-day nuclear pulse amplifiers use transistors as the active circuit elements. However, there are still many of the older valve-type amplifiers in use, and it is perhaps worth while at this stage to describe

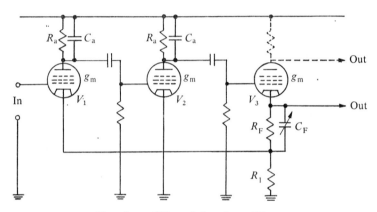

FIG. 6.29. 'Ring of three' amplifier.

briefly the feedback circuit that formed the basic building brick of so many of these early designs.

This circuit is known as the 'ring of three' and is shown in Fig. 6.29. It consists of two gain-producing stages, V_1 and V_2, the low-frequency voltage gain of each being $g_m R_a$ and each stage having a significant phase lag due to the time constant T_a formed by the anode load R_a and the stray capacitance C_a. The third valve, V_3, is a cathode follower, included primarily to isolate the feedback network $C_F R_F R_1$ from the anode of the second valve. The feedback signal is taken from the junction of R_F and R_1 and returned to the cathode of V_1, this being the circuit configuration that is necessary to make the feedback negative. As R_1 is then included in the cathode circuit of V_1 it will introduce local current feedback in this valve, effectively reducing its voltage gain by a factor $(1 + g_{mc} R_1)$ where g_{mc} is the mutual conductance in terms of the cathode current. The gain without feedback, A_0, and the feedback factor F will be similarly reduced, so there is a good case for making R_1 as small as possible. This in turn limits the

126

magnitude of the network impedance $R_F + R_1$ for a given gain with feedback, and is the prime reason why the feedback network must be isolated from the anode of V_2 to obtain maximum performance. The simplest way of securing this isolation is to introduce V_3 as a cathode follower. Because V_3 is functioning as a cathode follower, only a very small additional phase lag is introduced into the loop between its grid and cathode, and this can usually be neglected in comparison with the dominant lags due to the anode-circuit time constants of V_1 and V_2. Again, if V_3 is connected as a triode and has a load in the anode circuit of magnitude $R_F + R_1$, then a second output signal can be obtained from the anode, equal in magnitude to the signal on the cathode but of the opposite sign. This is a further advantage that comes from the inclusion of V_3.

Typical values for this circuit might be $g_m = 7\cdot5$ mA/V, $g_{mc} = 10$ mA/V, $R_a = 10$ kΩ, $C_a = 20$ pF, $R_F = 10$ kΩ, and $R_1 = 100\ \Omega$. The low-frequency voltage gain with feedback is then closely $1/B_0 = \{(R_F + R_1)/R_1\} \sim 100$. The low-frequency gain of V_1 is $\{g_m R_a/(1 + g_{mc} R_1)\} = 37\cdot5$, that of V_2 is 75 and that of V_3 approximately unity; thus $A_0 = 37\cdot5 \times 75$. The fraction fed back is $B_0 \sim 1/100$ so the overall feedback factor is $1 + (37\cdot5 \times 75)/100 = 29$. The value of T_F needed for critical compensation is $2T_a/\{1 + \sqrt{(29)}\} = 0\cdot062$ μs and for $R_F = 10$ kΩ this gives $C_F = 6\cdot2$ pF. Finally, the overall bandwidth is that due to two cascaded lag circuits, each having a time constant of $0\cdot2/\sqrt{(29)} = 0\cdot037$ μs and this corresponds to a 3 dB drop in amplitude at $2\cdot7$ MHz.

6.6.3. *Transistor amplifiers*

The design of transistor feedback amplifiers is complicated by the fact that several of the transistor parameters may vary both with temperature and between transistors of the same type. The two most important are the base-to-collector current gain β and the emitter resistance r_E. The current gain β may change markedly from one transistor to another, a maximum variation of ±50 per cent not being unreasonable. β is also temperature dependent, a representative figure being ±5 per cent for a $\pm5°$C temperature change. The emitter resistance r_E depends on both the emitter current and the temperature. If the steady operating current of the transistor has been stabilized, as is usual, then the principal variation in r_E is with temperature, and a typical figure here is $\pm1\cdot7$ per cent for a $\pm5°$C change in temperature.

Unfortunately, in the analysis of transistor feedback amplifiers, the feedback factor that is instrumental in minimizing the effects of changes in the current gain β is not the same as the feedback factor that operates in favour of changes in r_E. On the other hand, the change in gain resulting

127

from the temperature dependence of β is in the opposite direction to the change in gain resulting from the temperature dependence of r_E, so a measure of compensation is to be expected. In fact, by judicious choice of the circuit components, it is sometimes possible to balance one temperature effect completely by the other. If the transistor is physically replaced, however, then large changes in β are to be expected, whereas changes in r_E will remain small, particularly if the steady operating current through the transistor has been stabilized. For these reasons, the feedback factors calculated in the following paragraphs are those effective in stabilizing the feedback gain of the amplifier against changes in β. It is assumed that the effects of changes in r_E are likely to be less serious. This is frequently but not always the case.

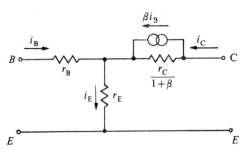

FIG. 6.30. Common-emitter low-frequency T equivalent circuit.

A transistor is basically a current-operated device, having a current gain α equal to the ratio of collector-to-emitter current, and a current gain β equal to the ratio of collector-to-base current. α is always slightly less than 1, and β, which is equal to $\alpha/(1-\alpha)$, usually lies in the range 20–200. Nuclear pulse amplifiers are, however, essentially voltage amplifiers and it is the voltage gain that usually has to be calculated and stabilized. In the following calculations, therefore, the emphasis is on evaluating the effective voltage gain of the circuit and the stabilization of this gain produced by the application of negative feedback.

The low-frequency equivalent T circuit of a transistor operating in the grounded emitter mode is shown in Fig. 6.30. If the collector load resistor R_L is small compared with the collector output impedance $r_c/(1+\beta)$ then the equivalent circuit can be simplified to the form shown in Fig. 6.31. In most of the following calculations this assumption will be valid. The circuit in Fig. 6.31 also shows an input signal V_S coming from a voltage source of impedance R_S; r_B is the ohmic base resistance and r_E is the emitter resistance which is equal to $(1/40I_E)$ at $300°K$ where I_E is the emitter current.

Analysis of this circuit gives the overall voltage gain as

$$V_{OC}/V_S = \frac{\dfrac{\beta R_L}{R_S + r_B + r_E}}{1 + \dfrac{\beta r_E}{R_S + r_B + r_E}} \tag{6.9}$$

and the input impedance at the base terminal as

$$Z_{IB} = r_B + (1 + \beta)r_E. \tag{6.10}$$

Eqn (6.9) is in the form $A_0/(1 + A_0 B_0)$, showing that the single-transistor circuit already has a small amount of built in negative feedback due to coupling between the output circuit (collector) and the input circuit (base) through the common resistance r_E. For normal circuit values the feedback factor is quite small, usually between 1 and 2, but it does increase the

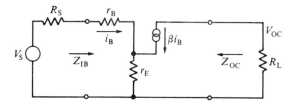

FIG. 6.31. Equivalent circuit of common emitter stage.

collector output impedance by the same factor and so helps to make the earlier assumption that R_L is small compared with the collector output impedance, more valid.

At higher frequencies the collector load R_L, shunted by stray capacitance C_L, will introduce an attenuation and phase shift in the output voltage, V_{OC}, just like that occurring in the anode circuit of a thermionic valve. A further property of the transistor is that its current gain α is frequency dependent, and this is usually described in terms of the cut-off frequency f_α, where the current gain has fallen by 3 dB from its low-frequency value. Phase shift also occurs as α falls, and although this may appear initially as if it were being caused by a single-lag circuit, it is much more complex fundamentally, and for low values of α it can significantly exceed 90°. If the transistor is being used to provide a base to collector current gain β, then β like α is frequency dependent, but starts falling at a much lower frequency; in fact it is 3 dB down at a frequency f_α/β. Here again the associated phase shift appears initially as if it were due to a single-lag circuit of time constant $\beta/2\pi f_\alpha$ but it too can exceed 90° for low values of β. This can be a complication in feedback amplifier design that cannot be allowed for easily.

129

The design technique is to assume that the β phase shift is in fact due to a single-lag circuit and to arrange that the loop gain will have fallen well below unity before this extra phase shift becomes serious. Thus in considering the behaviour of a transistor feedback amplifier it may be necessary to take into account two phase lags associated with each transistor, viz. that due to the load circuit time constant $C_L R_L$ and that due to the β cut-off frequency time constant $\beta/(2\pi f_\alpha)$.

The simplest feedback amplifiers are those involving a single transistor, and the properties of two common configurations will now be described. These circuits are important in their own right, and also as building bricks in multistage feedback amplifiers.

6.6.3.1. *Series feedback stage.* The low-frequency equivalent circuit is shown in Fig. 6.32.

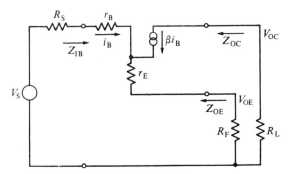

Fig. 6.32. Equivalent circuit of series feedback stage.

The input impedance at the base terminal is

$$Z_{IB} = r_B + (1 + \beta)(r_E + R_F). \tag{6.11}$$

Outputs can be taken from either the collector or emitter terminals. In the latter case the circuit is known as an emitter-follower.

The voltage gain to the collector is

$$V_{OC}/V_S = \dfrac{\dfrac{\beta R_L}{R_S + r_B + r_E + R_F}}{1 + \dfrac{\beta(r_E + R_F)}{R_S + r_B + r_E + R_F}} \tag{6.12}$$

$$\sim \dfrac{R_L}{r_E + R_F} \tag{6.13}$$

if the denominator of eqn (6.12) is $\geqslant 1$.

The collector output impedance Z_{OC} is this time very large compared with R_L due to the large feedback factor.

The voltage gain to the emitter is

$$V_{OE}/V_S = \frac{\dfrac{(1+\beta) R_F}{R_S + r_B}}{1 + \dfrac{(1+\beta)(r_E + R_F)}{R_S + r_B}} \tag{6.14}$$

$$\sim \frac{R_F}{r_E + R_F} \tag{6.15}$$

if the denominator of eqn (6.14) is $\gg 1$.

The emitter output impedance Z_{OE} is

$$Z_{OE} = r_E + \frac{R_S + r_B}{1 + \beta}. \tag{6.16}$$

Eqn (6.12) is again of the form $A_0/(1 + A_0 B_0)$ and the denominator can therefore be identified as the feedback factor F, which is instrumental in stabilizing variations in the output voltage V_{OC} against variations in the current gain β. Note that the feedback factor F diminishes as R_S increases, so in this circuit the feedback becomes non-existent for an infinite source impedance and the latter should therefore be kept as low as possible. A further deduction from eqn (6.12) is that the phase lag time constant associated with β is reduced F times by the feedback, while that due to stray capacitance across R_L is not, since R_L appears only in the numerator. This is of course consistent with very large Z_{OC}.

Considering now an output taken from the emitter, the denominator of eqn (6.14) is the feedback factor that is effective in stabilizing the gain against variation in the factor $(1 + \beta)$. Since $\beta \gg 1$, little error is involved if this is also taken as the feedback factor that is effective against changes in β alone. (The exact feedback factor is in fact $\dfrac{1 + \beta}{\beta}$ times greater.) As before, the phase lag time constant associated with β is also reduced by this same factor. Finally, any extra phase lag introduced by stray capacitance on the emitter output terminal will be small because of the very low output impedance at this point—see eqn (6.16).

6.6.3.2. *Shunt feedback stage.* The analysis of this circuit, Fig. 6.33, is considerably more complex than those for the two previous single-transistor circuits. The following approximate formulae are, however, of value in practical calculations.

The input impedance at the base terminal is

$$Z_{IB} \sim \left(r_E + \frac{r_B}{1 + \beta}\right)\left(\frac{R_F + R_L}{R_L}\right). \tag{6.17}$$

The output impedance at the collector terminal is

$$Z_{OC} \sim r_E + \frac{R_F}{1 + \dfrac{\beta R_S}{R_S + r_B + (1 + \beta)r_E}}. \tag{6.18}$$

The overall voltage gain is

$$\frac{V_{OC}}{V_S} \sim \frac{\beta R_F R_L}{(R_F + R_L)(R_S + r_B + r_E)} + 1 + \frac{\beta R_S R_L}{(R_F + R_L)(R_S + r_B + r_E)}. \tag{6.19}$$

$$\sim \frac{R_F}{R_S} \tag{6.20}$$

if the denominator of eqn (6.19) is $\gg 1$.

It should be noted that both the input and output impedances of this circuit are low. The denominator of eqn (6.19) is the feedback factor F

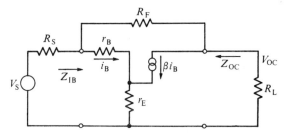

FIG. 6.33. Equivalent circuit of shunt feedback stage.

which is effective in minimizing changes in the voltage gain caused by changes in β. As before, the phase lag time constant associated with β will also be reduced F times, but any phase lag due to capacitance across R_F will not be affected by the feedback. Finally, stray capacitance on the output terminal will introduce only small additional phase lag because of the low output impedance at this point—see eqn (6.18).

6.6.3.3. *Multi-stage amplifiers.* In the following paragraphs the design data given in sections 6.6.1 and 6.6.3 are used to analyse the behaviour of three multi-stage feedback loops that have formed the building bricks of a number of nuclear pulse amplifiers.

The first circuit is shown in Fig. 6.34. Only the signal connections have been included; extra bias components have been deliberately left out. The circuit consists of two grounded emitter stages, J_1 and J_2, connected in cascade, with overall feedback returned from the collector of J_2 to the emitter of J_1. The important component values are marked on the circuit and it is assumed that both transistors have $\beta = 50$, $r_B = 50\ \Omega$, $f_\alpha = 100$

MHz and operate with 5 mA emitter current. The driving voltage source impedance is also assumed to be zero.

The gain of the first stage, with 250 Ω in the emitter and a collector load of 5 kΩ, is from eqn (6.12), equal to 19. The denominator of this equation also gives the local feedback factor of this stage as 43. This gain of 19 does not take account of the loading due to the second stage. This, however, is automatically allowed for if the gain of $J2$ is calculated for a source impedance of 5 kΩ. Using eqn (6.9), and taking the collector load of $J2$ as 2·5 kΩ (two 5 kΩ resistors in parallel) this works out to 23·6 with a local feedback factor of 1·05. Hence $A_0 = 19 \times 23\cdot6$, $B_0 = \frac{250}{5250}$ and this gives the overall gain with feedback as 20 and the overall feedback factor as 22·5.

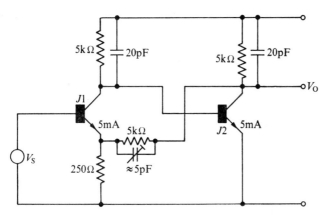

FIG. 6.34. Two-stage amplifier with feedback to input emitter.

In assessing the bandwidth of the amplifier, the various phase lags in the loop must first be quantified. The phase-lag time constant associated with the β of each transistor is $\beta/(2\pi f_\alpha) \sim 0\cdot08$ μs. For the first transistor this is reduced by the local feedback factor of 43 to approximately 0·002 μs. The next phase lag is that on the collector of $J1$. This is due to the 20 pF stray capacitance in shunt with both the 5 kΩ collector load of $J1$ and the input impedance of $J2$. The latter is 305 Ω from eqn (6.10). The time constant is thus 0·0058 μs. The lag time constant due to the β of $J2$ is only slightly reduced by the local feedback factor of 1·05 to 0·076 μs, and the final lag time constant is that due to the 20 pF strays on the collector of J_2, in shunt with 2·5 kΩ (two 5 kΩ resistors in parallel). This is 0·05 μs. The latter two time constants are taken as the dominant lags. The geometric mean is 0·066 μs and this divided by $\sqrt{(22\cdot5)}$ gives 0·014 μs. The overall bandwidth is then

$$\frac{0\cdot64}{2\pi \times 0\cdot014} \sim 7 \text{ MHz.}$$

Optimum pulse response will require a small compensating capacitor across the feedback resistor, the maximum value of this time constant being

$$\frac{2 \times 0\cdot66}{1 + \sqrt{(22\cdot5)}} = 0\cdot023 \ \mu s,$$

and for a 5 kΩ resistor this corresponds to a capacitance of 4·6 pF. In practice a slightly smaller value will be required as the two dominant lag time constants are not equal.

The second circuit is shown in Fig. 6.35. This consists of a grounded emitter stage J_1 feeding an emitter follower J_2 with feedback taken from

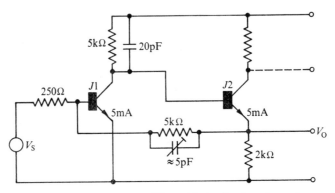

FIG. 6.35. Two-stage amplifier with feedback to input base.

the emitter of this stage and returned to the base of J_1. The input signal is also fed to the base of J_1 through an effective series resistance of 250 Ω, made up of the source resistance and, if necessary, added circuit resistance. The transistor parameters of β, r_B, and f_α have the same values as before, the emitter currents are again 5 mA each and the other main component values are marked on the circuit.

The gain of J_1 is computed from a factor $\frac{5000}{5250}$ due to the passive attenuation resulting from the source resistance and the overall feedback resistance, multiplied by the active gain of J_1 computed from eqn (6.9). The value for the source resistance used in eqn (6.9) is 240 Ω (250 Ω and 5 kΩ in parallel). This gives 440 with a local feedback factor of 1·85. The gain to the emitter of J_2 is calculated from eqn (6.14) taking the source resistance as 5 kΩ and the emitter resistance as 1430 Ω (2 kΩ and 5 kΩ resistors in parallel). This gives 0·93 with a local feedback factor of 15·5. Hence $A_0 = 440 \times 0\cdot93$, $B_0 = \frac{250}{5000}$ (i.e. R_S/R_F) and this gives the overall gain with feedback as 19 and the overall feedback factor as 21·5.

The β lag time constant is as before 0·08 μs, and for the first stage this is reduced by the local feedback factor of 1·85 to 0·043 μs. The lag time

134

constant on J_1 collector is due to the 20 pF strays in parallel with both 5 kΩ and the input resistance of J_2. The latter is calculated as 73 kΩ from eqn (6.11) and this gives a time-constant value of 0·094 μs. Finally the β lag time constant of J_2 is reduced by its local feedback factor of 15·5 to 0·0052 μs so the first two time constants are dominant in this example. The geometric mean is 0·063 μs and this divided by $\sqrt{(21·5)}$ gives 0·014 μs. The overall bandwidth is thus 7 MHz. Optimum pulse response will require a small compensating capacitor across the feedback resistor, the maximum value of the time constant being

$$\frac{2 \times 0·063}{1 + \sqrt{(21·5)}} = 0·023 \ \mu s$$

and for a 5 kΩ resistor this requires a capacitor of 4·6 pF. As before, a smaller value will probably suffice as the two dominant lag time constants are not equal.

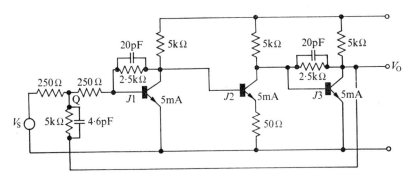

FIG. 6.36. Three-stage amplifier with local and overall feedback.

The third circuit is shown in Fig. 6.36. This consists of three stages, each with local feedback, connected in cascade and included in an overall feedback loop. This configuration has a number of advantages (Cherry and Hooper 1963). First, having local feedback on each stage confers a considerable degree of gain stabilization to the circuit, which can be further raised to a high value by a small additional feedback factor operating around the whole loop. Secondly, by using alternate series and shunt stages, interaction between the stages is reduced to a minimum. Finally, by placing a small capacitor across the feedback resistor of each shunt stage, this produces two dominant lags that are not, to a first order, dependent on the β of the transistors. The first and last advantages make it much easier to obtain a very large effective feedback factor on each transistor (i.e. combined local and overall) without the same danger of oscillation or pulse ringing as might occur in the previous circuits.

135

The component values are shown on the circuit diagram and as before it is assumed that each transistor has $\beta = 50$, $r_B = 50\ \Omega$, $f_\alpha = 100$ MHz and operates at an emitter current of 5 mA. Considering each stage separately, and taking the input to the first stage, before applying overall feedback, to be at point Q, the gain and local feedback factor of this stage are computed from eqn (6.19) and these are 9.7 and 28·4 respectively. The second stage gain and local feedback factor are computed from eqn (6.12) with $R_S \sim 100\ \Omega$ and these are 85 and 14·5 respectively. $R_S \sim 100\ \Omega$ is valid here since this stage is driven from the low output impedance of stage 1 (eqn. (6.18)). Finally, eqn (6.9) is again used to compute the gain and local feedback factor of stage 3 with $R_S = 5\ k\Omega$ and $R_L = 2\cdot5\ k\Omega$ (5 kΩ shunted by the 5 kΩ overall feedback resistor). These are 0·48 and 25·8 respectively.

Considering now the complete amplifier, the total loop gain A_0 is computed from a factor $\frac{240}{490}$ (see below) multiplied by the above three stage gains. This works out to 195. The factor $\frac{240}{490}$ represents the passive attenuation resulting from the source resistance (250 Ω) in conjunction with the overall feedback resistance (5 kΩ) in shunt with the impedance presented by the first stage at Q. The latter is approximately 250 Ω, giving a shunt combination of 240 Ω. B_0 is 1/20 so the gain with feedback is 18·1 and the overall feedback factor is 11. The effective feedback factors in this circuit are therefore 310 for stage 1, 160 for stage 2, and 280 for stage 3.

The bandwidth of the circuit is controlled by the dominant lag time constants of each shunt feedback stage, due to the 20 pF capacitors in parallel with the 2·5 kΩ feedback resistors. This time-constant value is 0·05 μs and reduces to 0·015 μs when divided by $\sqrt{(11)}$. The overall bandwidth is approximately 7 MHz, and the capacitor needed across the 5 kΩ feedback resistor to give optimum pulse compensation is again 4·6 pF.

The gain bandwidth product of this circuit is nearly the same as that of the previous two circuits, but in this case large feedback factors are operative on all transistors. The overall feedback factor on the other hand is quite small, and so the problem of preventing the multi-stage loop from oscillating or ringing on a fast pulse is much less formidable. This latter requirement is further helped by the two dominant lags being due essentially to simple CR time-constant circuits and not to pseudo-time-constant circuits that depend on the frequency behaviour of the β of the transistors. The circuit has a number of similarities to the 'ring of three' circuit using thermionic valves described in section 6.6.2. A further point of interest is that the star connection of resistors used at the input can be replaced by the equivalent delta configuration, and if the circuit is driven by a very low-impedance voltage source (as is assumed), then one resistor

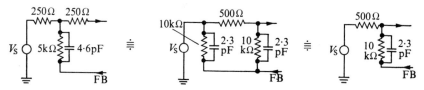

FIG. 6.37. Alternative input circuits for Fig. 6.36.

can be removed. This is shown in Fig. 6.37 and is perhaps the more common resistor arrangement for this circuit. The star connection is, however, helpful initially when analysing the circuit behaviour.

References

CHASE, R. L. and POULO, L. R. (1967) *I.E.E.E. Trans. Nucl. Sci.* **NS-14**, 83.

CHERRY, E. M. and HOOPER, D. E. (1963) *Proc. Instn elect. Engrs.* **110**, 375.

DUDLEY, R. A. and SCARPATETTI, R. (1964) *Nucl. Instrum. Meth.* **25**, 297.

FAIRSTEIN, E. (1956) *Rev. scient. Instrum.* **27**, 476.

—— and HAHN, J. (1965) *Nucleonics* **23**, 56.

GILLESPIE, A. B. (1953) *Signal, noise, and resolution in nuclear counter amplifiers*. Pergamon Press, London.

KANDIAH, K. (1966) Problems connected with high rates in pulse counting systems. *AERE Report R5302.*

—— STIRLING, A. and TROTMAN, D. L. (1968) A fast high resolution spectrometer for use with nuclear radiation detectors. *AERE Report R5852.*

LADD, J. A. and KENNEDY, J. M. (1961) A digital spectrum stabiliser for pulse analysing systems. *CREL-1063.*

NOWLIN, C. H., BLANKENSHIP, J. L., and BLALOCK, T. V. (1965) *Rev. scient. Instrum.* **38**, 1063.

ROBINSON, L. B. (1961) *Rev. scient. Instrum.* **32**, 1057.

GOULDING, F. S. and McNAUGHT, R. A. (1960) *Nucl. Instrum. Meth.* **8**, 282.

7 Low-Noise Head Amplifiers

By A. B. GILLESPIE

7.1. Introduction

DETERMINATION of the energy spectrum of a radioactive source by measuring the amplitude distribution of the pulses at the amplifier output is known as nuclear pulse spectrometry. It is in this field, rather than in the simpler measurement of source activity, that errors in pulse amplitude, caused by various random phenomena, can have their most serious impact. Such variations will change a line spectrum into one having peaks of a finite width, and so impose a limitation on the minimum energy separation between two peaks which can be satisfactorily resolved by the equipment. These random errors in pulse amplitude usually fall into one of two categories. These are:

(1) Errors that are a feature of the signal pulse train itself and arise from the pulse generation process in the detector or through the subsequent treatment of the pulses in the amplifier. Such errors only coexist with the signal pulses.

(2) Errors that arise from fundamental electrical and/or man-made noise disturbances in the detector and amplifier, combining with the signal pulses to produce a random spread of amplitude. In this case the distinguishing feature is that the noise disturbances causing the errors are present even in the absence of the signal pulses.

Pulse amplitude errors that arise from the pulse generation process in the detector are caused by the statistical uncertainty in the number of electron charges produced by the incident radiation. They are thus fundamental errors and cannot be reduced by any known technique. When a semiconductor or gas ionization detector is traversed by a high-speed charged particle (which may itself be the primary radiation or a secondary particle produced by the radiation) ionization is produced, and on the average one hole–electron pair or ion pair respectively is generated for every W electron volts of energy given up by the high-speed particle. W is approximately 3 eV for a semiconductor detector and approximately 30 eV for a gas ionization detector. If the energy of the incident radiation is E eV and all this energy is converted in the detector, then the average number of electronic charges produced is $\frac{E}{W}$. This number has a statistical uncertainty associated with it, but due to correlation between the ionizing events in the detector, the mean square fluctuation or variance is not normal and equal to $\dfrac{E}{W}$, as one might expect, but is equal to $\dfrac{E}{W} \times F$ where F is the Fano factor. This latter factor has been the object of much theoretical and experimental study, and for semiconductor and gas ionization detectors currently accepted values are 0·16 and 0·33 respectively (van Roosbroeck 1965, Fano 1947).

In the scintillation counter the energy-to-charge conversion process involves an additional step. First the high-speed charged particle produces photons in the phosphor and these are subsequently converted to photo-electrons by the light-sensitive cathode of the multiplier tube. The uncertainty is greatest in this latter stage, where the efficiency is around 10 per cent, and as a consequence the resulting spread in the number of photoelectrons produced is a normal one. The total energy required to produce one photoelectron can vary significantly with the type of phosphor in use, the type of radiation being measured, and the radiation energy, but for purposes of comparison a figure of 300 eV per photoelectron is assumed here.

Table 7.1 lists the magnitude of typical pulse generation errors in the three types of nuclear radiation detectors discussed above, for incident radiation energies of 100 keV up to 3 MeV. In addition to giving the r.m.s. error in the number of electronic charges in column 4, this is also converted to the corresponding full width at half maximum (fwhm) of a spectrum peak in column 5, and to the percentage width of the peak in column 6. For the scintillation counter the errors given refer to the electronic charge available at the multiplier tube photocathode. The percentage width errors at the output anode will be increased slightly, due to additional uncertainties in the dynode multiplication process, but the magnitudes given are still closely representative of the best that can be done.

A further source of error, which still depends on the presence of the signal pulses, can arise from the treatment that the pulses receive on their passage through the amplifier. Such errors arise primarily from the random time distribution of the nuclear events in the detector and the fact that the high-gain amplifier is usually a.c. coupled to prevent drift. Errors of this type have already been discussed in some detail in section 6.5 of Chapter 6, along with the techniques that can be applied to minimize them in comparison with the fundamental pulse-generation errors given in Table 7.1.

The remaining source of pulse amplitude error, and one that is not a function of the pulse train itself, is caused by fundamental electrical and/or man-made noise disturbances arising in the detector and the amplifier. Man-

139

made disturbances, like interference and pick-up, tend to be localized and to be a feature of the experimental environment and can usually be reduced to negligible proportions by the recognized techniques. Such errors will not be discussed further here. Errors due to the fundamental electrical noise in the detector and amplifier, however, constitute without doubt one of the

TABLE 7.1

Detector performance

Detector	Energy (MeV)	Total hole electron pairs	r.m.s. error hole–electron pairs	fwhm hole electron pairs	Width (%)
Semiconductor 3 eV per hole electron pair Fano factor = 0·16	3·0	10^6	400	940	0·094
	1·0	$3·3 \times 10^5$	230	540	0·165
	0·3	10^5	127	300	0·30
	0·1	$3·3 \times 10^4$	73	170	0·52

	Energy (MeV)	Total ion pairs	r.m.s. error ion pairs	fwhm ion pairs	Width (%)
Gas ionization 30 eV per ion pair Fano factor = 0·33	3·0	10^5	183	430	0·43
	1·0	$3·3 \times 10^4$	106	248	0·75
	0·3	10^4	58	136	1·36
	0·1	$3·3 \times 10^3$	33	77	2·35

	Energy (MeV)	Total photoelectrons	r.m.s. error photoelectrons	fwhm photoelectrons	Width (%)
Scintillation 300 eV per photo- electron	3·0	10^4	10^2	235	2·35
	1·0	$3·3 \times 10^3$	58	136	4·08
	0·3	10^3	32	74	7·4
	0·1	$3·3 \times 10^2$	18	43	13

major problems confronting the designer of high-performance nuclear pulse amplifiers for spectrometry applications, and this topic forms the subject matter of the remaining sections of this chapter.

To quantify the pulse amplitude errors produced by amplifier noise, it is customary to specify the noise in terms of an equivalent noise charge (ENC), this being a measure of the signal charge applied instantaneously at the amplifier–detector input that will produce a pulse signal at the ampli-fier output of amplitude equal to the r.m.s. noise. (This will be discussed in more detail in later sections.) At the time of writing the lowest equivalent noise charge for an amplifier with a thermionic valve as input element and with a total cold input capacitance of 50 pF is around 350 ion pairs.† With one or more cooled field effect transistors as input elements this can be reduced to some 200 ion pairs. Since these r.m.s. noise errors are not a function of the signal pulse amplitude, whereas the pulse generation errors are, then with a semiconductor radiation detector, it is clear from Table 7.1, column 4, that over the energy range 1–3 MeV both types of error may be significant. However, above 3 MeV the fundamental pulse generation errors are likely to predominate, whereas below 1 MeV the limit is likely to be set by the amplifier noise. A similar conclusion holds for the gas ionization type of detector, except that the transition region is between 3 MeV and 10 MeV. The scintillation counter is quite different in this respect, because all the figures in column 4 are increased by the gain of the photomultiplier before reaching the input of the amplifier. Photomultiplier gains are very high ($\approx 10^6$ upwards) and so the pulse generation errors will in this case always swamp any errors due to amplifier noise.

7.2. Some aspects of noise and signal-to-noise ratio

It has been shown in section 7.1 that for the lowest-noise head amplifiers presently available, the amplifier noise is a controlling factor in semiconductor detector measurements of particle or photon energies below 3 MeV and in gas ionization detector measurements of energies below 10 MeV. In most practical situations, however, the noise performance of the amplifier is not as good as that quoted in section 7.1 so, as a general rule, amplifier noise is usually always troublesome in experimental nuclear measurements, and methods of reducing the noise in relation to the signal amplitude must be explored. Signal-to-noise ratio is thus the important parameter.

In making calculations of the signal-to-noise ratio, it is convenient to assume an equivalent circuit for the amplifier and detector, and this is shown in Fig. 7.1. This is the same equivalent circuit as was assumed in Chapter 6. The input terminals are AA', $\sum C$ refers to the total cold input capacitance and includes contributions from the detector, the amplifier, and any connecting lead, and the bandwidth (or pulse-shaping network) is defined by the isolated time constants of differentiation and integration, T_1 and T_2 respectively. In the present section, discussion is confined to a pulse-shaping network of this type since it is probably the simplest network of all and

† A more general and accurate term would be 'electronic charges', but 'ion pairs' is now the well-established and understood method of defining the equivalent noise charge of a nuclear pulse amplifier.

11

the most commonly used. Other pulse-shaping networks are discussed in section 7.7. The noise contributions from the detector, from the active input element of the amplifier (valve, FET, etc.), and from the resistor R effectively shunting the detector are all considered to be effective at the input terminals AA'. The signal charge from the radiation detector is also effective at this point, but in making calculations of the signal-to-noise ratio, it is necessary to compare the maximum amplitude of the signal with the total noise at a point in the equivalent circuit where both have been limited by the finite bandwidth, for example at BB'.

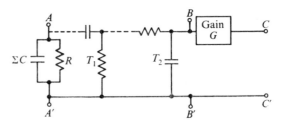

FIG. 7.1. Equivalent circuit of nuclear pulse amplifier.

For the present it is also proposed to describe the various noise contributions, other than thermal noise from resistors, in terms of the noises usually associated with thermionic valves. These are anode-current shot noise, flicker noise, and grid current noise. This is to some extent historical, but also quite logical, since these three types of noise have voltage spectral densities $v^2/\delta f$ at the amplifier input terminals AA' which are constant, vary as $1/f$ and $1/f^2$ respectively, and which, when classified in this way, are sufficient to encompass all additional noise contributions associated with the more recent advent of transistor amplifiers and semiconductor radiation detectors (Gillespie 1953). Put in another way, the transistor amplifier and the semiconductor radiation detector have not altered in any fundamental way the approach of the designer of high-performance nuclear pulse amplifiers to the basic problems of noise, signal-to-noise ratio, and time resolution in his designs, or the methods of optimization that are customarily used.

Considering first the thermal noise due to the resistor R in shunt with the detector, the r.m.s. noise voltage from this source reaching BB' depends on the value of R, and in particular is proportional to $\sqrt{R}$ for values of $\omega \sum CR \ll 1$ and is proportional to $\dfrac{1}{\sqrt{R}}$ for values $\omega \sum CR \gg 1$, where ω represents frequencies within the amplifier bandwidth. This suggests that R should be made sufficiently large for the noise from this source to become negligible compared with the other noises yet to be described, and this is invariably done in low-noise nuclear head amplifier technology. The magnitude of R needed to satisfy this condition will be deferred until the characteristics of grid current noise have been outlined. If v_t represents the r.m.s thermal noise voltage from R reaching BB' in the equivalent circuit, and $\omega \sum CR$ is much larger than unity for all frequencies within the amplifier pass band, then

$$v_t = \left\{ \frac{KT_e}{R(\sum C)^2} \frac{T_1^2}{T_1 + T_2} \right\}^{1/2}, \tag{7.1}$$

where $K = $ Boltzmann's constant $= 1\cdot37 \times 10^{-23}\ J/K$

and $T_e = $ room temperature $= 290$ K.

The anode-current shot noise is usually expressed in terms of an equivalent noise resistance R_n, which, if connected in the grid of the input valve, would, at room temperature, generate thermal noise equal in magnitude to the anode-current shot noise. The r.m.s. noise voltage reaching BB' from this source is

$$v_s = \left\{ KT_e \frac{2\cdot5}{g_m} \frac{T_1}{T_2(T_1 + T_2)} \right\}^{1/2}, \tag{7.2}$$

where $g_m = $ mutual conductance of the valve (A/V)

and $\dfrac{2\cdot5}{g_m} = $ equivalent noise resistance $R_n\ (\Omega)$.

The fundamental causes of flicker noise in valves are not well understood, but in low-power receiving valves measurements have shown that the noise magnitude does not vary markedly even between valves of different types, and the r.m.s. voltage contribution from this source at BB' is

$$v_f = \left\{ \frac{10^{-13}\ T_1}{T_1^2 - T_2^2} \ln(T_1/T_2) \right\}^{1/2}. \tag{7.3}$$

In this formula the constant 10^{-13} is a good average value obtained from experiment. Recent measurements, however, on very high slope valves used to obtain a low equivalent shot-noise resistance, do not agree well with eqn (7.3) and would suggest a value for the constant some two to five times larger. This uncertainty in the flicker noise magnitude must be accepted in theoretical calculations.

Grid current noise arises from the shot fluctuations in the steady grid current of the valve, generating a noise voltage across the input capacitance $\sum C$, and the r.m.s. magnitude of this noise voltage reaching BB' is

$$v_g = \left\{ \frac{eI_g}{2(\sum C)^2} \frac{T_2^2}{T_1 + T_2} \right\}^{1/2}, \tag{7.4}$$

where $e = $ Electronic charge $= 1\cdot6 \times 10^{-19}$ C

and $I_g = $ arithmetic sum of the positive and negative grid currents (A).

It will be noted that the frequency dependence of eqn (7.4) for grid current noise is identical to that of eqn (7.1) for thermal noise, so both noise magnitudes at BB' will always bear a fixed relation to one another. From this it follows that the value of R should be chosen to make the thermal noise component negligible, and this requires

$$R \gg \frac{0 \cdot 05}{I_g}. \tag{7.5}$$

A factor of 20 times or more would appear to be adequate as this will result in the thermal noise contribution at no time exceeding $2 \cdot 5$ per cent of the grid current noise.

The signal for comparison with the noise arises from charge collected in the detector, and for a charge Q_s collected instantaneously on $\sum C$ the signal voltage at BB' is

$$V = \frac{Q_s}{\sum C} \frac{T_1}{T_1 - T_2} \{\exp(-t/T_1) - \exp(-t/T_2)\}. \tag{7.6}$$

The signal-to-noise ratio at BB' is then equal to $V_m / \sqrt{(v_s^2 + v_f^2 + v_g^2)}$, where V_m is the maximum value of V. It has been shown that for a specified resolving time (i.e. pulse width) in the amplifier, the signal-to-noise ratio is an optimum if T_1 is made equal to T_2 (Gillespie 1953). Subject to this condition eqns (7.2), (7.3), (7.4), and (7.6) simplify to

$$v_s = \left(\frac{KT_e R_n}{2T_1}\right)^{1/2}, \tag{7.7}$$

$$v_f = \left(\frac{10^{-13}}{2}\right)^{1/2}, \tag{7.8}$$

$$v_g = \left\{\frac{eI_g T_1}{4(\sum C)^2}\right\}^{1/2}, \tag{7.9}$$

and

$$V = \frac{Q_s}{\sum C} \frac{t}{T_1} \exp(-t/T_1) \tag{7.10}$$

with

$$V_m = \frac{Q_s}{e \sum C}. \tag{7.11}$$

Other important conclusions are:

(a) The input valve should be a triode to prevent the appearance of a further noise known as partition noise. In a pentode valve this is generated by the probability distribution of cathode current electrons between the anode and the screen grid.

(b) $\sum C$ refers to the total *cold* input capacitance and not to some effective value produced by feedback.

(c) The signal-to-noise ratio is unaltered by feedback provided the bandwidth of the amplifier is not changed.

When using a triode input, however, thought must also be given to the following stage. For the input circuit shown in Fig. 7.2 the Miller capaci-

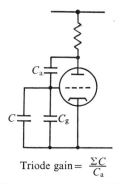

Triode gain $= \dfrac{\sum C}{C_a}$

FIG. 7.2. Triode input stage—Miller effect.

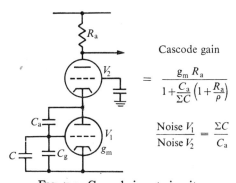

Cascode gain

$$= \frac{g_m R_a}{1 + \dfrac{C_a}{\sum C}\left(1 + \dfrac{R_a}{\rho}\right)}$$

$$\frac{\text{Noise } V_1}{\text{Noise } V_2} = \frac{\sum C}{C_a}$$

FIG. 7.3. Cascode input circuit.

tance limits the overall stage gain to $\sum C/C_a$, and if the detector capacitance is small this gain figure may only be 3 or 4. Thus a triode rather than a pentode may be necessary in the following stage to keep its added noise negligible. A circuit that has the same noise magnitudes as two cascaded triodes, but whose gain is less dependent on variation of the detector capacitance is known as the cascode and is shown in Fig. 7.3. The noise of V_1 is $\sum C/C_a$ times the noise of V_2, which is the same as two triodes in cascade, but the overall gain is

$$\frac{g_m R_a}{1 + \dfrac{C_a}{\sum C}\left(1 + \dfrac{R_a}{\rho}\right)}$$

145

where ρ is the anode impedance of each valve, and this is much less dependent on variation of the detector capacitance. The cascode is a popular input circuit with corresponding valve, transistor, and FET versions and appears also in hybrid forms, chiefly with a valve and transistor or with an FET and transistor.

7.3. Sensitivity

In nuclear pulse spectrometry, high signal-to-noise ratios are the rule rather than the exception, and the instrumental problem is usually that of distinguishing between pulses of nearly similar amplitude, corresponding to particles or photons differing only slightly in their energies. The sensitivity of an experimental nuclear pulse spectrometer may, therefore, be defined in terms of the minimum energy difference between two pulse amplitude groups that can be satisfactorily separated or resolved by a differential bias curve measurement. A convenient measure of this minimum difference is the full width at half maximum (fwhm) of a spectrum peak, and if the peak broadening or dispersion is of Gaussian shape then the fwhm is equal to 2·35 times the r.m.s. value. The experimental nuclear physicist usually quantifies this definition of the sensitivity of his equipment by stating the magnitude of the measured fwhm in terms of energy, referred to a particular type of radiation detector.

This overall sensitivity figure will almost certainly contain contributions due to the fundamental pulse generation errors discussed in section 7.1, the electrical noise from the amplifier–detector combination discussed in section 7.2, and others from the pulse rate effects discussed in section 6.5. In many cases the electrical noise from the amplifier–detector combination is the principal contributor, and this component has been singled out and is the one usually referred to by the designer of nuclear pulse amplifiers when specifying the sensitivity of his equipment.

Amplifier sensitivity, if only detector capacitance but not type is known, can not be expressed in terms of energy, and the practice here is to specify the sensitivity in terms of an equivalent noise charge (ENC), this being a charge of magnitude Q_n coulombs, which if applied instantaneously to the input capacitance $\sum C$ would give a maximum signal at BB' in the equivalent circuit just equal to the r.m.s. noise. A further step of convenience, and one that leads to the now standard method of quantifying ENC, is to divide Q_n by $1·6 \times 10^{-19}$, so converting it to a number of fundamental electronic charges or, to use the historically accepted term, ion pairs. It will be appreciated that both $\sum C$ and T_1 will also affect the sensitivity of the amplifier, so values for both must be stated along with the figure for ENC. From this more basic way of expressing sensitivity in ion pairs, it is a

146

simple matter for the experimentalist to calculate, should he wish, the amplifier sensitivity in terms of energy, by multiplying by the energy needed to produce one ion pair in the detector he is proposing to use and then multiplying by 2.35 to convert this energy figure to a fwhm.

It follows from eqns (7.7), (7.8), (7.9), and (7.11) and from the above definition of Q_n that

$$Q_n = e(\textstyle\sum C)\left\{\frac{KT_e R_n}{2T_1} + \frac{10^{-13}}{2} + \frac{eI_g T_1}{4(\sum C)^2}\right\}^{1/2}. \qquad (7.12)$$

Fig. 7.4 shows how the shot, flicker, and grid current components of Q_n vary with the amplifier-shaping time constant T_1. The interesting feature

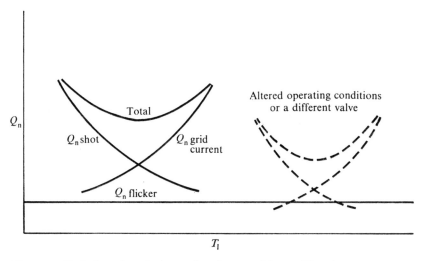

FIG. 7.4. Variation of equivalent noise charges with amplifier time-constants.

here is that Q_n (shot) varies as $1/\sqrt{T_1}$, Q_n (flicker) is constant, and Q_n (grid current) varies as $\sqrt{(T_1)}$. Clearly, Q_n (total) is a minimum when the shot and grid current components are equal and this defines a preferable shaping time constant to use in an experimental situation. In most cases, Q_n (flicker), although not negligible at the minimum point, is smaller than either of the other two components. The absolute values of the shot and grid current components of Q_n can also be varied by changing the operating conditions of the input valve, or for that matter the valve itself, to give an altered slope and grid current. Generally, lowering the slope of the valve reduces Q_n, if at the same time the pulse-shaping time constants are lengthened to make the shot and grid current components again equal (see Fig. 7.4) and this is the basis of the statement that in low-noise nuclear pulse amplifier design the ENC and resolving time are opposing factors; one can usually be improved at the expense of the other.

Typical figures for the optimized sensitivity or ENC of a good head amplifier using a medium slope input valve (<10 mA/V) and having a total cold input capacitance of some 50 pF are given in Table 7.2. It should be

TABLE 7.2

Head amplifier ENC

Shaping time constant, T_1 (μs)	ENC ion pairs
0·1	1600
1·0	700
10	400

noted that the flicker noise component for valves that agree with eqn (7.8) is 180 ion pairs at all time-constant settings.

Fig. 7.5 shows a cascode input stage employing two E83F valves. The circuit is representative of head amplifiers that were, and still are, used with

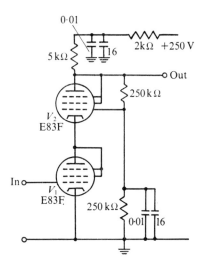

FIG. 7.5. Typical cascode input stage using valves.

gas ionization radiation detectors. An interesting feature of this circuit is the absence of a grid resistor in the bottom valve. This floating grid operation automatically sets the lower valve to the point on its grid base where the positive and negative grid current components are equal and where the I_g

value appropriate to eqn (7.9) is then twice the value of each component. This is a simple way of automatically stabilizing the bias of the input valve, without the danger of excessive grid current, and at the same time of removing completely the grid resistor and the attendant need to choose a value such that its thermal noise is insignificant. In this circuit both valves are operating with $g_m = 8$ mA/V (i.e. $R_n \sim 300\ \Omega$) and the effective value of I_g is 2×10^{-9} A. With a total cold input capacitance of 50 pF, the minimum ENC is approximately 500 ion pairs for shaping time constants of 5 μs. Theory and experiment are here in good agreement.

7.4. Amplifier sensitivity with semiconductor radiation detectors

Semiconductor radiation detectors (silicon and germanium) have undergone tremendous development in the past few years, and are now largely replacing gridded ionization chambers and scintillation counters for the accurate energy spectrometry of both highly energized particles and X- and γ-ray photons (Dearnaley and Northrop 1966, Goulding 1966). This impetus stems from the fact that only about 3 eV of energy (2·9 Ge, 3·5 Si) are needed to generate one hole–electron pair in a semiconductor detector, compared with some 30 eV for an ion pair in a gas ionization detector. As a consequence the charge available from the semiconductor detector is nearly ten times greater than that from the ionization chamber, and for the same detector capacitance, will produce a correspondingly larger signal voltage. This substantial improvement in the signal amplitude is, however, not fully effective, because the semiconductor detector also introduces some further components of electrical noise that will be discussed later in this section. In addition, the consequential increase in the pulse generation errors must also be considered, so the effects of using a semiconductor radiation detector on the definitions of spectrometer sensitivity and amplifier sensitivity given in section 7.3 are as follows:

(a) Spectrometer sensitivity, or fwhm, expressed in units of energy will get less by a factor usually between $\sqrt{10}$ and 10 times. This reflects the overall improvement stemming from the large increase in the signal.

(b) Amplifier sensitivity, or ENC, expressed in ion pairs will increase by an amount equal to the additional electrical noise. This reflects the worsening of the amplifier performance due to the extra noise from the detector.

The first use of semiconductor radiation detectors was in the measurement of energetic short-range particles like protons, alpha particles, and fission fragments. Such detectors consist essentially of a reversed biased

149

semiconductor junction, usually between a heavily doped n-material and a more lightly doped p-material. The active depth of the detector is set by the thickness of the junction depletion layer, and although this can be controlled within certain limits by the applied reverse bias, depletion layers are usually quite thin, a representative figure for silicon with 300 V bias being some 0·02 cm. The collected charge from the active volume of the detector forms a voltage pulse across the depletion layer capacitance (50 pF in the above example for each cm² of surface area), but as this capacitance depends on the layer thickness, the voltage pulse for a given collected charge will vary with the bias applied to the detector. This complication, not met when using gas ionization detectors, can be got around fairly simply with a slightly different head-amplifier configuration—see section 7.5.

A further complication, which adversely affects the signal-to-electrical noise ratio and thus the sensitivity of the amplifier–detector combination, is the leakage current through the semiconductor detector due to the reverse bias. This current stems primarily from the thermal generation of hole-electron pairs in the depletion layer volume, and will have associated with it a shot fluctuation that will generate a noise voltage across the detector capacitance, in an identical way to the noise voltage generated by the grid current of the input valve. In fact, when calculating this type of noise, the detector leakage current and the input valve grid current are simply added to give the appropriate value for I_g in eqn (7.9).

At room temperature, leakage currents in silicon radiation detectors usually lie in the range 10^{-8}–10^{-6} A. This is much larger than the grid current of a good valve, and so the sensitivity or ENC of the amplifier will be that much worse. For example, assuming a detector leakage current of 10^{-7} A, and an input valve having a $g_m = 8\,\mathrm{mA/V}$ ($R_n \sim 300\,\Omega$) the minimum ENC of the amplifier, for a total cold input capacitance $\sum C$ of 50 pF and for 0·7 μs shaping time constants is 1200 ion pairs. The minimum value occurs here at much shorter shaping time constants than before, because reducing the time-constant value below 5 μs in the first instance reduces the greater noise due to the detector leakage current, but only until the shot noise of the input valve again becomes comparable (see Fig. 7.4).

In a practical situation, the ENC may be greater than the figure given above because of a further component of detector leakage current. This is current leaking across the surface of the junction, which gives rise to a noise voltage across the detector having a spectral power density that varies approximately as $1/f$. This noise voltage behaves in an identical way to flicker noise in valves, in so far as the amplifier-shaping time constants are concerned, and likewise is no better understood or calculable. Surface leakage currents are very variable, depending critically on the surface conditions and its preparation, and generally extreme care is taken in the fabrication of the detector to try and make this noise component insignificant. Using surface passivation and guard ring techniques, noise from this source can be made negligible, but should the measured ENC of the amplifier turn out to be much larger than expected from calculation then this source must always be suspect (Goulding 1966).

The semiconductor radiation detectors just described are not suitable for X- and γ-ray spectrometry because of their small active volume. Using modern lithium drifting techniques, however, it is now possible to produce detectors with depletion layers of 1–$1 \cdot 5$ cm thick, and proportionately lower capacitance (Goulding 1966). In addition, cooling of silicon detectors to 220–$170°$ K can reduce very markedly the bulk leakage current to around 10^{-9} A or less. Such detectors now find wide application in the measurement of particles with energies up to 100 MeV and also in the measurement of low energy γ-rays.

Germanium detectors are attractive for γ-ray spectrometry because of the higher atomic number ($Z = 32$) and also for the lower number of electron volts needed to generate a hole–electron pair ($W = 2 \cdot 9$ eV). Germanium detectors, however, cannot be used at room temperature as the bulk leakage current is several orders of magnitude greater than that of silicon, but by cooling to liquid nitrogen temperature ($77°$ K) bulk leakage currents of 10^{-9} A or less can be realized.

Hence, by cooling semiconductor radiation detectors the minimum ENC of the amplifier for an E83F input valve once again approaches 500 ion pairs for $\sum C = 50$ pF and shaping time constants of 5 μs. This figure can be reduced still further by taking advantage of high slope valves that have been developed in recent years. For example, the E810F valve can operate with a $g_m = 25$ mA/V ($R_n \sim 100$ Ω) without the grid current of selected valves exceeding 10^{-9} A. In some cases the flicker noise may be higher than expected from eqn (7.8) and further selection may be necessary. Under favourable conditions a minimum ENC of some 380 ion pairs is possible with $\sum C = 50$ pF and shaping time constants of $2 \cdot 5$ μS. A lower value for $\sum C$ could reduce this still more.

7.5. The charge-sensitive amplifier configuration

In the discussion so far, it has been assumed that the signal charge Q_s develops a voltage across the total cold input capacitance $\sum C$, and that this signal is subsequently amplified by an amplifier having a very stable voltage gain. With a semiconductor radiation detector, however, the capacitance depends on the applied bias voltage, so for a given input signal charge the output signal amplitude will likewise depend on bias volts. This is most unsatisfactory in nuclear pulse spectrometry, where a known and stable relationship between output signal amplitude and radiation energy is required. What is needed is a head-amplifier configuration that supplies a constant output signal amplitude for a given input charge, irrespective of the value of the detector capacitance. This is not perfectly realizable, but can be closely approached by the application of negative feedback. Amplifiers having this configuration are known as 'charge sensitive'.

A block schematic of the arrangement is shown in Fig. 7.6. Negative feedback is applied from the output of the head amplifier to its input through a small feedback capacitance C_f. It will be appreciated that C_f now adds to the total cold input capacitance, so in the diagram $\sum C = C + C_f$. The effective input capacitance of the circuit when operating is $C + (A + 1)C_f = \sum C + AC_f$, the input voltage is $Q_s/\{\sum (C) + AC_f\}$ and

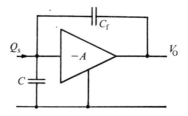

FIG. 7.6. Schematic of charge-sensitive amplifier.

the output voltage $V_0 = AQ_s/\{\sum (C) + AC_f\}$. If $AC_f \gg \sum C$, then the output voltage V_0 approaches Q_s/C_f showing that the circuit is behaving as if all the signal charge developed its voltage across C_f and all this voltage in turn appeared at the output. To determine the exact stability of this charge to voltage conversion process, differentiation of the above expression for V_0 gives

$$\frac{dV_0/V_0}{d \sum C/\sum C} = \frac{-1}{1 + \dfrac{AC_f}{\sum C}}, \qquad (7.13)$$

showing, as one might expect, that the fractional change in the output voltage resulting from a fractional change in the total cold input capacitance is reduced by the feedback factor $1 + AC_f/\sum C$. Thus if $\sum C = 50$ pF and $C_f = 2 \cdot 5$ pF then A must be 1000 in order that a 10 per cent change in $\sum C$ (i.e. 5 pF) should not affect the output signal amplitude by more than 0·2 per cent.

Because one has to add C_f to the cold input capacitance in order to apply the feedback, then fundamentally the ENC of a charge-sensitive type of amplifier must be greater than that of the corresponding voltage sensitive type. However, the added capacitance is usually only a few picofarads, so the deterioration in the ENC is very small. Most low-noise head amplifiers for use with semiconductor radiation detectors now use this charge-sensitive configuration.

The charge-sensitive feedback ring for a head amplifier using a hybrid cascode input stage consisting of a valve and transistor is shown in Fig. 7.7. The valve is the E810F operating with with $g_m = 25$ mA/V at $I_a = 10$

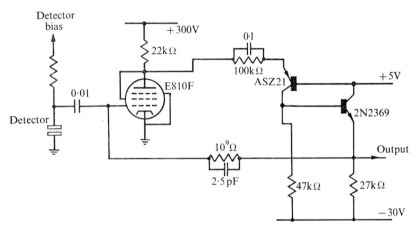

FIG. 7.7. Charge-sensitive amplifier with hybrid cascode input stage.

mA and having $I_a \sim 10^{-9}$ A. This circuit can give, under favourable conditions, the minimum ENC figures quoted in section 7.4.

7.6. Amplifiers with transistor and FET input stages

Most present-day nuclear-pulse head amplifiers use transistors in all the stages following the input. A thermionic valve as an input element is inconvenient, because of the need for both a heater supply and an H.T. supply of some 200–300 V with currents up to 50 mA. It is thus natural to enquire if there is a semiconductor element that could replace the thermionic valve and at the same time lead to an improvement, or at least no worsening of the noise performance.

Bipolar transistors are not suitable in this respect, because the base current generates noise across the cold input capacitance, just like the grid current of a valve, only base currents have to be reckoned in terms of many microamperes compared with 10^{-9} A for the grid current of a good valve. However, the bulk field effect transistor (FET) is now attracting a lot of attention, and it is likely that FETs, designed with this application in mind, will completely surpass the thermionic valve.

153

The FET consists basically of a rod or channel of semiconducting material (n- or p-type) with an electrode called the source at one end and a second electrode called the drain at the other. A voltage applied between these two electrodes will cause a current to flow in the body, its magnitude depending on the resistance of the material. For the case of an n-type body, a third electrode called the gate consists of a p-type region diffused on the side of the rod to form a p–n junction. Reverse biasing of the junction then causes the associated depletion layer to extend into the body of the device and to modulate the cross-sectional area and consequently the resistance of the channel. The FET is essentially a voltage-operated device, having a high-impedance control terminal (the gate) since only the current of the reverse biased junction flows in this circuit. Typical currents are in the region 10^{-9}–10^{-11} A. In this respect the FET is much more like a valve, and the effect of the gate voltage on the channel current is likewise described in terms of a mutual conductance g_m.

The noise contributions from an FET fall into the same three spectral power density categories as do the contributions from a thermionic valve. The shot noise of the current flowing in the channel can be represented by an equivalent shot-noise resistance R_n in the gate circuit, and it has been shown that R_n for an FET is $0.7/g_m$ (Van der Ziel 1962). Thus an FET having $g_m = 7$ mA/V should generate no more shot noise than the E810F valve operating with $g_m = 25$ mA/V. Early FETs had mutual conductances generally less than 2 mA/V but with improved understanding of their operation and manufacture, FETs are now available with mutual conductances of 10–15 mA/V. In this respect the FET is thus the equal, if not better than any available thermionic valve in so far as shot noise is concerned.

Gate current noise of the FET is identical to grid current noise of the valve, and here again, as typical gate currents are 10^{-9} A or less, the FET is superior to most valves capable of providing, at the same time, low shot noise.

The remaining component is that having a $1/f$ power density, and it is perhaps true to say that at present even less is known fundamentally about this source in FETs than in valves or semiconductor radiation detectors. The belief is that it is largely associated with the fabrication of the FET and that careful study and attention to detail in manufacture will ultimately minimize this noise component. Recent work along these lines is encouraging. A skeleton circuit for a charge-sensitive feedback ring using a hybrid cascode input stage consisting of an FET and a transistor is shown in Fig. 7.8. The FET quoted is one presently under development, having a g_m of some 10–15 mA/V and aiming to meet a specification where the $1/f$ noise is negligible compared with the shot- and gate-current components. Expected ENC figures for this amplifier are around 120 ion pairs for zero external capacitance, 250 ion pairs for $\sum C = 50$ pF, and 400 ion pairs for $\sum C = 100$ pF. Optimum pulse-shaping time constants are in the range 1–10 μs.

A further very interesting feature of the FET is the increase in the mutual conductance that can be gained by cooling. This stems from the increased

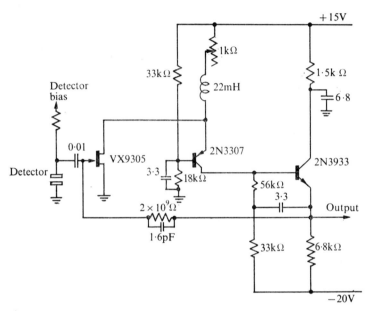

FIG. 7.8. Charge-sensitive amplifier with FET input stage.

mobility of the carriers. In addition, the thermally generated bulk leakage current of the reverse biased junction will get less, so all round it would appear that quite significant improvements in noise can be gained by low-temperature operation. Experimental work on various selected FETs cooled to liquid nitrogen temperature would indicate that a 40 per cent reduction in ENC is possible.

7.7. Amplifier sensitivity with other linear shaping networks

All the figures for ENC quoted so far refer to the simplest and most commonly used shaping network consisting of equal differentiating and integrating time constants. Other pulse-shaping networks using a combination of delay lines and/or a multiplicity of equal time constants were discussed in Chapter 6 in the context of amplifier behaviour at high counting rates. The behaviour of these networks in regard to noise and time resolution

155

will now be considered. In this comparison it is convenient to use a shorthand method of identifying the different networks and typical examples are:

$CR.CR$—equal CR differentiator and integrator,
$CR.(CR)^2$—equal CR differentiator and two integrators, and
$DL.CR$—delay line differentiator and CR integrator.

It is assumed that only components of shot type noise and grid current type noise are present, and that the voltage spectral density of the combined noise at the amplifier input terminals AA' is $v^2/\delta f = a^2 + g^2/\omega^2$ where $a^2 = 4KT_e R_n$ and $g^2 = 2eI_g/(\sum C)^2$. The minimum ENC occurs when both noise components are equal at BB' in the equivalent circuit and for the $CR.CR$ filter, eqn (7.12) shows this to occur when $T_1 = a/g$ with a corresponding equivalent noise charge $Q_n = 1\cdot36 \sum C\sqrt{(ag)}$ (neglecting flicker noise).

It has been shown theoretically that Q_n can never be less than $\sum C\sqrt{(ag)}$ so it is appropriate to compare all the linear filters in this section with this theoretical ideal (Baldinger and Franzen 1956). A simple linear filter to realize this ideal is not practically possible, but if it were, then the output pulse would have the cusp shape shown in entry 1, Table 7.3. The remaining entries (2–9) are for the other most practically useful linear filters, and the minimum value of Q_n for each is given in column 3. This minimum value for Q_n must also be supplemented by the delay and/or time-constant values for each filter and these are given in column 2. To complete the comparison the effective pulse width must be given in each case, as this defines the resolving time T_r and consequently the maximum counting rate capability of the system. This is shown in column 4, the effective pulse width for a unipolar pulse being defined as $\dfrac{\text{area}}{\text{maximum amplitude}}$ and that for a bipolar pulse as $\dfrac{+\text{ve area}}{+\text{ve amplitude}} + \dfrac{-\text{ve area}}{-\text{ve amplitude}}$. The noise integrals required for Table 7.3 are straightforward but their working can be quite laborious (Tsukuda 1961). A good table of standard integral forms is here very useful (Dwight 1947).

Of the filters listed in Table 7.3, it is interesting to note that Q_n for the simple equal time-constant filter $CR.CR$ is only 36 per cent greater than the best possible value. Additional integration leads to an improvement in Q_n and T_r, for example filter 3, $CR.(CR)^2$ with two integrations, and filter 4, $CR.(CR)^4$ with four integrations, but the adoption of bipolar pulse shaping, filters 5, 6, and 8, leads to no improvement—generally to a worsening of Q_n and a considerable worsening of the resolving time T_r. Filter 7 is an interesting one involving a single delay-line shaper and a single

TABLE 7.3

Noise performance of various filters

Filter	Conditions	Equiv. noise charge Q_n in units of $\sum c\sqrt{(ag)}$ coulombs	Resolving time T_r in units of $\dfrac{a}{g}$ seconds
1. To give a cusp-shaped pulse	Rise and fall time constants $= \dfrac{a}{g}$ Time to peak $= \dfrac{2a}{g}$	1·0	2·0
2. $CR \cdot CR$ T_1	$T_1 = T_2 = \dfrac{a}{g}$	1·36	2·72
3. $CR \cdot (CR)^2$ $2T_1$	$T_1 = T_2 = \dfrac{1}{\sqrt{3}}\dfrac{a}{g}$	1·215	2·13
4. $CR \cdot (CR^4)$ $4T_1$	$T_1 = T_2 = \dfrac{1}{\sqrt{7}}\dfrac{a}{g}$	1·16	1·93
5. $(CR)^2 \cdot CR$ $3\cdot4\,T_1$ $0\cdot6\,T_1$	$T_1 = T_2 = \sqrt{3}\dfrac{a}{g}$	1·43	7·93
6. $(CR)^2 \cdot (CR)^2$ $4\cdot7\,T_1$ $1\cdot3\,T_1$	$T_1 = T_2 = \dfrac{a}{g}$	1·36	5·62
7. $DL \cdot CR$ T_d	$T_d = 1\cdot036\,T_2$ $T_2 = 1\cdot29\dfrac{a}{g}$	1·096	2·07

TABLE 7.3—*cont.*

Filter	Conditions	Equiv. noise charge Q_n in units of $\sum c\sqrt{(ag)}$ coulombs	Resolving time T_r in units of $\dfrac{a}{g}$ seconds
8. $(DL)^2 \cdot CR$	$T_d = T_2$ $T_2 = 2 \cdot 2\dfrac{a}{g}$	1·37	4·5
9. $(DL)^2 \cdot CR$	$T_d = \sqrt{3}\dfrac{a}{g}$ $T_2 \to \infty$ (true integration)	1·075	1·73

CR integrator. Provided the integrator time constant is almost equal to the square pulse duration T_d, this filter gives a very good Q_n and also a good T_r. The pulse shape has, however, no flat region at the top and is not very useful in practice. Filter 9 uses double delay-line shaping to provide initially a bipolar square pulse, but this is then integrated with a CR integrator having a very large time constant (approaching ∞ for perfect integration) or with an operational amplifier integrator to give a symmetrical unipolar triangular pulse (Nowlin *et al.* 1965). This has the best Q_n and T_r of all the filters listed in the table, but again has a pointed top. Note that as Q_n and T_r improve, the pulse top becomes more pointed, tending ultimately to the cusp shape.

7.8. Amplifier sensitivity with gated shaping networks

The technique of rendering operative for a finite period, all or a part of the pulse-shaping filter, only on the arrival of a signal, can lead to improvements in the ENC and/or the resolving time, because of the absence of stored noise energy in the filter prior to its activation by the gate waveform and the possibility of reducing the signal waveform to zero very quickly after the closure of the gate. Two interesting applications of this technique, relate to amplifiers with double delay-line shaping (filter 8, Table 7.3) and to amplifiers containing an active integrating element (Blalock 1965, Kandiah 1965, Deighton 1968).

In the former case the second delay-line shaper is gated on for a time

T_d, coincident with and equal to the pulse of duration T_d emerging from the first delay-line shaper (time-constant integration comes after the second shaper). In this way the signal-to-noise ratio is not worsened by noise stored in the second delay line, and the latter only performs the function of inverting the positive lobe to generate a bipolar pulse. The equivalent noise charge Q_n is then the same as that for single delay-line shaping (filter 7, Table 7.3) with a relative minimum value of 1·096. Moreover the conditions for minimum Q_n are those given in column 2 for the $DL.CR$ filter, so the resolving time T_r is likewise reduced in the ratio of the integrating time constants, i.e. 1·7 times. Thus the advantages of a bipolar pulse, in respect of base-line shift at high rates, are retained, with, at the same time, the good Q_n of a unipolar pulse and a significant improvement in the resolving time.

In the case of amplifiers using an active integrator the equivalent circuit is shown in Fig. 7.9. This is the same as that previously used and shown

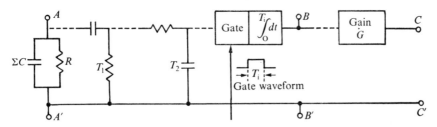

FIG. 7.9. Equivalent circuit for amplifier with active integrator.

in Fig. 7.1, with the addition of the active integrator circuit. This latter circuit is normally held with its output voltage zero, until the arrival of a signal pulse, when it is rendered active by the gating waveform for an integrating time T_1. At the end of the integrating period the voltage output from the integrator is held for a short time, just long enough for subsequent instruments to measure the pulse amplitude, and then the integrator is quickly discharged and the output voltage returned to zero. A comprehensive analysis of the noise performance of this circuit is given by Deighton (1968) and two interesting examples are singled out here for discussion.

The first corresponds to the conditions $T_1 = T_i = 1·29a/g$ and $T_2 = 0$, where a/g is the value of the time constant that gives the minimum Q_n for the basic equal time-constant amplifier. The condition $T_2 = 0$ implies an infinitely wideband amplifier but it has been shown that if T_2 does not exceed 0·129a/g the effect of this on the ENC is less than 2·5 per cent. The above ideal conditions give a relative minimum value for Q_n of 1·096, which is identical to that of a single delay-line shaper. The simple definition of the

159

resolving time T_r used in Table 7.3 is not applicable to gated integrator systems; in the analysis given by Deighton (1968) the resolving time is defined as the time separation between two identical signals, such that the amplitude measurement of the second is in error by 1 per cent because of residual voltages from the first. This gives a resolving time for the above gated-integrator case of $6(a/g)$. If this definition is applied to the single delay-line shaper, $DL.CR$ in Table 7.3, the resolving time is also equal to $6(a/g)$. Thus the gated-integrator amplifier with $T_1 = T_i = 1\cdot29(a/g)$ and $T_2 = 0$ has an identical performance, in respect of Q_n and T_r, to a conventional amplifier using a single delay-line shaper. The practical advantages of the gated integrator are, however, very significant, as there is no longer a need for high quality delay lines, and the complication of changing lines for different delay times can be replaced by simply varying the duration T_i of the integrator-gate waveform.

The second example is for the conditions $T_1 = T_2 = 0\cdot17(a/g)$ and $T_i = 0\cdot85(a/g)$. This gives a relative minimum value for Q_n of $1\cdot36$, which is the same as that of the conventional equal time-constant amplifier but has a resolving time (new definition) of $1\cdot13(a/g)$ compared with $6\cdot5(a/g)$ for the equal time-constant amplifier. This is a very significant improvement.

It should be remembered that no system can ever produce a minimum value for Q_n less than $\sum C\sqrt{(ag)}$, and this is only 26 per cent better than the conventional equal time-constant amplifier, but potentially large improvements in resolving time are possible, since the pulse amplitude information has been obtained very shortly after the pulse reaches its maximum value. It would appear that gated integrator systems, with the addition of time-constant switching and d.c. coupling in the amplifier, might enable the theoretical ultimate in Q_n and T_r to be very closely approached, even at very high pulse rates (Deighton 1968, Kandiah 1968).

References

BALDINGER, E. and FRANZEN, W. (1956) *Advances in electronics and electron physics*. Academic Press, New York.

BLALOCK, T. V. (1965) *Rev. scient. Instrum.* **36**, 1448.

DEIGHTON, M. O. (1968) *Nucl. Instrum. Meth.* **58**, 201.

DEARNALEY, G. and NORTHROP, D. C. (1966) *Semiconductor counters for nuclear radiation*. E. and F. Spon, London.

DWIGHT, H. B. (1947) *Tables of integrals and other mathematical data*. Macmillan, New York.

FANO, U. (1947) *Phys. Rev.* **72**, 26.

GILLESPIE, A. B. (1953) *Signal, noise and resolution in nuclear counter amplifiers*. Pergamon Press, London.

GOULDING, F. S. (1966) *Nucl. Instrum. Meth.* **43**, 1.

KANDIAH, K. (1965) Active integrators in spectrometry with radiation detectors. *AERE Report R*5019.

—— (1968) A fast high-resolution spectrometer for use with nuclear-radiation detectors. *AERE Report R*5852.

NOWLIN, C. H., BLANKENSHIP, C. L., and BLALOCK, T. V. (1965) *Rev. scient. Instrum.* **38**, 1063.

TSUKUDA, M. (1961) *Nucl. Instrum. Meth.* **14**, 241.

VAN DER ZIEL, A. (1962) *Proc. Instn elect. Engrs* **50**, 1808.

VAN ROOSBROECK, W. (1965) *Phys. Rev.* **139**, 1702.

8 *Pulse Amplitude Discriminators*

By L. J. HERBST

A PULSE amplitude discriminator is required to give an output when the input signal exceeds a threshold value that is usually adjustable. The output constitutes digital intelligence and is usually applied to a scaler or coincidence unit. It may also be fed to a special module like a time-to-amplitude converter for timing measurements. The discriminator output should be a shaped digital-coupling pulse of fixed amplitude and duration. Its rise time should be very short for accurate timing information and the time displacement between input and output should be as independent of input amplitude as possible. The change in said time displacement with input amplitude is called 'time walk' or 'time jitter'. In an ideal discriminator the time walk will be equal to the rise time of the input signal. This can be seen from Fig. 8.1 where the large signal might have an amplitude many times (50

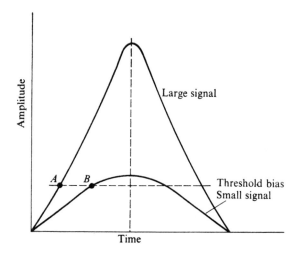

FIG. 8.1. Discriminator triggering.

or more) that of the small signal. A further requirement in discriminators is that threshold sensitivity should be independent of signal rate. For that reason the circuit between the input and the discriminator element should be d.c.-coupled throughout. An alternative is to have a.c. coupling and shape the input signal into a bipolar pulse prior to reaching the first d.c. blocking capacitor. The former method is preferred.

162

A discriminator contains the various blocks shown in Fig. 8.2. The amplifier, often a buffer with unity current gain, feeds the discriminator element whose output drives the pulse former. A bias circuit supplies the threshold bias to the discriminator. The next section on tunnel diode

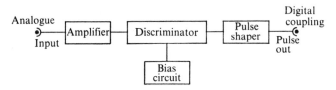

FIG. 8.2. Discriminator schematic.

discrimination constitutes a major part of this chapter. The tunnel diode is an ideal element for discriminators in nuclear instrumentation, especially at high pulse rates and where the time walk should be as small as possible.

8.1. Tunnel-diode discriminator

8.1.1. *The tunnel diode*

The tunnel diode (R.C.A. 1963) is a p–n junction diode with doping levels of 10^{19} to 10^{20} atoms/cm^3, i.e. about 1000 times higher than used for conventional diodes. This results in an extremely thin depletion layer, about 100 Å, compared with 10^4 Å for a normal junction. The i/v characteristic is due to a quantum-mechanical tunnelling effect that permits a current carrier to pass through a potential barrier considered by classical physical analysis to be too high for the carrier to surmount. The practical requirements for this to be possible are heavy doping and a very narrow junction; junction areas are of the order 0·025 in^2. Since the tunnelling action is due to majority carriers, the frequency response of such devices is virtually equal to that of a normal conductor up to the 'valley' point V in Fig. 8.3. At greater voltages, the junction barrier potential is reduced to a level that permits conventional current flow. Charge storage in that region may be significant, particularly when switching the diode back from the high to the low voltage state.

It is possible to reduce the peak (P in Fig. 8.3) in the forward region to negligible proportions (0·05–1 mA) resulting in the back diode, also known as tunnel rectifier. The construction of back and tunnel diodes is basically the same, but the doping level is reduced for back diodes (though it is still very much larger than for conventional junction diodes). A typical back-diode characteristic is shown in Fig. 8.4; the maximum current at P is 0·2 mA for the BD3 (General Electric) back diode used extensively. (The measured average value of ten samples chosen at random was 96 μA.) The

163

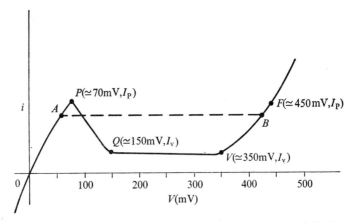

FIG. 8.3. Typical i/v characteristic for germanium tunnel diode.

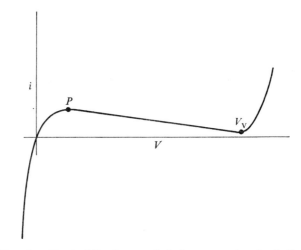

FIG. 8.4. Typical i/v characteristic for germanium back diode.

back diode switches from the OFF region ($0 < V < V_v$) to the negative region in Fig. 8.4 with a speed comparable to that of the tunnel diode. It provides a means of coupling between tunnel diodes and of buffering tunnel and back diodes from one another. An important application of this will be mentioned later. Silicon and gallium arsenide are alternative materials, but are inferior to germanium in respect of speed (silicon) and reliability (gallium arsenide).

8.1.2. *Tunnel-diode behaviour*

Let us now look into tunnel-diode switching speed and also discriminator sensitivity as a function of pulse shape.

164

The switching-speed calculations below apply for a step-function input with negligible rise time. The computations are based on the equivalent circuit of Fig. 8.5. C is the sum of diode plus circuit capacitance. The

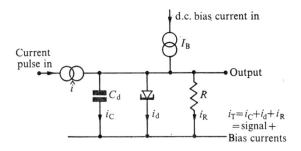

FIG. 8.5. Basic tunnel-diode switch—equivalent circuit.

equation for Fig. 8.5 is

$$\hat{\imath} + I_B = i_c + i_d + i_R \qquad (8.1)$$

where $\hat{\imath}$ = signal current,

I_B = bias current through diode,

i_c = current flowing through capacitor,

i_d = diode current, and

i_R = current through load R.

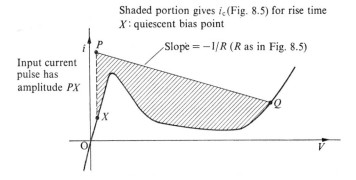

FIG. 8.6. Tunnel-diode characteristic—rise time.

Figs. 8.6 and 8.7 show charge and discharge currents flowing into and out of C when the diode is switched from the low to the high voltage stage (Fig. 8.6) and back (Fig. 8.7). Switching speed can be calculated to a very close approximation by the method of piece-wise linear analysis. The tunnel

165

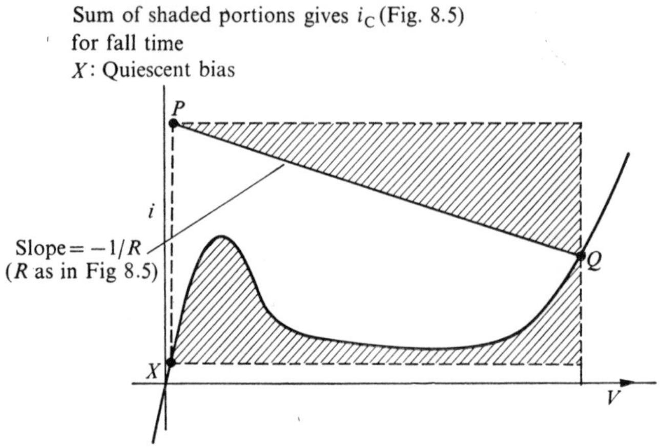

Sum of shaded portions gives i_C (Fig. 8.5)
for fall time
X: Quiescent bias

FIG. 8.7. Tunnel-diode characteristic—fall time.

diode is quiescently biased to X and switched to the high-voltage state. Rise and fall times are obtained by integrating

$$t = \int \frac{C dV}{i_c} \qquad (8.2)$$

over the appropriate interval. The following assumptions are made:

(1) Input-signal rise and fall times are very short compared to tunnel-diode rise and fall times.
(2) The input pulse width exceeds the sum of t_r and t_f.
(3) The effects of diode and passive-component lead inductances may be ignored.
(4) C is independent of voltage. (The theoretical relationship over the range $0 < V < V_v$ is $C \propto (\phi - V)^{-1/2}$ where ϕ is 0·6 for germanium and 1·1 for silicon.

The tunnel-diode i/v characteristic is approximated by Figs. 8.8 and 8.9, which correspond to Figs. 8.6 and 8.7.

For the unloaded diode i_R is zero; in practice the diode is often only lightly loaded so that i_R can be neglected. Since $\hat{i}$ and I_B are constant over the rise-time interval, differentiation of (8.1) gives

$$d i_c = -d i_d. \qquad (8.3)$$

Consider interval BC in Fig. 8.8 as example. Over BC

$$i_d = \{I_p - (V - V_B)/R_d\} \qquad (8.4)$$

where R_d, the modulus of the slope resistance of BC, is given by

$$R_d = (V_c - V_p)/(I_p - I_v). \qquad (8.5)$$

166

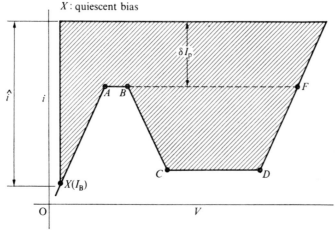

FIG. 8.8. Approximate i/v tunnel-diode characteristic—rise-time calculation.

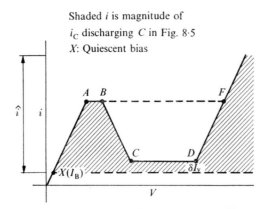

FIG. 8.9. Approximate i/v tunnel-diode characteristic—fall-time calculation.

From (8.3), $dV = -R_d \times di_c$, and (8.2) becomes

$$t_{BC} = CR_d \int_{i_c \text{ at } B}^{i_c \text{ at } C} \frac{di_c}{i_c}$$

$$= CR_d \ln \{(I_B + \hat{\imath} - I_v)/\delta I_p\}. \qquad (8.6)$$

δI_p is the amount by which the peak current excursion exceeds I_p, i.e.

$$\delta I_p = I_B + \hat{\imath} - I_p. \qquad (8.7)$$

167

Another convenient parameter, this time for calculating switching time for resetting the diode from the high- to the low-voltage state, is δI_V, the excess of peak current below I_V.

$$\delta I_v = I_v - I_B. \tag{8.8}$$

The complete expressions for rise and fall times in Figs. 8.8 and 8.9 are

$$t_r = t_{XA} + t_{AB} + t_{BC} + t_{CD} + t_{DF}, \tag{8.9}$$

where

$$t_{XA} = (CV_p/I_p)\ln(\hat{\imath}/\delta I_p), \tag{8.10}$$

$$t_{AB} = C(V_B - V_A)/\delta I_p, \tag{8.11}$$

$$t_{BC} = CR_{BC}\ln\{(I_B + \hat{\imath} - I_v)/\delta I_p\}, \tag{8.12}$$

and

$$R_{BC} = (V_c - V_p)/(I_p - I_v). \tag{8.13}$$

R_{BC} has been used because, in the absence of published i/v characteristics; it is closely given by the empirical formula

$$R_{BC} = \tfrac{1}{2}(V_v - V_p)/(I_p - I_v), \tag{8.14}$$

$$t_{CD} = C(V_v - V_c)/\{\hat{\imath} + I_B - I_v\}, \tag{8.15}$$

$$t_{DF} = \frac{C(V_F - V_v)}{(I_p - I_v)}\ln\left\{\frac{\hat{\imath} + I_B - I_v}{\hat{\imath} + I_B - I_p}\right\}. \tag{8.16}$$

Switching times are calculated to point F in Fig. 8.8 because the circuit driven by the diode has usually been actuated by then.

Inspection of the above equations shows that the switching times are largely dependent on C/I_p, which is the figure of merit for a tunnel diode. C/I_p appears directly in the expressions for t_{XA} and t_{AB}, and indirectly in t_{BC} and t_{DF} (remembering that $I_p \gg I_v$).

We can now calculate switching times for the types of tunnel diodes used in our instrumentation. The parameters of the diodes are listed in Table 8.1. C/I_p is not the only criterion determining choice of a diode. In many instances the diode remains in the high-voltage state where the dissipation will be relatively high for long periods. It is necessary to choose a diode capable of withstanding this dissipation. The point is stressed because tunnel diodes are available with lower C/I_p ratios than those for the devices listed in this chapter, but with insufficient anode dissipation to permit operation in the high-voltage state. Typical figures for the various voltages in Fig. 8.8 are $V_A = 60$ mV, $V_B = 80$ mV, $V_C = 150$ mV, $V_D = 350$ mV ($=V_v$), and $V_F = 520$ mV.

Calculated rise times for the IN 3713 (operating with zero bias) and the

TABLE 8.1

Tunnel-diode parameters

| Type No. | I_p (mA) | C (pF) | | I_p/C (mA/pF) |
		typ.	max.	typ.
IN 3713	1	3·5	5	0·29
IN 3715	2·2	7	10	0·31
IN 3717	4·7	13	25	0·36
IN 3128	5	7	15	0·71
IN 3129	20	10	20	1·0
IN 3857	5	6	8	0·85
IN 3858	10	6	8	1·67
IN 3859	20	8	10	2·50
40566	5	11	15	0·45
40571	5	6	8	0·85
40572	10	6	8	1·67
40573	20	8	10	2·50

IN 3858 (biased forward by 8 mA) are plotted against signal amplitude in Figs. 8.10 and 8.11. The standard discriminators in use here operate with the above bias conditions when the threshold is set for maximum sensitivity. These indicate time walks of the order 7 and 1 ns respectively, results that agree with observed behaviour.

Fall times can be calculated from Fig. 8.9 in a manner similar to the rise-time evaluations based on Fig. 8.8. In the faster discriminators operating

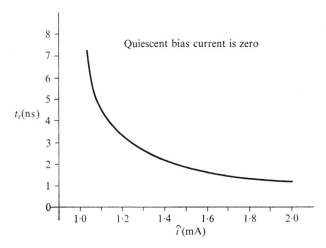

FIG. 8.10. Calculated t_r for IN 3713 tunnel diode.

at high peak rates the tunnel diode is reset by a current generated in the reset circuit. Fall time is thus determined by the reset circuit, whereas rise time is determined by the input signal.

An approximation often quoted for t_r is

$$t_r \sim C(V_v - V_p)/(I_p - I_v), \qquad (8.17)$$

$$\sim C(V_v - V_p)/0{\cdot}9I_p. \qquad (8.17a)$$

Eqn (8.17a) applies for signals substantially above the threshold $(\delta I_p/I_p > 0{\cdot}2)$ and implies that the rise time consists largely of the interval

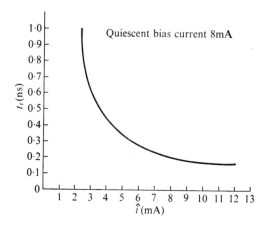

FIG. 8.11. Calculated t_r for IN 3858 tunnel diode.

needed to traverse $BCDF$ in Fig. 8.8. Calculated rise times based on (8.17a) are given in Table 8.2. Though approximate, they give some idea of the relative speeds obtainable.

Circuit capacitance is readily allowed for by adding it to C in Table 8.1. Some of the diodes are capable of switching at clock rates in excess of 100 MHz. Taking the IN 3859, putting t_f equal to $2t_r$, allowing 10 pF for circuit capacitance and 2 ns for propagation delays in the external circuit, the maximum repetition rate $1/(t_r + t_f + 2)$ comes to 312 megapulses/s.

Actual signal pulses are not step functions but near-triangular in shape. For satisfactory operation the tunnel diode should be triggered by section OP of the signal in Fig. 8.12. Consider the case of a diode biased in the low-voltage state close to the peak and switched to the high-voltage state by section OP of the input in Fig. 8.12(b). Assuming the diode to be switched to V_v for satisfactory triggering,

$$(\hat{\imath} \times t_r)/2 \gg (C_d + C_s)(V_v - V_p), \qquad (8.18)$$

TABLE 8.2

Approximate rise times for tunnel diodes
(Values as given in Table 8.1 have been taken for C)

Type No.	I_p (mA)	t_r (ns)	
		typ.	max.
IN 3713	1	1·4	2·1
IN 3715	2·2	1·3	1·9
IN 3717	4·7	1·1	2·5
IN 3128	5	0·6	1·2
IN 3129	20	0·2	0·4
IN 3857	5	0·5	0·7
IN 3858	10	0·3	0·4
IN 3859	20	0·2	0·2
40566	5	0·9	1·2
40571	5	0·5	0·7
40572	10	0·2	0·3
40573	20	0·2	0·2

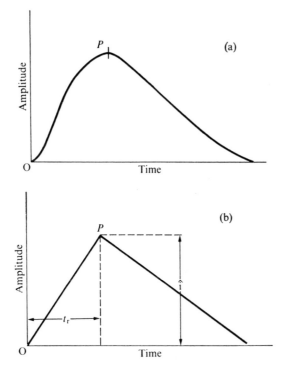

FIG. 8.12. Linear approximation of signal. (a) Actual signal. (b) Approximation.

171

where C_d is the diode and C_s the circuit capacitance. In practice the maximum forward bias for stable operation is $0.8I_p$. For a signal lifting the diode 10 per cent above I_p, $\hat{i}$ then equals $(0.2I_p + 0.1I_p) = 0.3I_p$, and $(V_v - V_p) \simeq 0.3$ V. Hence (8.18) reduces to

$$0.5I_p \times t_r > (C_d + C_s). \tag{8.19}$$

Hence the minimum rise time for satisfactory triggering is given by

$$t_r(\min) = 2(C_d + C_s)/I_p \tag{8.20}$$

for signals of amplitude $\hat{i} = 0.3I_p$.

Table 8.3 gives $t_r(\min)$, calculated from (8.20).

TABLE 8.3

Minimum rise time for satisfactory triggering

Type	I_p (mA)	i (mA)	$t_r(\min)$ (ns)
IN 3713	1	0.3	17
IN 3715	2.2	0.7	11
IN 3717	4.7	1.5	8
IN 3128	5	1.6	5
IN 3129	20	6.6	1.5
IN 3857	5	1.6	4.4
IN 3858	10	3.3	2.2
IN 3859	20	6.6	1.3
40566	5	1.6	6.4
40571	5	1.6	4.4
40572	10	3.3	2.2
40573	20	6.6	1.3

Experimental evidence supports Table 8.3 as regards $t_r(\min)$ though i tends to be $0.2I_p$ rather than $0.3I_p$. Tests with a photomultiplier simulator giving near-triangular pulses with 2 ns rise time and 5 ns fall time showed that the IN 3858, IN 3129, and IN 3959 performed well, whereas the IN 3128 and IN 3857 did not trigger decisively.

8.1.3. *Operation with quiescent bias below I_v*

A typical circuit is shown in Fig. 8.13 where the signal is fed to the diode via a common-base amplifier J_1, which offers a high source impedance to the tunnel diode. R_1 in the emitter path matches the input impedance to the coaxial cable (50 or 100 Ω) connecting photomultiplier and discriminator.

$J2/J3$ is an emitter-coupled pair with $J2$ OFF and $J3$ ON, the V_{EB} drop for $J2$ being set by $RV1$ to about 150 mV below conduction value (0·3 V for germanium, 0·7 V for silicon). The tunnel diode, when triggered, will divert current from $J3$ into $J2$ and this condition will be reversed again when the diode returns to the low-voltage state. This monostable-trigger circuit is only practicable with the diode biased below the valley current (in practice this means reverse bias) and the maximum discriminator sensitivity will then be I_p.

An important advantage of the tunnel diode over other discriminator elements is the inherent hysteresis of the device. When a signal has just

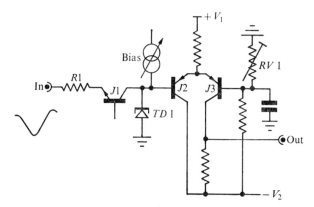

Fig. 8.13. Discriminator for bias below I_v.

exceeded P in Fig. 8.3 the diode will remain in the high-voltage state until the signal decreases by an amount of about I_p or more. In other discriminators the element would revert to the original state for a slight decrease in signal under the above conditions. Hysteresis gives more decisive triggering and for that reason is provided in most discriminators by adding special circuits. No such addition is needed for the tunnel diode; the hysteresis holds for all discriminator circuits described in this chapter. It is also an important feature of the synchronizing discriminator in Chapter 10.

8.1.4. *Operation with quiescent bias between I_v and I_p*

Discriminator sensitivity can be increased by forward biasing the diode to somewhere along OP in Fig. 8.3. Practical considerations, namely worst case design allowing for tolerance and temperature dependence of I_p, limit the maximum forward bias to about 0·8I_p.

Such a forward-biased trigger circuit differs from that of the previous section in two respects. Suppose the diode is biased at a point A in Fig.

8.3. A triggering pulse will switch the diode to its high-voltage state and the diode will remain there on removal of the pulse, providing the shunt load across it is not too great, because it now has an alternative stable quiescent point along VF (B in the case of negligible loading). Note that no such point existed in the case of quiescent static bias below I_v. Also in the absence of signals a surge occurring when the power supplies are switched on, or a transient break-through on one of the power lines, may switch the diode to the high-voltage state, rendering the discriminator inoperative.

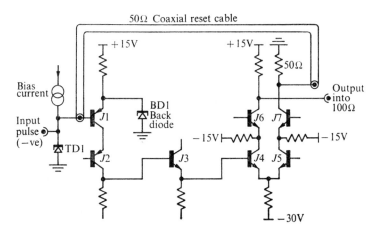

FIG. 8.14. Ultra-fast pulse former.

The techniques employed in one of our fast discriminators and sketched out in Fig. 8.14 provides both a reset pulse and a guard against accidental triggering.

The essence of the circuit is the d.c. loop of the path tunnel-diode–trigger-circuit reset cable–tunnel diode. When triggered $TD1$ will turn on $J1$ diverting the standing current through the back diode $BD1$ into $J1$. This current is then passed via the common-base amplifier $J2$ to the emitter-follower $J3$, which triggers the emitter coupled pair $J4/J5$ turning $J4$ ON and $J5$ OFF. $J2$, $J6$, and $J7$ are common-base amplifiers which remain in conduction throughout the entire switching cycle. They present low impedances to the collectors of $J1$, $J4$, and $J5$, thereby increasing their switching speed, and pass the current pulses with only very small increase in rise- and fall-time to their respective collectors. The pulse at the collector of $J7$ resets the tunnel diode via the reset cable whose length, together with the constant circuit delay, determines the width of the output pulse. Consequently the user can adjust the width of the output pulse by choosing

174

the appropriate length of reset cable. Note that the output pulse ($J6$ collector) is directly coupled to the output socket and based at zero d.c. level. This particular pulse former has a maximum rate of about 125 megapulses/s.

An improved pulse former operating similarly to Fig. 8.14 and now in extensive use is shown in Fig. 8.15 (Herbst 1965, 1969). $J4$ and $J5$ are common-base amplifiers arranged to be in conduction throughout the entire cycle, and the npn–pnp combination in each branch is a simple method of obtaining the required d.c. reference levels for reset and output pulses. The reset cable is now in the emitter path of $J4$ and is matched by $R1$. The current through $R2$ provides zero base line for the output pulse;

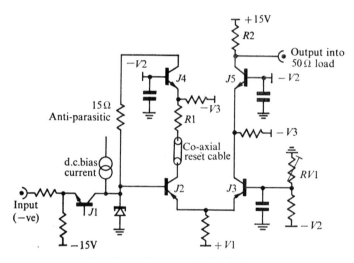

FIG. 8.15. Improved pulse former.

without it the base line would be negative by an amount $50 \times I$ volts, I being the quiescent collector current (usually a few milliamperes) of $J5$. The bias of $J2$ is adjusted by $RV1$ so that the tunnel diode will cause $J2$, $J3$ to change over when switching to the high-voltage state (in practice, V_{eb} of $J2$ is forward-biased to about 450 mV for silicon planar transistors).

The main advantages of the circuit, apart from its simplicity, are:

(a) The common-base amplifier is the fastest configuration available; a typical propagation delay is 0·3 ns at 10 mA signal level.

(b) The entire current steered from $J3$ into $J2$ is available for resetting the tunnel diode, compared with only half this amount in the circuit of Fig. 8.14. Consequently, for a given tunnel diode, only half the current of the emitter-coupled pair in Fig. 8.14 will have to be steered by the corresponding transistors of Fig. 8.15. When aiming at the highest possible pulse rates

175

this is important, because the switching speed of $J2$ and $J3$ increases with decreasing current levels for a given tunnel-diode drive.

Pulse rates of 100 megapulses/s have been obtained consistently using the circuit shown in Fig. 8.15.

8.1.5. *Drive circuit for tunnel diode*

The input circuit generally chosen, specially for ultra-fast operation, consists of the cascaded npn–pnp common-base amplifiers $J1$ and $J2$ in Fig. 8.16 where a negative input signal is assumed. The order of amplifiers and

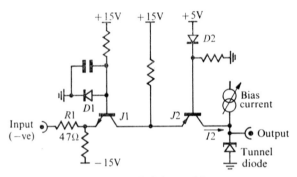

FIG. 8.16. Drive amplifier.

the polarity of the tunnel diode would be reversed for a positive signal. $J1$ stands in a quiescent current of a few milliamperes, the level being determined to give a good h_{fe} and f_t. $R1$ matches the amplifier to the characteristic impedance of the transmission system, 50 Ω in the case of Fig. 8.16. The input resistance is given by

$$r_{in} = R1 + r_e + r_b/(1 + h_{fe}) \tag{8.21}$$

$$= R1 + \{26/I_e(\text{mA})\} + r_b/(1 + h_{fe}). \tag{8.21a}$$

For u.h.f. transistors $r_b \sim 30$ Ω, $h_{fe} \sim 20$, and I_e will be about 5 mA giving $r_{in} \sim (R1 + 7)$ Ω. Correct matching for 50 Ω demands $R1$ equal to 43 Ω. Precise matching is not necessary and a 39 or 47 Ω resistor will be acceptable in practice. $D1$ and $D2$ are compensating diodes which maintain the collector currents of $J1$ and $J2$ virtually independent of temperature. The diode compensation maintains constant emitter current. Collector and emitter current are related by

$$I_c = \frac{h_{fe}}{1 + h_{fe}} \times I_e \tag{8.22}$$

Differentiating (8.22),

$$\frac{dI_c/I_c}{dt} = \frac{1}{1 + h_{fe}} \times \frac{dh_{fe}/h_{fe}}{dt}. \tag{8.23}$$

In silicon planar transistors, $(dh_{fe}/h_{fe})/dt$ lies between 0·5 and 2 per cent/°C. If we take a value of 1 per cent and put h_{fe} equal to 20 $(dI_c/I_c)/dt$ ~ 0·05 per cent/°C. For I_c equal to 5 mA the drift in I_c will then be 2·5 μA/°C. The circuit in Fig. 8.16 compensates to some extent for this because the increase in $J1$ collector current decreases the emitter current of $J2$. The above calculations apply to quiescent d.c. conditions. Signal analysis is more complex; the temperature dependence of r_e (which is directly proportional to absolute temperature) has to be allowed for. Calculations indicate perfect compensation when $R1$ has a value between 30 and 100 Ω. Experimentally we have found excellent temperature stability with this circuit. $J2$ limits the signals to a peak amplitude I_2. This limiter is very efficient in protecting the tunnel diode from large photomultiplier surges and recovers quickly from overload. The circuit is d.c. coupled and hence rate-independent.

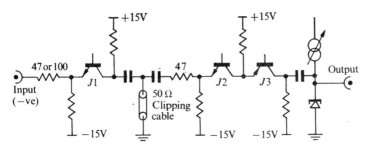

FIG. 8.17. Amplifier with clip.

Sometimes there is a need to shorten input pulses by clipping, for example to prevent multiple firing of the pulse former by wide signals. A suitable input circuit is outlined in Fig. 8.17 where $J1$ and $J3$ perform similarly to $J1$ and $J2$ in Fig. 8.16. The clipping cable is connected in the emitter circuit of $J2$ and d.c. blocking capacitors can now be used because the bipolar pulses produced by the clipping ensure rate-independent performance. $J2$ limits the positive and $J3$ the negative section of the bipolar pulse.

8.1.6. *Bias circuit*

A circuit developed to provide variable threshold bias for the tunnel diode will now be described by referring to Fig. 8.18, the complete bias arrangement for one of our discriminators. The diode current has to be proportional to the setting of the helical potentiometer $RV2$ and must also be independent of ambient temperature.

In this discriminator, the bias of the 5 mA tunnel diode $TD1$ ranges from 4 mA forward (i.e. within 1 mA of peak current) to 3 mA reverse. Remembering that the quiescent collector current I_1 of the drive transistor flows through TD_1, the bias current I_B has to range from 10 to 17 mA.

Consider first transistors $J2$ and $J3$. A voltage change δV_{in} at the base of $J2$ will result in an equal voltage change at $J3$ emitter due to the emitter-follower action of these transistors. $J1$ present a very high emitter load for $J2$ resulting in very good linearity for $\delta V_e/\delta V_{in}$. The current change δI_2 is then given by $(\delta V_{in}/R1)$ and this equals δI_B (assuming equal collector and emitter currents for $J3$). Temperature compensation is ensured by the pnp–npn combination $J2$, $J3$. Zener diodes $ZD1$ and $ZD2$ stabilize the negative supply voltage. To obtain good temperature stability we use Zener diodes with a breakdown voltage round 5·6 V. Zener breakdown has a

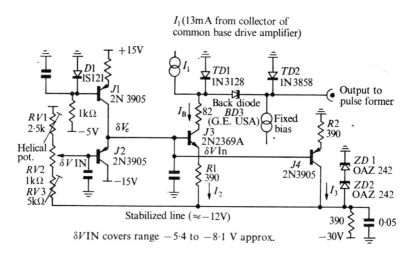

FIG. 8.18. Discriminator bias circuit.

negative and avalanche breakdown a positive temperature coefficient, resulting in very good temperature stability in the transition region (around 5·6 V) between these two mechanisms. In 5·6 V Zener diodes there is a typical temperature coefficient of 0·005 per cent/°C at currents of a few mA. The more expensive alternative is to use a reference element containing a Zener diode, constructed for avalanche breakdown, in series with an ordinary junction diode that has a negative temperature coefficient. Numerous such devices are available: the IN 825 with 6·2 V and a maximum temperature coefficient of 0·002 per cent/°C at 7·5 mA is a representative example. In each case the current through the Zener diode(s) must remain constant, i.e. independent of bias setting ($RV2$) to realize the above temperature stability. This is achieved by $J4$. A change δV_e increases I_2 by $\delta V_e/R1$ and

178

decreases I_3 by an equal amount $\delta V_e/R2$ (since $R1$ and $R2$ are equal) maintaining $I_2 + I_3$ constant.

To obtain the best possible stability for ambient thermal gradients, $J2$ and $J3$ should be contained in the same envelope. Suitable dual npn–pnp transistors are available for this purpose.

8.1.7. *Tunnel-diode–back-diode–tunnel-diode coupling*

In many discriminators the threshold bias extends from somewhere near I_p into the reverse region. A circuit suitable for these conditions and providing an output pulse of constant width is shown in Fig. 8.19. Fig. 8.20 gives

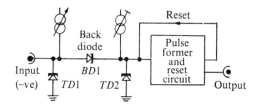

FIG. 8.19. Tunnel-diode–back-diode–tunnel-diode coupling.

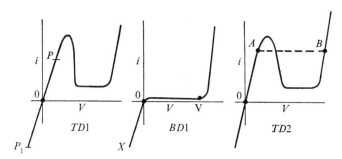

FIG. 8.20. Characteristics relevant to Fig. 8.19.

the i/v characteristics of the various diodes. The bias range of $TD1$, the discriminator diode, extends from P to P_1. $TD2$ is biased to A near the peak. Normally $TD2$ has a peak current twice that of $TD1$. This is not a necessary condition for circuit operation but is preferred practice. Quiescently $BD1$ will be biased to zero ($TD1$ at P) or at a point along OV close to O ($TD1$ at P_1). A negative input signal will switch $TD1$ to the high-voltage state. $TD1$ will switch $TD2$ via $BD1$, now in the high-conductance region OX. When $TD2$ switches to B, $BD1$ reverts to a low conductance. The circuit driven by $TD2$ will generate a reset pulse of fixed duration as described in section 8.1.4. This reset will return $TD2$ to the low-voltage

179

state and will act likewise on $TD1$, if necessary, via $BD1$. There will be two possible conditions:

(i) $TD1$ is biased above I_v. It will then be in the high-voltage state on removal of the signal and will be returned to its bias point by the reset pulse.

(ii) $TD1$ is biased below I_v. It will then return to its bias point on removal of the signal, but $TD2$ will remain in the high-voltage state being buffered by $BD1$. Some of the reset current will be applied to $TD1$ via $BD1$ momentarily biasing $TD1$ more reverse. This will have no adverse effect on discriminator operation.

An important advantage of the circuit is the provision of a powerful drive for the pulse former. The coupling from $TD1$ to $TD2$ gives significant current gain.

8.2. Pulse formers

Fig. 8.15 is a standard pulse former for our fast discriminators giving a constant dead time equal to twice the output pulse width. It is capable of minimum clock rates of 100 MHz at an output of 14 mA into 50 Ω in production. The rise time of the output can be calculated from Roehr (1963),

$$t_r = 0 \cdot 8 \times \frac{I_c}{E_1} \times \left\{ \frac{1}{2\pi f_1} + R_1 \times C' \right\} /(R_s + 2r_b'),\qquad(8.24)$$

which applies to $J2$ and $J3$ in Fig. 8.15 working into a collector load R_1 via the common-base amplifiers $J4$ and $J5$. I_c is the switched current and E_1 the amplitude of the switching signal above the conduction threshold of $J2$ base. C' is the sum of transistor output (C_{ob}) and circuit capacitance, and R_s the resistance of the source driving $J2$ base, whose base spreading resistance is r_b'. Fig. 8.15 is similar to the cascode amplifier. The common-base amplifiers presents a very low-input impedance to $J2$ and $J3$. The effect of capacitive collector-base feedback for $J2$ and $J3$ (which was not taken into account in the treatment yielding (8.24)) is thereby greatly reduced, because voltage swings at the collectors of $J2$ and $J3$ are very small. The common-base amplifier has a bandwidth

$$B = 1/(1/f_1 + 2\pi R_1 \times C')\qquad(8.25)$$

and its rise time is related to B by (Lewis and Wells 1959)

$$t_r(\text{C.B.}) = 0 \cdot 5/B.\qquad(8.26)$$

In eqn (8.26) the factor $0 \cdot 5$ is an empirical preference to the theoretical $0 \cdot 31$.

180

Typical values for the fast transistors used (type 2N918 or similar) are $f_1 = 900$ MHz, $C_{ob} = 1 \cdot 5$ pF, $R_1 = 50$ Ω, and $r_b' = 30$ Ω. Allowing $1 \cdot 5$ pF for circuit capacitance, $C' = 3$ pF. Substituting in (8.25) and (8.26) gives t_r(C.B.) equal to $1 \cdot 2$ ns.

Reverting to (8.24) we calculate the performance for a pulse-former like Fig. 8.15 driven by the circuit of Fig. 8.19 in which $TD2$ is a 10 mA diode. I_c is chosen to be 14 mA, R_s put at 10 Ω, and E_1 at 240 mV. It is difficult to quote a precise value for E_1. The figure we have given is arrived at as follows: $RV1$ in Fig. 8.15 sets the V_{EB} of the OFF transistor $J2$ forward to 500 mV. The tunnel diode will change $J2$ base from -60 mV to -500 mV when switching to the high-voltage states. If we take 700 mV to be the threshold value mentioned below (8.24), E_1 equals 440 mV (swing at

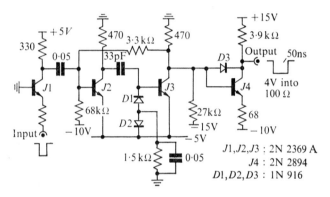

Fig. 8.21. Pulse former giving A.E.R.E. type 2 output.

$J2$ base) less 200 mV (amount of swing needed to bring base from 500 to 700 mV), i.e. 240 mV. From (8.24) t_r comes to $1 \cdot 0$ ns.

The overall rise time T_r is related to t_r and t_r(C.B.) by

$$T_r = \{(t_r)^2 + (t_r(\text{C.B.}))^2\}^{1/2}, \qquad (8.27)$$

giving T_r equal to $1 \cdot 6$ ns, a value in good agreement with observed rise times between $1 \cdot 5$ and $1 \cdot 8$ ns on production models.

Slower discriminators require pulse formers with wider outputs. Fig. 8.15 is not suitable for pulse widths exceeding 40 ns; the length of coaxial cable needed becomes excessive. Monostable multivibrators with CR timing are used and examples of these are given in Figs. 8.21–8.23. The recovery times are about one-third of the pulse width in each case. Fig. 8.21 gives an output conforming to the A.E.R.E. 2000 series type 2 specification. The output is -4 V into 100 Ω load with a rise time of 5 ns. The emitter-follower output stage in all the pulse formers shown has the following features:

181

(i) The output is d.c. coupled to the load with a d.c. quiescent level within ±70 mV of zero.

(ii) An un-bypassed 68 Ω collector resistor protects the emitter-follower from excessive surges in case of accidental grounding of the output.

(iii) A diode connected between base and emitter protects the transistor from surges that exceed the reverse V_{eb} rating. Such damage may occur with the output a long way from the module, connected to it via a coaxial cable, and short-circuited by accident.

Figs. 8.22 and 8.23 give outputs conforming to the NIMS specifications of the U.S.A. Atomic Energy Commission. All pulse formers shown are

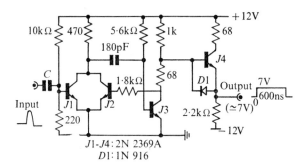

FIG. 8.22. NIMS pulse former—slow.

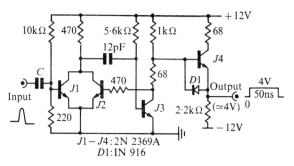

FIG. 8.23. NIMS pulse former—fast.

saturated monostables using npn transistors. At present npn devices are superior to pnp for fast saturated switching because they have much smaller storage times for excess minority carrier storage in the base. We have found it advisable to have buffer transistors between the trigger and the monostable. Some designers omit these on grounds of economy. In Figs. 8.22 and 8.23 positive trigger inputs are applied via C, which is sometimes chosen to differentiate the input signal. The buffer transistors are readily adapted for

182

negative trigger signals by grounding the base and applying the input to the emitter.

8.3. Performance of standard discriminators

The performance of two standard discriminators in general use at A.E.R.E. is now summarized.

	Type A	*Type B*
Discriminator element	Tunnel diode IN 3713. Threshold 1·3–10·2 mA for input pulse with 7 ns rise- and fall-times	Tunnel diode IN 3858. Threshold 2–20 mA for input pulses with 1·5 ns rise- and fall-times
Input impedance	50 or 100 Ω	50 Ω
Output(s)	A.E.R.E. type 2 digital coupling pulse. −3 V into 100 Ω. Rise time 10 ns. Dead time 100 ns	(i) Digital coupling level level 2 ns rise time and 3 ns fall time, amplitude of 28 mA into 50 Ω. Pulse width determined by chosen length of coaxial cable plugged into sockets provided. Minimum pulse width at 10 per cent peak amplitude 5 ns (giving minimum dead time of 10 ns). (ii) A.E.R.E. type 2 digital coupling pulse triggered by (i) above: −4 V into 100 Ω. Rise time 5 ns.
Time walk	$(4 + t_r)$ ns for input signals that rise to a level greater than 3 per cent above threshold; t_r is rise time of input pulse	$(1 + t_r)$ ns for conditions defined in entry under type A
Threshold temperature stability over range 25–45 °C.	Sensitivity decreases by about 0·1 mA	Sensitivity decreases by about 2·5 per cent

8.4. Constant dead-time generation

The pulse former in Fig. 8.15 has a dead time independent of signal amplitude and width (provided this is less than the width of the pulse-former output). The same cannot be said of the monostables in Figs. 8.21–8.23. There the recovery time is a function of the CR timing constant and the

183

monostable dead time depends therefore on the amplitude of the trigger signal. The pulse former is usually fed from an emitter-coupled current switch driven by a tunnel-diode discriminator. The output amplitude of this switch depends somewhat on signal amplitude, especially when the

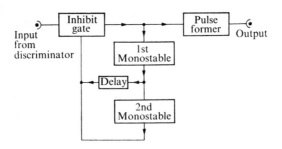

FIG. 8.24. Schematic—constant dead-time generator.

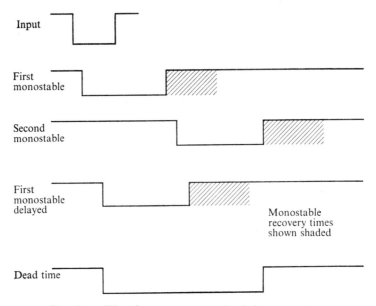

FIG. 8.25. Waveforms—constant dead-time generator.

latter is a fast pulse that does not trigger the tunnel diode decisively. What this amounts to is that while the tunnel diode is the major discriminating element, the subsequent stages perform some discrimination.

Constant dead time is ensured by using two monostables as shown in the block schematic of Fig. 8.24. Fig. 8.25 gives the relevant waveforms. The output of the first monostable is delayed so that the signal cannot break through the inevitable gap between the two monostable outputs. The above

184

arrangement is a convenient method of providing a variable dead-time discriminator.

8.5. IC discriminators

A standard IC discriminator is sketched in Fig. 8.26 (S. G. S. Fairchild 1967). J_1 and J_2 constitute a differential amplifier with a constant current source J_3. An input change of 2 mV is sufficient to cause a change in the output by saturating amplifier A. In the Type 710 discriminator, an IC element originated by Fairchild and made by many other manufacturers, output levels are -0.5 V and $+3$ V with a time walk of about 30 ns for conditions as defined in section 8.3. This element is much slower than the

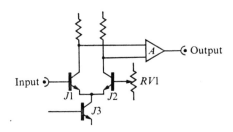

Fig. 8.26. Schematic—IC discriminator.

discriminators mentioned in this chapter. This, together with the need of a pulse former to give a specified digital coupling pulse, means that the Type 710 IC is not generally suitable for fast nuclear instrumentation. However, discrete component discriminators are again being built with a front end similar to Fig. 8.26 now that very fast transistors are freely available. The advantages over the tunnel diode are:

(i) excellent threshold stability with temperature. The coefficient is about 3.5 μV/°C—two orders of magnitude better than the coefficients quoted in section 8.3;

(ii) wide dynamic range from a few millivolts to about 5 V;

(iii) easy polarity reversal. Threshold polarity is reversed by interchanging the circuits at the bases of J_1 and J_2 in Fig. 8.26;

(iv) increased reliability. The tunnel diode is a relatively delicate device.

For slower working with discriminator dead times of 100 ns or higher, transistor discriminators on the lines of Fig. 8.26 are likely to supersede the tunnel-diode discriminator which remains the best configuration for fast working.

References

HERBST, L. J. (1965) *Electron. Lett.* **1**, 43.

—— (1969) *Nucl. Instrum. Meth.* **70**, 189.

LEWIS, I. A. D. and WELLS, F. H. (1959) *Millimicrosecond pulse techniques*, 2nd edn, p. 7.

ROEHR, W. (ed.) *Switching transistor handbook*, p. 238. Motorola Inc., Arizona, U.S.A.

R.C.A. Technical manual TD-30, tunnel diodes. R.C.A., U.S.A., 1963.

FAIRCHILD, S. G. S. *The application of linear microcircuits*, 2nd issue. 1967.

9 *Counting Circuits*

By F. H. HALE

9.1. Introduction

THE most commonly used counting instruments are usually referred to as scalers. This name probably stems from the days when scaling became necessary to reduce the increasing count-rates to a frequency acceptable by the then only easily available counter—the Post Office mechanical register. It was a small step to add indicators such as miniature neon lamps to show the state of the scaling circuits. With the advent of transistors their economy of power and space, together with the very fast switching attainable, led to the exclusion of the Post Office register and the original reason for scaling. The name has so far remained, but there are indications that the correct term, counting register, is becoming more widely adopted. Scalers and counting registers employ the same basic circuits, but whereas scalers need neither display of, nor access to, the stored information, counting registers always have one and/or the other and often scaling facilities as well.

9.1.1. *Dead time, recovery time, and resolving time*

During the time an electronic circuit is processing an accepted signal it is *dead* to further inputs. This *dead time* is normally followed by a further period, known as the *recovery time*, during which the circuit can be operated with gradually increasing sensitivity from zero to normal. Since it is accepted practice to drive counting circuits by standardized signals there will be some point during the recovery time when the circuit will respond to another input. The dead time plus the non-responsive part of the recovery time is the *resolving time* of the circuit for the particular input signal. Any attempt to drive a circuit with input signals separated by less than its resolving time will cause it to miss counts. With evenly spaced inputs at frequencies up to twice the maximum acceptable without loss it will miss alternate inputs and divide, or scale, by 2. With still higher frequencies it will divide by 3, then by 4, and so on. While this principle has been used—particularly in 'blocking oscillator' form—for scaling up to factors of 5 or so, it is unreliable unless component tolerances and supply voltages are rigidly controlled; and at still higher frequencies it is difficult to prevent a factor of, say, 10 drifting to 9 or 11, or even further. Clearly, scaling in this manner

187

is possible only near to a fixed-input frequency and it is generally referred to as frequency division.

9.1.2. *Pulses occurring randomly in time*

The nature of nuclear radiation, its random occurrence, the detection, amplification, and selection processes have all been dealt with in other chapters. In particular, it has been shown that the true mean rate N_m is related to the observed rate N_o and the resolving (dead) time τ by

$$N_m = \frac{N_o}{(1 - N_o \tau)}. \tag{9.1}$$

If $N_o \tau$ can be kept small it may be permissible to omit resolving-time corrections at the end of a count. For instance, if $N_o \tau < \dfrac{1}{100}$, N_o approximates to N_m to within 1 per cent. In other words, a circuit that can accept uniformly-spaced input signals up to a frequency f can be used to count randomly spaced signals with a mean frequency of only $f/100$ if 1 per cent accuracy is required without correction.

It is usual when dealing with randomly spaced events to express the accuracy of any particular count in terms of the standard deviation, which is defined as the square root of the count, T, i.e.

$$\text{Standard deviation} = \sqrt{T}. \tag{9.2}$$

This is sometimes more conveniently expressed as a percentage of the total count.

$$\text{Percentage standard deviation} = \frac{\sqrt{(T)} \times 100}{T} = \frac{100}{\sqrt{T}}. \tag{9.3}$$

The percentage deviations of various counts are tabulated thus:

100	1000	10 000	100 000	Total counts
10%	3%	1%	0·3%	Standard deviation

Note that the larger the count the higher the accuracy. This is one of the basic laws of statistics. Its significance to the experimenter is that there is almost a two-to-one chance that the recorded total of random counts will be within plus or minus the standard deviation of the true figure. More detailed information may be obtained from standard textbooks on statistics.

9.1.3. *The input signal*

The input signal should be standardized and supplied by an associated unit such as a discriminator, pulse-amplitude selector (window discriminator), coincidence unit, analogue to digital converter, etc. Some counting instruments accept analogue and/or non-standardized inputs and include built-in standardizing circuits.

Input signals should preferably be d.c. coupled, since a.c. couplings often cause level shift and possible miscounting of closely spaced signals unless some form of d.c. restorer, or level clamp, is used. The latter is difficult to achieve in the 50-Ω and 100-Ω input signal lines normally used.

9.2. The technique of counting

9.2.1. *Binary counting*

Electronic scalers and counting registers are all based on some form of two-state electronic switch, or binary element. Examples are the classical Eccles–Jordan trigger circuit, square-loop ferromagnetic cores, cold-cathode trigger tubes, tunnel diodes, etc. It is desirable that these switches toggle on a series of identical input signals and many arrangements employ a pair of two-state elements to achieve this. These are usually connected symmetrically, giving the desirable features of equal sensitivity in each direction and complementary outputs. Switching occurs very shortly after the input's transition to its significant state. There is invariably a minimum duration specified for the input signal to ensure that the circuit latches in its new state before the input is removed.

Arrangements outlined above switch alternately between their two states and, since their complementary outputs change state at half the input rate, divide by 2. If the transition from ON to OFF (or OFF to ON) of one of the outputs is used as the input to a second identical circuit the latter will divide the original input by 4, and so on.

The two states of a binary element's output(s) are referred to as 0 and 1. If a number of elements are suitably connected in cascade and all set with their outputs at 0, the first input will put the first output to the 1 state, while the second input will return it to zero which will set the second element's output to 1. If these binary elements have indicators to show when they are in the 1 state the indication will be 1, 0, which in binary notation is 2 (not 10). Similarly, the next input produces 1, 1 (decimal 3), while the fourth will give 1, 0, 0 (decimal 4). This is the start of the binary series which is increasing powers of 2. The first ten terms are given in Table 9.1 with the least significant figure on the right—in conformance with decimal notation.

The binary system with its radix of 2 (the decimal system has a radix of 10) has the feature that it uses only two numbers, 0 and 1. These can both be indicated by a single indicator lamp by its being off or alight, respectively; or by the absence or presence of a signal level on a line. This

14

two-state signal gives considerable economy of display and interconnection since all other radices require one lamp or line per number.

The total of the decimal equivalents in the table below is 1023 which in binary notation is 1 111 111 111. The decimal number 999 is 24 less which requires the deletion of 16 and 8 leaving 1 111 100 111. This is difficult to interpret visually, but it uses the most precise and fastest of all counting circuits and, what is most attractive in data transmission, any number up to 999 can be conveyed on only ten lines (or in ten 'bits') against thirty lines necessary with decimal notation. Due to the difficulty in visual

TABLE 9.1

The binary series

Element (or term)	10th	9th	8th	7th	6th	5th	4th	3rd	2nd	1st
Power of 2	2^9	2^8	2^7	2^6	2^5	2^4	2^3	2^2	2^1	2^0
Decimal equivalent	512	256	128	64	32	16	8	4	2	1

comprehension of binary notation a compromise decimal code based on binary code has been evolved. This is known as 'Binary Coded Decimal', usually shortened to B.C.D.

9.2.2. *Binary coded decimal counting*

The most convenient and generally used binary decimal code is based on the first four terms of the binary series. These are given in full in Table 9.2 with the value, or weight, of each bit in brackets. The table follows the numerical convention of the least significant bit (or figure) on the right.

The sixteenth input results in the 0 state for every bit and, as would be expected, the four bits produce a scale or radix of 16. But B.C.D. scaling circuits detect the number 9, 1001, and arrange that the tenth input produces all zeros instead of 1010. Thus it becomes a scale of 10 instead of 16. Under these conditions the fourth bit transition from 1 to 0 can be used as a carry signal to operate a subsequent decade. In B.C.D. notation 999 becomes 1001,1001,1001, requiring twelve bits against the previously explained ten for binary and thirty for decimal systems. It will be observed that each of the ten numbers (0–9) in a decade are represented by a combination of four binary bits, and one quickly acquires the ability to decode these visually.

Scalers and counting registers frequency have a maximum store of six decades (999 999). For this figure the binary counter requires only twenty bits, while the B.C.D. and decimal systems need twenty-four and sixty respectively. Therefore, where only electronic read-out is used pure binary counting is the best choice, but if both visual and electronic read-out are required the binary decimal code is usually preferred. On the other hand, registers for visual reading only are tending to decimal display despite greater complication and cost. This is partly due to the greater ease of reading and partly due to the fact that a lamp and/or line failure is self-evident because of the absence of information. It will be appreciated that

TABLE 9.2

Decimal equivalent of all four-bit binary numbers

Number	Bits			
	4th(8)	3rd(4)	2nd(2)	1st(1)
0	0	0	0	0
1	0	0	0	1
2	0	0	1	0
3	0	0	1	1
4	0	1	0	0
5	0	1	0	1
6	0	1	1	0
7	0	1	1	1
8	1	0	0	0
9	1	0	0	1
10	1	0	1	0
11	1	0	1	1
12	1	1	0	0
13	1	1	0	1
14	1	1	1	0
15	1	1	1	1
(16) = 0				

the failure of the fourth lamp in the B.C.D. number 9 (1001) will be innocently—and correctly—read as 1 (0001). Where this situation must be avoided there are various ways round it such as using the complements of bits in the zero state. Thus the signal present level for $\overline{8}$ (not 8) would be used in preference to no signal present on the 8 line. Alternatively, parity check techniques can be employed. These offer a degree of protection by generating an additional parity 'bit' when the number of 'ones' in the number ('word') is odd or even, depending on the system. These safeguards always add considerably to cost and sometimes reduce operating speeds.

9.3. Binary counting circuits

Any bi-stable circuit can be used for counting, but a symmetrical type is preferable since it usually toggles one way and then the other way with equal sensitivity and speed on receipt of a single unidirectional train of input signals. Such circuits can be assembled with thermionic tubes, cold-cathode tubes, semiconductors, etc.; but since modern instruments almost invariably use semiconductors these notes will be confined to the basic principles of transistor circuits in the discrete component and integrated circuit forms.

9.3.1. *The Eccles–Jordan circuit*

The best known circuit is the Eccles–Jordan circuit. It is a classical bi-stable circuit using two transistors in a cross-coupled ring-of-two configuration.

Fig. 9.1 gives the basic circuit and there are many variations and elaborations of it. It is shown using npn transistors since, at present, these are more common and cheaper than pnp types.

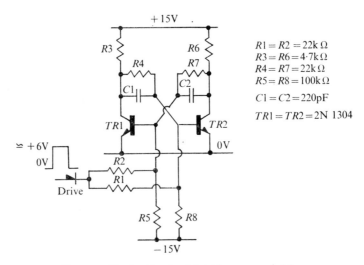

$$R1 = R2 = 22\text{k}\,\Omega$$
$$R3 = R6 = 4\cdot7\text{k}\,\Omega$$
$$R4 = R7 = 22\text{k}\,\Omega$$
$$R5 = R8 = 100\text{k}\,\Omega$$

$$C1 = C2 = 220\text{pF}$$

$$TR1 = TR2 = 2\text{N }1304$$

FIG. 9.1. Eccles–Jordan bistable—normal drive.

The two resistive potentiometers $R3$, $R4$, $R8$ and $R6$, $R7$, $R5$ are arranged so that, in the absence of the two transistors, the junctions of $R5$, $R7$ and $R8$, $R4$ would be appreciably positive with respect to the OV line. The addition of the npn transistors with their emitters at OV and their bases to these positive points means that they will both be initially turned ON when the supply voltages are switched on. The two bases would then be clamped to within a fraction of a volt of the OV line by the emitter/base junctions.

192

If it is assumed that each transistor has a common-emitter current gain of 50, then a random increase in current in the base of $TR1$ would result in a 50 times larger current at its collector, which accordingly would move negatively. The consequent reduction in base current of $TR2$ would appear as a fifty times greater reduction in its collector current and in a corresponding increase in collector potential. This would have the effect of increasing the base current of $TR1$ and, in fact, reinforcing the random increase responsible for this chain of events. Since the reinforcement is the result of the original change amplified by 50^2, or 2500 times, it constitutes heavy positive feedback. The action is cumulative and rapidly turns on so much current in $TR1$ that it 'bottoms', that is, the voltage drop across $R3$ due to the collector current of $TR1$, plus the current in $R4 + R8$, almost equals the positive supply voltage and there is only about $+0.25$ V left at the collector of $TR1$. $R4$ and $R8$ now form a potentiometer across the -15 V supply so that $TR2$ base is held at about -2.5 V and this transistor is turned OFF. However, $R6$, $R7$, and $R5$ are still across both supplies and with the base of $TR1$ clamping the $R5$, $R7$ junction at about $+0.6$ V. The base current of $TR1$ is defined by $R6 + R7$ less the small drain through $R5$. Thus the circuit has flipped into a state with $TR1$ hard ON and 'bottomed' and $TR2$ turned OFF. This is a very stable condition which the circuit will hold indefinitely. It is important to note that the voltage across $C1$ (about 3 V) is relatively small relative to that across $C2$ (about 12 V).

The circuit is toggled by the application of a positive step of 6 V or so to the drive point. This input is split by two identical resistors each connected to the base of a transistor. Due to the clamping effect of the emitter-base junction of $TR1$ this point cannot move and the input merely increases the base current. The base of $TR2$ is at about -2.5 V with a capacitor $C1$ effectively between it and OV via the bottomed collector of $TR1$. Therefore the input signal has to supply charge to $C1$ in order to raise the base potential of $TR2$ to the point of conduction. $R1$ and $C1$ form an inconvenient integrating time-constant which delays the turn-on of $TR2$. The main outcome of this is that the input signal must be maintained at its $+6$ V level for several $C1R1$ time constants. Eventually there will be only a little voltage across $C1$, and $TR2$ will start to conduct. Its collector will fall in potential and, due to the gain of $TR2$, start turning $TR1$ OFF despite the turning-ON effort of the input via $R2$. The cumulative cycle described before will take place very rapidly, but this time in the opposite direction. It is here that $C1$ and $C2$, which are often referred to as 'memory' capacitors, play an important part. As $TR1$ turns OFF the rapid rise of its collector is communicated via $C1$ (with no volts across it) directly to $TR2$ base, thus assisting the input via $R1$ and turning that transistor ON

193

very hard. $TR2$ collector therefore bottoms extremely rapidly setting at about $+0.25$ V. Now there were about 12 V across $C2$ so that $TR1$ base side of $C2$ will go 12 V negative and ensure that this transistor is cut off. In this second stable state $C2$, like $C1$ previously, will stabilize with about 3 V across it and the time for this will be determined mainly by the time constant $C2R7$. $C2$ will, of course, charge up to about 12 V on the much shorter time constant $C1R3$. Thus the memory capacitors have ensured that the circuit has changed state with no indecision.

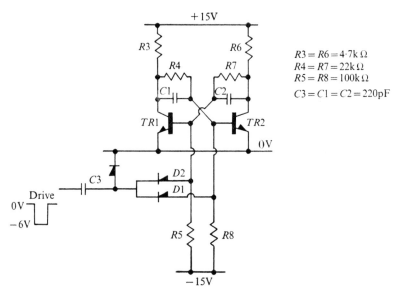

$$R3 = R6 = 4.7\text{k}\,\Omega$$
$$R4 = R7 = 22\text{k}\,\Omega$$
$$R5 = R8 = 100\text{k}\,\Omega$$
$$C3 = C1 = C2 = 220\text{pF}$$

FIG. 9.2. Eccles–Jordan bistable—inverted drive.

The above description covers the fundamental circuit operation. In fact the action of the circuit is very much more complex. The time constant of the memories is related to the switching speed of the transistors and the rise time, duration, and spacing of the input signals. Stray circuit capacitances and the loading effects of associated gates, indicators, succeeding binaries, etc., all tend to disturb the circuit symmetry and modify the operation. The circuit as given can operate with a resolution around 1 μs, but by changing transistors and the component values it can operate up to several hundred MHz.

By substituting two diodes for the resistors $R1$ and $R2$ the circuit can be driven by negative-going instead of positive-going input signals. Fig. 9.2 shows this alternative arrangement. On the assumption that the circuit has flipped with $TR1$ ON and $TR2$ OFF the bases will be as before at approximately $+0.6$ V and -2.5 V and, in particular, there will be 3 V across

194

C_1 and 12 V across C_2. A fast edge in excess of 3 V at the drive input will open diodes D_1 and D_2 and take their anodes and the bases of the transistors to some negative potential, say −6 V. This will turn both transistors OFF by the same amount. The input capacitor C_3 has a similar value to C_1 and C_2 and the series combination C_1, C_3, with only a small charge (3 V) on C_1, will charge positively faster than the combination C_2, C_3 with the result that the base of TR_2 will be brought into conduction before the base of TR_1. Again this is a simplified description and similar modifying conditions to those outlined for the previous case still exist.

The circuits in Fig. 9.1 and 9.2 are shown with high resistance 'tails' for compatibility with the classical Eccles–Jordan circuit using thermionic tubes. As such they are equally suitable when used with germanium or

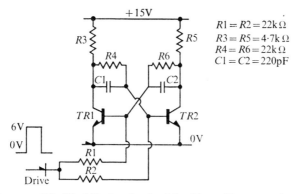

$$R1 = R2 = 22k\,\Omega$$
$$R3 = R5 = 4 \cdot 7k\,\Omega$$
$$R4 = R6 = 22k\,\Omega$$
$$C1 = C2 = 220pF$$

FIG. 9.3. Modified Eccles–Jordan bistable—silicon transistors.

silicon transistors. However, silicon transistors bottom with a much smaller residual voltage than is required to turn ON their bases and it is therefore possible to design silicon transistor switches without returning the bases to a negative supply. This is nowadays normal practice with both integrated and discrete-component circuits. The basic arrangement is given in Fig. 9.3. If TR_1 is bottomed its collector will be within a fraction of a volt ($\simeq 0 \cdot 2$ V) of OV and well below the cutting OFF voltage ($\simeq 0 \cdot 5$ V) of TR_2 base. Hence, there will be no TR_2 collector current and the base current of TR_1 will be defined by R_5 and $R6$. Since there is no base current in TR_2 there will be no current in R_4 and no volts across C_1, but there will be about 12 V across C_2. The operation of this circuit is as for that of Fig. 9.1 and the same comments apply.

9.3.2. *A tunnel-diode binary circuit*

The tunnel diode has two stable states but it is not easy to design a satisfactory binary with a single diode and practical designs usually employ two. The reason for using tunnel diodes rather than transistors is that very fast switching, and therefore very high count-rates are attainable, but silicon monolithic integrated circuits now achieve guaranteed minimum bistable clock-rates of up to 300 MHz and the tunnel diode is likely to be superceded.

The circuit given in Fig. 9.4 and described below can record input signals separated by 10 ns and register regularly spaced signals at 10^8 Hz. $R4$ and $R5$ are load resistors for the two tunnel-diodes $TD1$ and $TD2$. $R1$ is chosen such that only one of the diodes can be in the high-voltage low-current

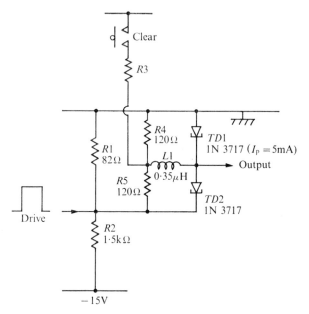

FIG. 9.4. Tunnel-diode bistable.

state while the other is in the low-voltage high-current state, the difference current passing through the inductance $L1$. The time constant of $L1$ and the effective resistance presented by $R1$, $R4$, and $R5$ is longer than the duration of the drive signal. In the o state of the binary $TD1$ is in the low-voltage condition. The drive signal takes all the current from $R2$, about 10 mA, leaving none for the diodes, but a current then flows in L_1 in the reverse direction to that flowing before the drive signal arrived. The combined effect of this current and that supplied by $R2$ at the cessation of the drive signal is to set and hold $TD1$ in the high-voltage low-current, or 1, state. The next drive signal has the opposite effect and returns the circuit to its original o state with a low voltage across $TD1$. The circuit is preset in the o state by the application of a positive voltage to the junction of $R4$ and $R5$ via $R3$.

196

This type of circuit cannot be cascaded directly and very fast non-inverting buffer amplifiers are needed between successive binaries.

9.4. Logic gates

With the advent of integrated circuits and the great increase in data handling the emphasis in instrument design has shifted from the circuit diagram to the logic diagram. With integrated circuits one must know precisely their characteristics and then set about interconnecting them to do what is required.

9.4.1. *Basic gates*

Referring back to Fig. 9.1 the transistors are required to act solely as switches in which their collectors are either open-circuited or short-circuited to the OV line. Fig. 9.5(a) shows the basic arrangement. If the input

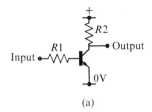

(a)

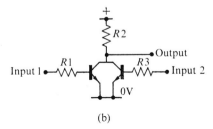

(b)

FIG. 9.5. Basic-gate arrangement. (a) Switch. (b) Two-input switch.

terminal is at OV the transistor is turned OFF and, as there is no current in R_2, the output point is at the supply potential. If the input terminal is taken to the supply rail the transistor bottoms and the output is virtually at OV. The significant point is that the input and output signal amplitudes are of equal but opposite polarity, and that similar circuits can be directly coupled in cascade indefinitely without loss of d.c. levels. If two such circuits are connected with a common collector-load resistor the arrangement of Fig. 9.5(b) results. Here, if one *or* the other input (or both) is taken to the supply rail the output will always be OV (binary o); but they *both* must be taken to OV for a high level at the output (binary 1).

9.4.2. *Basic gates and the Eccles–Jordan bistable*

Fig. 9.6 shows the circuit of Fig. 9.3 redrawn using simple two-input gates instead of single transistors. This has the advantage that a separate transistor is available each side for the trigger input. The operation of the circuit is substantially the same as that in Fig. 9.3. The two gates would normally be in a single integrated circuit element and the only conventional components would be C_1, C_2 and R_4, R_6.

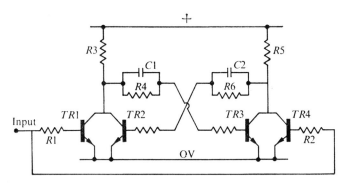

FIG. 9.6. Bistable with trigger transistors.

9.5. Integrated circuit counting arrangements

From here on the following conventions have been adopted:

(a) The circuits are referred to as counters, not scalers.

(b) Since, in both binary and decimal notation, the least significant figures are conventionally written on the right, the diagrams are presented with the inputs on the right and the signal path from right to left.

(c) The binary 'zero' and 'one' states are written as o and 1 respectively.

(d) The commercial convention of allocating to the high signal level the significance of binary 1 has been adopted, but where possible preference is given to 'high' and 'low'.

(e) Where the counting elements are flip-flops they are referred to as FF_1, FF_2, FF_4, and FF_8 from the input-signal end. Their corresponding Q and $\bar{Q}$ outputs are designated Q_1, $\bar{Q}_1$, Q_2, $\bar{Q}_2$, etc.

(f) It is assumed that connecting both the J and K of a flip-flop to a high level 'enables' it to be toggled, while a low level 'disenables' or 'inhibits' toggling. It should be appreciated that this is not a universal rule.

(g) It is assumed that J–K flip-flops may have up to three J and K inputs. For either side to be 'enabled' all its three inputs must be high, or at 1.

If any one of the three is low that function is inhibited. Thus all six must be high for toggle action. However, only the number of inputs necessary for circuit requirements are shown on the diagrams in the interests of clarity and it is assumed that surplus inputs are either taken permanently to 1 or paralleled with one of the used inputs. A D-type flip-flop toggles when the D terminal is connected to its $\bar{Q}$ output.

9.5.1. *Serial, or ripple-through, binary counters*

As explained earlier, electronic switches which will toggle on successive inputs can usually be cascaded to form a binary counting chain. The

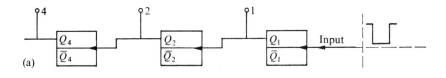

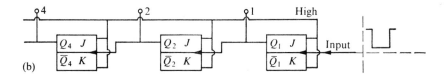

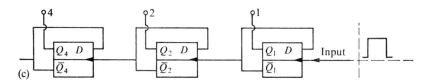

FIG. 9.7. Ripple-through binary counters. (a) Eccles–Jordan bistable. (b) *J–K* flip-flop. (c) *D* flip-flop.

switches can be of the simple Eccles–Jordan type or one of its modern derivatives, the *J–K*, or *D* flip-flop. The simplest arrangement is where the elements trigger one another.

Fig. 9.7 shows three such arrangements all designed to count 'up', or additively. The sequence of operation of these circuits is given in column 2 of Table 9.3.

Column 2 shows that the transition for 1 to 0 of the Q output of each binary operates its successor. This, of course, cannot produce subtraction; but since the elements shown are symmetrical they all have complementary

199

outputs and if the elements in each case are operated from the $\bar{Q}$ of their predecessors, the sequence of column 4 results. The complementary Q states are given in column 3, which will be seen to be in the reverse direction to column 2.

TABLE 9.3

Counting sequence for three cascaded binaries—scale of eight

Column 1 Input	Column 2 (addition)			Column 3 (subtraction)			Column 4 (subtraction)		
	$Q4$	$Q2$	$Q1$	$Q4$	$Q2$	$Q1$	$\bar{Q}4$	$\bar{Q}2$	$\bar{Q}1$
0	0	0	0	0	0	0	1	1	1
1	0	0	1	1	1	1	0	0	0
2	0	1	0	1	1	0	0	0	1
3	0	1	1	1	0	1	0	1	0
4	1	0	0	1	0	0	0	1	1
5	1	0	1	0	1	1	1	0	0
6	1	1	0	0	1	0	1	0	1
7	1	1	1	0	0	1	1	1	0
8 or 0	0	0	0	0	0	0	1	1	1

Serial binary counters will operate satisfactorily at the maximum toggle capability of the first flip-flop and in long cascades can accept a second input before the first has been propagated to the far end. This 'propagation delay' can cause embarrassment in certain situations, such as when the counter's state is in constant comparison with a pre-set number, since anomalous parity might be achieved during propagation. Under such circumstances synchronous counters are necessary in which all the binaries toggle together from the input signal.

9.5.2. *Synchronous binary counters*

Fig. 9.8 shows a synchronous binary counter using J–K flip-flops. The J and K terminals of the first element are taken permanently to a high signal level and it therefore toggles on successive input signals. If the series is initially cleared with all its Qs at 0 the first input will operate only $FF1$. The resulting 1 at $Q1$ will enable $FF2$ which can switch in synchronism with $FF1$ at the next input. But for this input $FF4$ is still inhibited at its second J and K by the 0 from $Q2$. It is clear that the normal counting sequence given in column 2 of Table 9.3 will be achieved, and also that transferring all the Js and Ks to their preceding $\bar{Q}$s will result again in a count-down.

This arrangement is just as fast as the ripple-through counter, but two difficulties are that successive elements need progressively more J and K

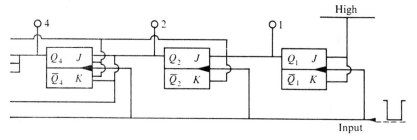

FIG. 9.8. Synchronous binary counter with parallel 'carry'.

terminals, and the loading on the earlier Qs becomes excessive. For instance, a twenty-stage counter would require 19J and 19K inputs for the last flip-flop and a fan-out capability of 38 for $Q1$. The first requirement could be accommodated by using separate multi-input AND gates, but the second remains.

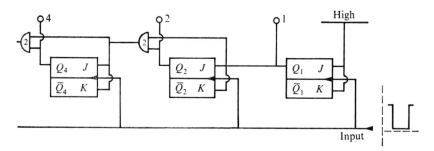

FIG. 9.9. Synchronous binary counter with serial 'carry'.

The situation is eased if a serial 'carry' system is used while retaining the parallel triggering. Fig. 9.9 shows how this can be done. Here the necessary gating information is passed to the J and K of a stage and AND-ed with its output to gate its successor. This is relatively economical and avoids the fan-out problem, but it introduces a progressive delay through the series combination of AND gates. This delay, while less than that through the same number of flip-flops, requires that an input signal must not be closer to its predecessor than the propagation time through the counter, with the result that the maximum count-rate capability decreases as more stages are added.

201

9.5.3. *Reversible binary counters*

It has been shown that transferring a serial counter's carry, or a synchronous counter's J and K gating, from the Q to $\bar{Q}$ of its flip-flops reverses the direction of counting. This is shown, using a multi-pole change-over switch, in Fig. 9.10. However, mechanical switches are unattractive for

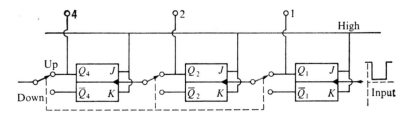

FIG. 9.10. Up/down (reversible) binary counter.

reasons of size, reliability, and flexibility of layout—particularly in the larger assemblies—and electronic switching is preferred.

The integrated circuit element most suited to this function is shown in Fig. 9.11 and is usually referred to as an 'exclusive OR' although it is sometimes available in a slightly modified form as a 'half adder'. Assuming the gates are T^2L or MDTL a low signal level at any one of the four inputs

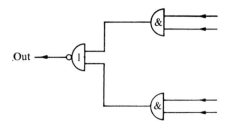

FIG. 9.11. Exclusive OR.

will disenable the associated AND gate and a signal to the other input terminal of that gate becomes ineffective. Similarly, if one of the inputs to the other AND gate is fed with a high signal-level that gate is enabled and can respond to signals at its other input. Reversing these high and low 'control' levels transfers the signal response to the first AND gate. The AND gates' outputs are OR-ed in the output gate. In an UP/DOWN counter, this circuit can be used in place of the single pole two-position switches of Fig. 9.10 and this is shown in Fig. 9.12. If the down-line is low and the up-line high the signals from the Qs will be ineffective, but those from the $\bar{Q}$s will be fed to trigger the succeeding binaries, and vice versa. (It should

202

be noted, in passing, that signal inversion takes place in the exclusive OR which is why the UP 'enables' are derived from the $\bar{Q}$s instead of the Qs as in the earlier simpler circuits.) Hence a pair of control buses carrying complementary signal levels can be supplied by a switch, or flip-flop, and routed round a complex layout to control UP and DOWN counting by selecting one or other of signal input pairs fed to a large number of exclusive ORs. Changing over the UP and DOWN line levels reverses the direction of count.

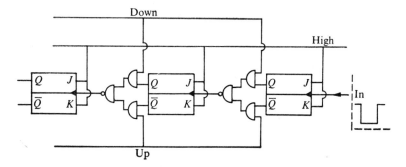

FIG. 9.12. Up/down binary counter.

For simplicity this technique has been described for a serial counter, but it is equally applicable to any of the other arrangements given. Fig. 9.13 shows the serial-carry, parallel-triggered counter of Fig. 9.9 modified for UP/DOWN operation.

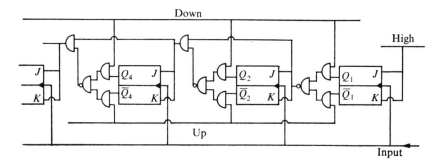

FIG. 9.13. Serial-carry synchronous reversible binary counter.

One advantage of the electronic over the mechanical switch is that no 'count' signals flow in the control circuits that carry only steady d.c. levels. A considerable disadvantage is the longer propagation time through the additional gates, resulting in a reduced maximum count-rate.

203

9.6. Counting in other than binary code

In binary counting with only two numbers, o and 1, there is a radix of 2; in octal code employing eight numbers, o–7, the radix is 8, and so on. To count in a radix it is necessary to cascade series of scales to that radix so that, as each one 'overflows', it operates its successor. An octal counter can consist of three stages of binary counting since $2^3 = 8$ and it follows that any counter with a power of 2 as its radix can be assembled with equal ease. The usual method of counting to a non-binary radix is to cascade sufficient binary elements to exceed the radix, detect when the radix number has been registered in binary form, and arrange for the next input signal to set all binaries to zero. However, there are exceptions and an original decade due to Cooke-Yarborough (1964), has been used almost exclusively for many years in scalers of the Harwell 2000 series.

9.6.1. *A seven-transistor decade counter*

The novelty of this design is the use of the three-transistor six-state circuit in which only one state needs suppression to achieve a scale-of-five.

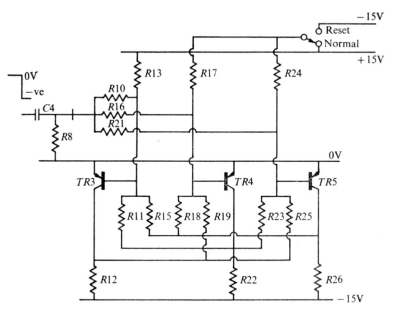

FIG. 9.14. D.c. couplings in scale-of-five.

This is preceded by a conventional binary element giving an overall scale of ten. Discrete components and pnp transistors are used.

The d.c. arrangement of the three-state circuit is given in Fig. 9.14. The resistor values are such that two of the three transistors will always be ON

204

(bottomed) while the third is biased OFF. The collector of each transistor is d.c. coupled to the bases of the other two and the component values associated with one transistor are repeated for the other two. If TR_3 happens to be in the biased OFF condition then the other two will be ON with their collectors within about 0·25 V of the 0V line and TR_3 will be held OFF by the resulting positive voltage at the junction of R_{13}, R_{11}, and R_{15}. Since TR_3 is passing no current its collector load R_{12} and the two parallel branches $(R_{17} + R_{19})$ and $(R_{24} + R_{25})$ form a potentiometer network in which the lower ends of R_{17} and R_{24} would be negative with respect to 0V were they not virtually clamped to within 0·3 V of 0V by the conducting

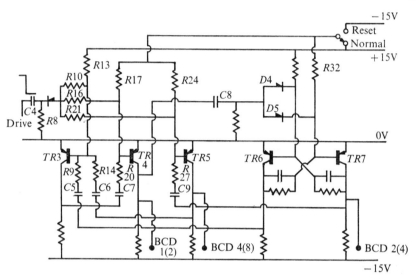

FIG. 9.15. A.c. couplings in scale-of-five. Figures in parenthesis are preceded by a scale-of-two.

emitter/base junctions of TR_4 and TR_5. These two transistors are thus held in the bottomed condition.

Fig. 9.15 shows the a.c. couplings of the complete scale-of-five. These are the switching impulse paths. The d.c. circuits hold the scale-of-five indefinitely in any state to which the a.c. couplings have set it. The circuit is 'cleared', or set to the 0 condition by switching the top ends of R_{17}, R_{24}, and R_{32} to −15 V for not less than 30 μs. This switches ON and bottoms TR_4, TR_5, and TR_7. The input signal is a negative-going edge, which is differentiated by C_4R_8 with a time constant long compared with the circuit's switching time. The counting sequence is as follows:

(a) The first signal cannot turn TR_4 and TR_5 ON any more than their already bottomed condition, but it turns ON TR_3, which amplifies and inverts it to turn OFF TR_4, despite the turning ON input also applied to its base. D_4 and D_5 isolate the binary TR_5 and TR_6 from the resulting negative excursion at TR_4 collector.

(b) The second input signal is amplified and inverted by TR_4, which bottoms, and the positive voltage change at TR_4 collector, coupled through C_8 and D_5, switches the TR_6/TR_7 binary to its opposite state. The positive edge at TR_6 collector coupled through C_5 and R_9 causes TR_3 to be turned OFF despite the turning ON input signal at its base and the transistors have reverted to their original state with TR_3 OFF and TR_4 and TR_5 ON. However the TR_6/TR_7 binary has changed state.

(c) The third input signal has the same effect as the first, i.e. it is amplified and inverted by TR_3 and fed to turn OFF TR_4 despite the turning-on input signal at its base.

(d) The fourth input is amplified and inverted by TR_4 and the positive edge at TR_4 collector is fed via C_8 to switch the TR_6/TR_7 binary back to its original state. The positive edge at TR_7 collector coupled through C_9 and R_{27} turns off TR_5 despite the turning ON input signal at its base.

(e) The fifth input signal is amplified and inverted by TR_5, and fed via C_6 and R_{14} to turn OFF TR_3 despite the turning ON input signal at its base. The initial state is again achieved but this time for all five transistors.

Adopting the convention that a bottomed collector represents binary 0 and that the collector voltage (-12 V) of a biased OFF transistor represents binary 1, it will be found that the states of the collectors of TR_5, TR_7, and TR_4 will be as shown in Table 9.4 for successive input signals.

TABLE 9.4

Switching sequence for three-transistor ring-of-five

Input signal	Collector outputs (see Fig. 9.15)		
	TR_5	TR_7	TR_4
0 (clear)	0	0	0
1	0	0	1
2	0	1	0
3	0	1	1
4	1	0	0
5	0	0	0

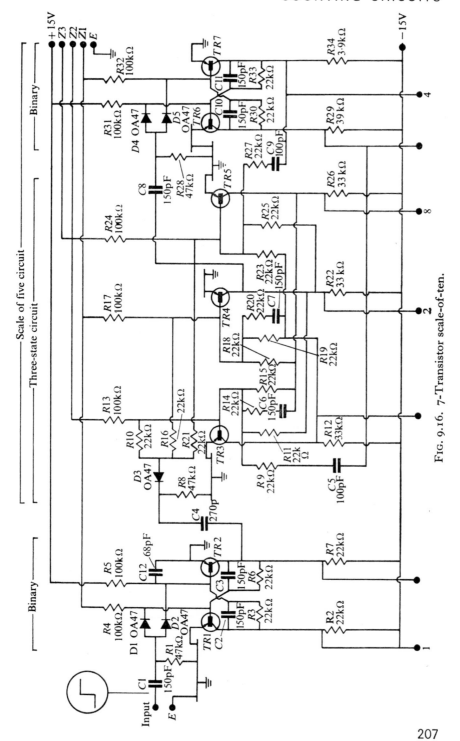

FIG. 9.16. 7-Transistor scale-of-ten.

Thus the circuit has divided by 5, and if it is preceded by a single divide by 2 binary the combination will divide by 10. In the scale-of-ten the digit 8 comes from $TR5$ and is terminated as a positive edge by the tenth input signal to the full decade. In order that decades may be cascaded without auxiliary coupling-circuits the input scale-of-two must operate off this edge and for this reason operates in the 'turn-off' mode described earlier. Fig. 9.16 gives the full circuit, which operates up to 500 kHz.

9.6.2. *Decade counters based on binary elements*

These usually consist of a four-binary scale of 16 with six of the states suppressed, or skipped. There is a large number of ways of doing this, but modern design tends to favour the first ten states of pure binary giving

TABLE 9.5

1–2–4–8 BCD counting sequences

Input No.	Count up 8–4–2–1				Count down 8–4–2–1			
Clear = 0	o	o	o	o	o	o	o	o
1	o	o	o	I	I	o	o	I
2	o	o	I	o	I	o	o	o
3	o	o	I	I	o	I	I	I
4	o	I	o	o	o	I	I	o
5	o	I	o	I	o	I	o	I
6	o	I	I	o	o	I	o	o
7	o	I	I	I	o	o	I	I
8	I	o	o	o	o	o	I	o
9	I	o	o	I	o	o	o	I
10	o	o	o	o	o	o	o	o
11	o	o	o	I	I	o	o	I
12	o	o	I	o	I	o	o	o
13	o	o	I	I	o	I	I	I

1–2–4–8 BCD. However, any code of four numbers that can be assembled in part, or *in toto*, to give 1 to 9 can be used, i.e. 1–2–2–4, 1–2–4–2, 1–2–3–3, 1–2–3–4, 1–2–3–5, etc. 1–2–2–4 and 1–2–4–2 have been used quite extensively in the past. Only the better known 1–2–4–8 BCD circuits are dealt with from now on.

Table 9.5 gives the counting sequence for 1–2–4–8 BCD in the count-up and count-down modes.

In either case the number above the dotted line has to be followed by a number out of binary sequence. In the count-up mode this next number would normally be binary $10 = 1010$, while in the count-down mode it would be binary $15 = 1111$. Also, in the count-down mode the change in sequence must occur at the first input instead of at the tenth as in the case for count-up, because the first number when counting from 9 to 0 is 9, just as zero is the first number when counting from 0 to 9. What is required is that the state above the dotted lines must be detected and used to modify the sequence to achieve the desired result in each case.

9.6.3. *Serial, or ripple-through, decade counters*

Serial decade counting in the count-up mode can be achieved easily by a means that has been exploited since BCD counting was first thought of. It consists of using the 1 at $Q8$ to divert the outputs of $FF1$ from $FF2$ directly to $FF8$, so that the tenth input sets both the 1 and 8 to zero which, together with the undisturbed 2 and 4, give the desired result, 0000. Fig. 9.17 shows how this is achieved. It is drawn with non-inverting gates since the early circuits nearly always used diodes.

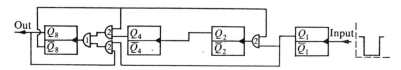

FIG. 9.17. Serial 1–2–4–8 BCD counter.

In the cleared state all the Qs are at 0 and the 1 at $\bar{Q}8$ enables the AND gates between $FF1$ and $FF2$, and $FF4$ and $FF8$, while the zero at $Q8$ inhibits the AND gate between $FF1$ and $FF8$, so that the four elements are connected for normal serial binary counting. Table 9.5 shows that on the eighth count $Q1$, $Q2$, and $Q4$ all go to 0 while $Q8$ goes to 1. This latter state inhibits the AND gate between $FF1$ and $FF2$ and enables that between $FF1$ and $FF8$. The ninth input sets $FF1$ to 1 and the tenth returns it to 0, which transition also sets $Q8$ to 0. This returns all the gates to their initial state and, since the flip-flops are also all at 0, the sequence can repeat.

The same basic idea has been used in some modern integrated circuit decades except that it has been simplified by changing the last binary to an edge triggered set–reset flip-flop. Only one AND gate is now needed—between $FF1$ and $FF2$. Fig. 9.18 shows a schematic of this arrangement. Normal serial counting proceeds until the eighth count has been registered, when the resulting 0 at $\bar{Q}8$ disenables the gate between the first two flip-flops. Again the ninth input sets $Q1$ to 1 and the tenth returns it to 0, which transition also switches the set–reset flip-flop back to 0. Previous 1 to 0 transitions at the second, fourth, sixth, and eighth counts had merely tried to switch $FF4$ to the state it was already in, and since the eighth input arrives undelayed compared with that from $Q4$ the latter is able to switch $FF8$ after its propagation delay through the first three flip-flops.

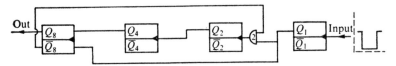

FIG. 9.18. IC serial 1–2–4–8 BCD counter.

Fig. 9.19 shows a modern decade using four J-K flip-flops. The first flip-flop operates in the normal toggle mode with its J and K both taken to a high level. The second and third flip-flops are driven normally from the outputs of their predecessors and count in binary up to 7 since $J2$ is held

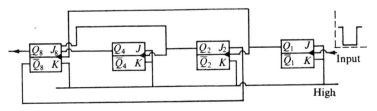

FIG. 9.19. Count-up decade counter.

at 1 by $\bar{Q}8$. At count 7, $J8$ is enabled by $Q2$ and $Q4$ so that the eighth input can switch $FF8$ producing a zero at $\bar{Q}8$ and hence at $J2$. This prevents further counting of $FF2$ and $FF4$, the Qs of which are at 0 and both connected to $J8$. Therefore the tenth input will switch $Q8$ to 0 and the complementary 1 at $\bar{Q}8$ will permit $FF2$ to toggle again and the full sequence can repeat.

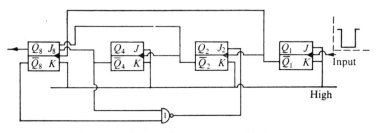

FIG. 9.20. Count-down version of Fig. 9.19.

Fig. 9.20 shows a count-down version of Fig. 9.19. Apart from the NOR gate—from $\bar{Q}4$ and $\bar{Q}8$ to $J2$—in place of the single $Q8$ to $J2$ link in Fig. 9.19, and the driving of the flip-flops from the preceding $\bar{Q}$s instead of the

Qs, the circuits are the same. The effect of the gate is to feed a 1 to $J2$ only when there is a 1 at $Q4$ or $Q8$, which, as examination of Table 9.5 will show, is the requirement. The sequence starts with all Qs at 0 and the first input will operate only $FF1$ and $FF8$ (1001 = 9). $J2$ will now be enabled and normal binary count-down takes place until 0011 (3) is registered. The 1 at $\bar{Q}8$ (or $\bar{Q}4$) will result in a 0 at $J2$ so that when the ninth input switches $Q1$ to 1 the resultant transition of $\bar{Q}1$ switches the 0 at $J2$ into $Q2$ giving 0001. The tenth input clears $Q1$ to 0 and the sequence can repeat.

9.6.4. *Synchronous, or parallel-clocked, decade counters*

Synchronous counters have come into general use with the advent of clocked flip-flops—particularly J–K and D types. Fig. 9.21 shows the

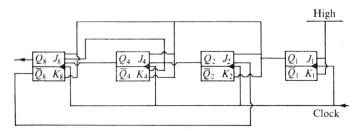

FIG. 9.21. Synchronous count-up decade.

synchronous binary counter of Fig. 9.8 modified to a radix of 10. Up to the count of 8 the operation is as in Fig. 9.8 since $J2$ is enabled by $\bar{Q}8$. But after the eighth count has been registered the 0 at $\bar{Q}8$ inhibits $J2$. The ninth count is registered in $FF1$ only and the resultant 1 at $Q1$ enables

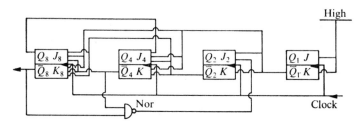

FIG. 9.22. Synchronous count-down decade.

$K8$ only, since the 0s at $Q2$ and $Q4$ are fed to $J8$. Hence the tenth input, which sets $Q1$ to 0, will set $\bar{Q}8$ to 1, or $Q8$ to 0—and the correct state 0000 obtains.

Fig. 9.22 gives the count-down version of Fig. 9.21. As before, to achieve a count-down the carry gating is derived from the $\bar{Q}$s. At the start, when the counter is registering all 0s, $Q8$ inhibits $J4$ while $\bar{Q}8$ and/or $\bar{Q}4$ inhibit

211

(via the NOR gate) $J2$. The first input will therefore switch only $FF1$ and $FF8$ to 1 giving the correct result, 1001 = 9. The second input returns $FF1$ to its initial state leaving 1000 = 8. The 1 at $Q8$ now no longer inhibits $J4$ while the 0 (low) at $\bar{Q}$ enables $J2$ via the NOR gate. The third input will switch all flip-flops, resulting in 0111 (7). $\bar{Q}4$, instead of $\bar{Q}8$ now holds $J2$ enabled via the NOR gate while the low at $\bar{Q}4$ will inhibit $FF8$ and the normal count-down with parallel carry sequence can proceed until 0000 is achieved when the whole sequence repeats.

9.6.5. *Parallel-clocked serial-carry counters*

The decades shown in Figs. 9.21 and 9.22 have minimum propagation delays, but, when cascaded, cannot be clocked in parallel from a common

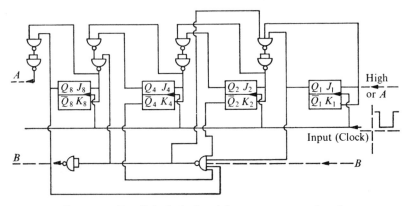

FIG. 9.23. Parallel-clocked serial-carry count-up decade.

line without complex multi-input carry AND gates. They can be cascaded easily by using $Q8$ or $\bar{Q}8$, as appropriate, to trigger each decade's successor, but the feature of synchronous switching is lost. However, this may be retained if serial carry with its increased propagation delay is acceptable. Fig. 9.23 shows such an arrangement in the count-up mode. Here the carry gates between the flip-flops are series-connected pairs in order to avoid inversion and to obtain the requisite NAND/NOR functions. Apart from this the circuit operates similarly to that of Fig. 9.9. The four-input gate detects only the state of 9 (1001) being fed with $1\bar{0}\bar{0}1$ (which is 1111). This results in a low signal-level at its output which is fed to the first gate of the carry-pair between $FF1$ and $FF2$, and also to the centre of the carry-pair between $FF4$ and $FF8$. Thus the second flip-flop is inhibited while the last is enabled so that the tenth input signal can switch only $FF1$ and $FF8$ setting their Qs to 0 giving 0000 as required. Such a system is also applicable to up-down operation, but with additional slowing down.

212

Fig. 9.24 gives the same circuit in a count-down mode. The carry-gate inputs are transferred from the Qs to the $\bar{Q}$s and the four-input gate is

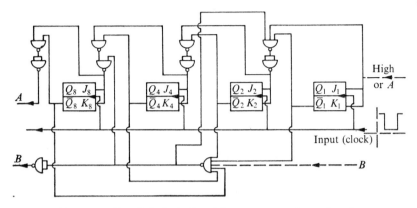

FIG. 9.24. Parallel-clocked serial-carry count-down decade.

connected to detect 0000. The latter is achieved by feeding it with $\bar{o}$ $\bar{o}$ $\bar{o}$ $\bar{o}$ (or 1111) and leaving its output connected as before, thereby ensuring that only $FF1$ and $FF8$ will switch on the first input signal and achieve the required 9, 1001. Thereafter normal count-down proceeds until 0000 is registered when the four-input gate will again interfere and cause 9 to come up.

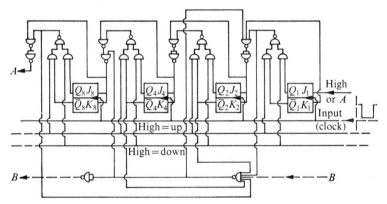

FIG. 9.25. Parallel-clocked serial-carry decade count up/down.

Fig. 9.25 is a combination of Figs. 9.23 and 9.24 to obtain up/down operation. The selection of the carry signals from the Qs or $\bar{Q}$s is achieved as described earlier by exclusive-OR circuits. Since the exclusive-OR gates invert, the UP direction line is associated with the Qs and vice versa. The dotted lines are serial carries to and from associated decades.

213

9.6.6. *Ring counters*

This category includes multi-electrode tubes like dekatrons and trocho-trons, and circuits employing rings of trigger elements such as thyratrons, cold-cathode tubes, tunnel diodes, and square-loop ferromagnetic cores. Thermionic hard-valve circuits have also been used, but these are more complex than circuits using two-state devices and although capable of fast counting they were never widely used.

Multi-electrode tubes usually have ten stable states (hence the name dekatron) although a version with twelve stable states was produced. Other ring-counting circuits can be assembled with any number of elements, but it is rare to exceed ten since circuit tolerances become critical. Early scale-of-ten circuits usually employed a ring-of-five driven by a binary (ring-of-two). Even then tolerance problems are more severe than for decades based on cascaded binaries.

9.6.6.1. *The dekatron.* This is the simplest and most widely used multi-electrode counting tube (Bacon and Pollard 1950). It is a self-displaying glow discharge tube with its anode in the form of a metal disc around which are positioned a ring of 30(36) symmetrically spaced cathode wires. A group of 10(12) wires consisting of every third wire constitute the main counting and display cathodes. The remaining wires are internally connected in two similar rings and are known as the transfer, or guide cathodes (see Fig. 9.26). The tube contains an inert gas—usually neon.

The ten stable states of the tube consist of a glow resting on one of the cathodes, K_0–K_9. The addition of a digit requires the transfer of the glow in a clockwise direction and is achieved by the application of a staggered pair of negative pulses to the guide rings, G_1 and G_2. The duration and voltage amplitude of these pulses must be great enough to ensure satis-factory transfer from a cathode to the adjacent G_1 and further transfer to G_2 after which, on the cessation of the second pulse of the staggered pair, the glow will transfer to the adjacent (and next) display cathode.

Dekatrons are simple and cheap, but although economical in current they require power supplies of several hundreds of volts which is inconvenient in transistorized equipment. They are also very slow, being incapable of operating much above 6 kHz. Later developments of the tube achieved single pulse transfer by the use of shaped guides giving asymmetrical potential gradients, but all tubes of this kind have been rendered virtually obsolete by the overwhelming advantage of transistor circuits.

9.6.6.2. *Counting rings.* The form of counting ring generally encountered is a variation of the shift register based on D or J–K flip-flops. If a ten-bit shift register containing a single 1 has its output connected to its input the 1 can be 'clocked' round and round and detected each time it passes a

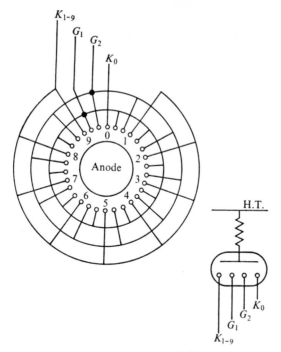

FIG. 9.26. Arrangement of dekatron.

particular point, resulting in a single decade counter. However, ten flip-flops to count up to 10 is uneconomical and a neat variation due to Johnson and called the 'twisted ring' uses only five flip-flops. Fig. 9.27(a) shows the Johnson twisted-ring arrangement. In the 'cleared' state all Qs are at 0 and

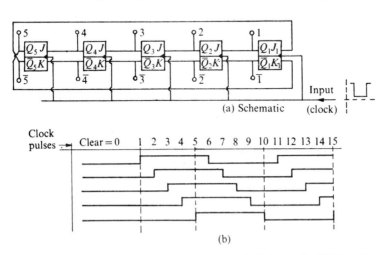

FIG. 9.27. Johnson twisted-ring decade counter. (a) Schematic. (b) Waveforms.

215

since J_1 is connected to $\bar{Q}$ (=1) of FF_5 the first five input pulses will progressively fill the register with 1's. When Q_5 goes to 1 its complement $\bar{Q}_5$ will go to 0 so that the next five input pulses will progressively fill the register with 0s and the initial 'cleared' state obtains. Obviously this is a scale of 10 and any one Q is high for five counts and low for the next five, according to Fig. 9.27(b). An attraction of the Johnson twisted-ring counter is the high count-rate attainable, since each flip-flop works at only one-fifth the input rate. However, the realization of a high rate depends on layout and the type of flip-flop used. A decimal display is easily achieved by using ten two-input AND gates according to Table 9.6, which can be checked against the waveforms of Fig. 9.27(b).

TABLE 9.6

Decimal decode of Johnson twisted-ring counter

Decimal number	0	1	2	3	4	5	6	7	8	9
AND-ed Outputs	$\dfrac{\bar{1}}{\bar{5}}$	$\dfrac{1}{\bar{2}}$	$\dfrac{2}{3}$	$\dfrac{3}{4}$	$\dfrac{4}{5}$	$\dfrac{5}{1}$	$\dfrac{2}{\bar{1}}$	$\dfrac{3}{\bar{2}}$	$\dfrac{4}{\bar{3}}$	$\dfrac{5}{\bar{4}}$

Notice that each output is used only twice so that fan-out, or loading problems are slight. It is cumbersome to derive BCD code from this counter. As with any ring counter a spurious 1 will cycle round and give anomalous results. This can be corrected by AND-ing $\bar{Q}_5$ and $\bar{Q}_4$ into J_1.

9.7. Derandomization

Modern detectors and transducers can be so fast that random events can produce input signals well inside the resolving time of many data-handling equipments, even when the mean count-rate is quite low. To overcome this derandomizing techniques have been evolved. Two methods will be described which are quite different approaches to the same problem.

Fig. 9.28 illustrates the first arrangement. The two binary counters (of

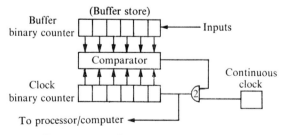

FIG. 9.28. Buffer store derandomizer.

four or five bits each) must be as fast as possible and must be parallel clocked. The comparator must be equally fast and seeks parity between corresponding bits of each counter. Whenever parity exists the comparator closes the gate in the clock output line. Hence at the start if both counters

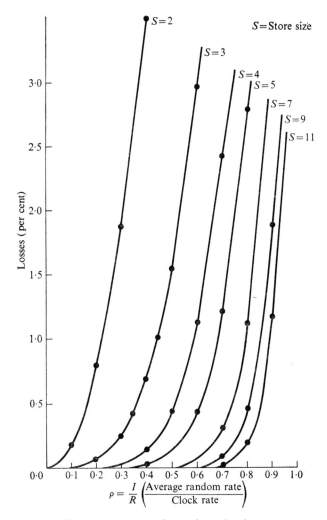

FIG. 9.29. ρ versus loss—derandomizer.

are cleared parity exists and the gate is shut. It is necessary for the interval between clock signals to be slightly longer than the resolving time of the associated processor, computer, or counter. Also the input signal mean rate must be less than the operating rate of the clock. As soon as the buffer counter registers an input, parity is destroyed and the clock gate opens. If only one

217

input is registered the first clock signal passed by the gate restores parity and the gate shuts, but of course the processor has received one input signal too. If a quick succession of input signals is recorded by the buffer counter it might take some time for the clock counter to catch up, but it will do so because the rate of its inputs is faster than the mean rate of those fed to its companion. The efficiency of derandomization depends on the number

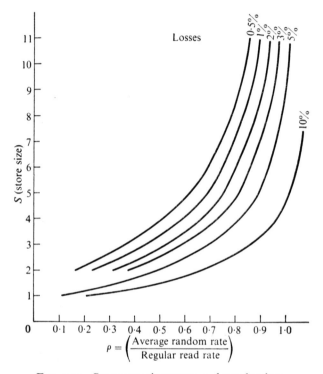

FIG. 9.30. Store capacity versus ρ-derandomizer.

of bits of the stores and the ratio ρ of the input's mean count-rate to the clock rate. Alexander, Rederring, and Kennedy (1959) calculated results using a Poisson distribution of incoming signals for ρ varying between 0·1 and 1·0 and a buffer store capacity s from 1 to 19 bits. Fig. 9.29 shows a family of curves for the percentage loss as a function of ρ, while Fig. 9.30 gives the store capacity required for various percentage losses as a function of ρ. The advantage of derandomization is apparent immediately from these curves. For example, if losses of 1 per cent are acceptable an experiment can be carried out with

$$\rho = 0\text{·}68 \text{ and a buffer store of 5 bits, or}$$

$$\rho = 0\text{·}88 \text{ and a buffer store of 10 bits.}$$

This should be compared with a ρ of o·oi necessary without derandomization.

The second method of derandomization, due to Orman (1967) is illustrated in Fig. 9.31. Input signals, of a few nanoseconds duration each, are fed via an OR gate C into a delay line. If the output monostable circuit at the other end of the delay line is quiescent, gate A is open and gate B closed. The first input signal will trigger the output monostable, which is designed to generate a signal lasting long enough to bridge the resolving time of the receiving processor. During this time gate A is shut and gate B recirculates via gate C any inputs that arrive. Regeneration takes place at

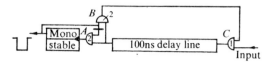

FIG. 9.31. Delay-line derandomizer.

gate B or gate C by circuitry not shown in Fig. 9.31, thereby ensuring that the signals are not degraded by re-circulation. As soon as the output monostable resets it takes the next available signal from the delay line and the operation is repeated. As before the gates must be very fast and tunnel diodes have been used successfully. Again the signal input mean rate must be less than the extraction rate.

9.8. Access, or read-out

Large arrays of counters, such as might be used in nuclear physics experiments, are normally read automatically into a computer or other processor. However, there are also frequent requirements to read-out the contents of a small number of counters (for instance, an UP/DOWN counter operating from a shaft encoder). Two methods of read-out are available. The first is to transfer the stored counts from a non-access counter into one with access facilities. The other is to use only counters with parallel access to the stored bits. The latter has always been preferred and, since the advent of integrated circuits, is almost universal.

One technique for transferring from a non-access to an access counter is to feed a train of signals into both and use the overflow of the non-access counter (which contains the required information) to inhibit both. Fig. 9.32 shows this arrangement. Two major disadvantages of this method are the complementary form of the transfer and the loss of the original information. These can be avoided by the use of a further 'transfer' counter having the same storage capacity as the non-access counter. All three counters are fed from a common transfer-signal generator, but at the start of a transfer the transfer signals are withheld from the access counter by the absence of an overflow signal from the non-access counter. The overflow from the transfer counter is used to inhibit transfer signals to all counters and since it has

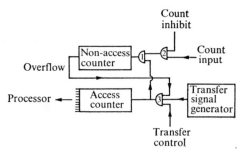

FIG. 9.32. Simple transfer from non-access to access counter.

the same storage capacity as the non-access counter the latter will have cycled once and finish with the same stored number as it started with. However, its overflow allows transfer signals into the access scaler so that this finishes with the same number as was originally in the non-access counter. Fig. 9.33 illustrates such an arrangement.

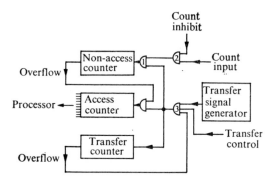

FIG. 9.33. Non-destructive transfer from non-access to access counter.

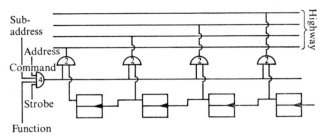

FIG. 9.34. Parallel transfer.

9.8.1. *Parallel transfer*

Fig. 9.34 gives a typical arrangement using integrated circuits. All counters are connected in parallel onto a common 'highway' and each has an associated gate (which could be integral with it) to switch its stored information onto the appropriate line of the highway on receipt of a command signal. Typically the command could be the ANDed result of address, sub-address, function, and strobe signals.

9.9. Design trends

Early instruments were an assembly of units in a large box—usually of standard 19-in width. For instance, the A.E.R.E. 1009 scalers comprised, besides the counting circuits, a complete stabilized mains power supply and an input discriminator with adjustable dead time. The modern approach is to build these units individually as plug-in modules and in the Harwell 2000 series the power unit, discriminator, and counter are each individual modules that plug into a standard shelf-unit. However, there are certain facilities in addition to counting which are conveniently included in a counting module. These are:

(1) *An input circuit.* This is a circuit that ensures that the input signal— even if defined as in the 2000 system—is suitable to operate the first decade. It should also remain 'dead' to further inputs from the time a count is accepted until it has been completely processed by the counter. It may include one or more individual inhibit inputs operating through OR logic for remote and/or automatic control purposes, and also facilities for injecting local or remotely generated test signals behind the inhibit gate.

(2) *A test oscillator.* This is an internally generated source of test signals variable in frequency, often by the scaling factor switch, to enable each decade to be tested at about two counts per second for easy visual check. Sophisticated assemblies of counters often include a test routine applied periodically by the processor or computer.

(3) *Reset circuit.* This accepts a range of control levels and converts them all to a suitable form to clear all decades, overflows, etc., to 0.

(4) *Scaling factor switch.* This and its associated circuits select the scaling factor at which a carry output signal is generated. In its simplest form it can connect the output-signal generating circuit directly to the appropriate decade, but in more refined arrangements it works on an 'all nines' fast-carry circuit, as in Fig. 9.35. In this particular example a progressive shorting switch is used, and is shown connecting, through OR diodes, the output of the input-circuit plus digits 1 and 8 (9) of the first decade. Examination of the waveforms in Fig. 9.35(b) shows that the only time the output line

will not be held at o by one or other of the diodes is during the count of 9, at the end of which it can fall to 1. It returns to o at the transition of 9 to o.

Rotation of the switch clockwise progressively adds higher-order decades. These will reach the releasing state of 9 in descending order and it will always be the input circuit that releases the output line and returns it to o. Hence, no matter how many decades are incorporated, the scaled output

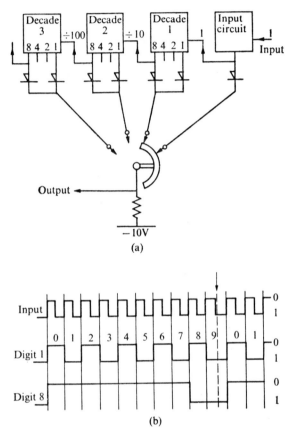

FIG. 9.35. 'All-nines' fast-carry circuit. (a) Schematic. (b) Waveforms.

will always be synchronized with the relevant input signal and the propagation delay through cascaded decades thereby avoided. This is of considerable importance when scaling is included in a system.

(5) *Output circuits.* It is desirable for a scaler to generate two output signals. The first should be a standard brief-carry, or digital-coupling signal and the second a standard binary-control level. The former is best generated by a monostable and the latter by an *R–S* flip-flop. Such an arrangement will produce a carry signal every time the total selected by the scaling factor

switch is passed, but the level will be generated the first time this occurs and persist until it is cancelled. Cancellation is usually effected when the counter is cleared, or reset. The binary level can be fed back into the scaler's own inhibit gate so that it will stop itself on achieving the scaling-factor total. Due to the fast-carry circuits described above this can be achieved within the resolution of a 10 MHz scaler, i.e. in less than 0·1 μs. It is advantageous to precede the R–S flip-flop by two cascaded binary flip-flops and an extension of the fast-carry circuit so that the level is available at one, two, or four times the scaling factor.

(6) *Overflow circuit*. If the total store of a counter is exceeded the most significant digits are lost. Therefore the last decade, on switching to 0000, should set an R–S flip-flop which operates a warning device such as a lamp. The warning remains in operation no matter how many times the total store is exceeded and the operator, or processor, is informed that the available information is invalid. The overflow flip-flop is cleared with the counting circuits.

9.9.1. *Pre-set counters*

There is a demand for scalers which will scale by every number from one up to the maximum of the store's capacity. Extending the fast-carry technique and fitting binary-coded decimal switches makes this possible and such an instrument is available in the A.E.R.E. 2000 series (pre-set scaler, type 2166). This type of scaler, when used with a crystal-controlled signal generator, enables accurate digital-gate and delay systems to be assembled.

9.10. Counting ratemeters

If a constant charge is drawn each time an input signal arrives, the mean current, which can be displayed on a meter, is proportional to the mean rate of the input signals. The diode pump circuit of Fig. 9.36 is the best

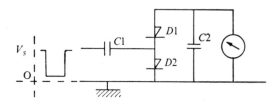

FIG. 9.36. Basic diode pump ratemeter.

known technique for doing this. Between inputs $C1$ will rapidly charge through $D2$ to V_s. The input signal in effect grounds the input terminal and the voltage across $C1$ will be redistributed via $D1$ between $C1$ and $C2$.

223

For satisfactory operation $C2$ must be much larger than $C1$ and the time constant of the meter resistance and $C2$ should be longer than the mean interval between inputs. If $C2$ is 100 times as large as $C1$ there will be slightly less than 1 per cent of V_s across $C2$ after the first input. This small voltage will drive a small and diminishing current through the meter. The charge of the next input will add to what is left of the first charge and a slightly larger current will flow through the meter. Eventually, as $C1$ pumps charge into $C2$, the voltage across $C2$ will reach a value at which it drives a current through the meter which is equal to the mean rate of arrival of current out of $C1$. Thus a state of equilibrium obtains. The mean voltage across $C2$ can become an appreciable fraction of V_s and the indication is non-linear since the charge per event is $(V_s - V_t)C$, where V_t is the voltage on $C2$. This feature has been exploited in some instruments, particularly hand-portables, to produce an approximately logarithmic scale over two or three decades.

9.10.1. *Linear ratemeters*

When a linear scale is required it is possible to use an elaboration of Fig. 9.36 wherein the voltage at the top of $C2$ does not change appreciably. The basic scheme is shown in Fig. 9.37. $C2$ no longer feeds the indicating meter,

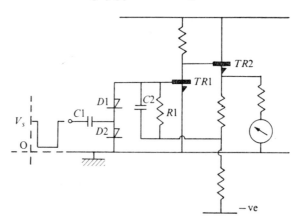

FIG. 9.37. Linear-diode pump ratemeter.

but is connected across a high-gain phase-inverting d.c. amplifier that has its input referred to ground. The indicating meter is connected across the output of the amplifier and $C2$ is loaded by $R1$, which can be much higher than the resistance of a meter, so that reasonable time constants are possible without inconveniently large values of $C2$. Because of the gain A of the amplifier the voltage excursion at its input is only $\{1/(1 + A)\}V_t$ where V_t is the voltage across $C2$.

An alternative linear ratemeter circuit can be obtained very easily by substituting a transistor for $D1$ in Fig. 9.37. This exploits the transistor's constant current collector characteristic and is shown in Fig. 9.38. Here $C1$ charges and discharges virtually between two opposed diodes connected to ground, and the charge per input is constant irrespective of the voltage across $C2$. In fact the voltage across $C2$ can be several times greater than V_s without non-linearity being introduced. Hence the combined resistance of the meter and $R1$ can be quite high, again permitting reasonable time constants without excessive values of $C2$.

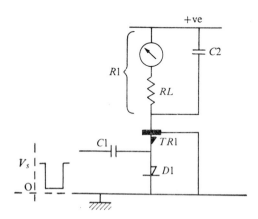

FIG. 9.38. Linear-diode transistor-pump ratemeter.

In the circuits of Figs. 9.36–9.38 the voltage across $C2$ is proportional to $NV_sC_1R_1$, where N is the mean input-rate, $C1$ is the value of the input capacitor, V_s the voltage to which it charges between inputs, and $R1$ the resistance across $C2$ (the meter resistance in the case of Fig. 9.36). It is significant that the value of $C2$ does not affect the meter reading, but only its steadiness. It is therefore possible by switching the value of $C2$ to vary the response speed and the steadiness of the reading.

In all cases for regularly spaced inputs the value of $C2$ determines the time taken to achieve stability and the degree of variation about the mean level as each charge arrives. Large values of $C2$ give small variations but slow response speeds, and vice versa. For random inputs these conditions still obtain, but superimposed on them is the random-arrival characteristic of the input-signal train.

It can be shown that the expected standard deviation of any single reading from a ratemeter will be given by (Taylor 1951)

$$\text{standard deviation} = (2NCR)^{1/2}, \tag{9.4}$$

225

where N is the count-rate and C and R are in farads and ohms respectively. The corresponding standard deviation for a counter was shown in eqn (9.2) to be $T^{1/2}$, where T is the total count, or the count-rate N multiplied by the counting time t, i.e. $T = Nt$.

For a common standard deviation, the expressions of eqns (9.2) and (9.4) are equal: $Nt = 2NCR$ or $t = 2CR$, which indicates that a single reading of a ratemeter of time constant CR will have the same expected accuracy as the result from a count lasting $2CR$. However, a word of warning is necessary since the comparison assumes that the ratemeter has already reached stability. In fact it takes $2 \cdot 3CR$ to achieve 90 per cent of the true reading when starting from zero and $4.6CR$ to achieve 99 per cent.

An interesting variation of Fig. 9.37 omits $R1$, which turns the circuit into a linear integrator.

9.10.2. *Logarithmic ratemeters*

As already explained, the circuit of Fig. 9.36 can be used to obtain a near logarithmic scale. A more precise and elaborate system employs a number of individual diode pump circuits with feed capacitors increasing by a factor of 10 (Cooke-Yarborough and Pulsford 1951). If the voltages across all these circuits are summed linearly it can be shown that the output is approximately a logarithm of the input rate.

Other systems employ the logarithmic I_a/V_a characteristics of a diode, but this is confined to currents of a few microamps, although extending over several decades. Later developments use the same characteristic of the emitter-base diode of a transistor and very satisfactory results can be obtained by this method when used as the non-linear feedback component with an integrated circuit operational amplifier.

References

ALEXANDER, T. K., REDDERING, H. G., and KENNEDY, J. M. (1954) Transistorized counting systems. *Chalk River Report CREL*-779, p. 3.

BACON, R. C. and POLLARD, J. R. (1950) *Electron. Engng* **22**, 173.

COOKE-YARBOROUGH, E. H. (1964) *Nucl. Instrum. Meth.* **30**, 106.

—— and PULSFORD, E. W. (1951) *Proc. Instn elect. Engrs* **98**, Part 2, 196.

ORMAN, P. R. (1967) Private communication.

TAYLOR, D. T. (1951) *The measurement of radio isotopes*, p. 48. Methuen, London.

10 Coincidence and Time-spectrometer Circuits

By P. R. ORMAN

10.1. Introduction

THE electrical pulses from a radiation detector give information on at least four important points. Their frequency of occurrence gives a measure of the probability of a particular event or the yield in a nuclear reaction. The amplitude of a pulse may be related to the energy absorbed in the detector. The shape or 'waveform' of the pulse may be characteristic of a particular type of interaction and so may be used as a means of distinguishing between particular particles or radiations. A fourth piece of information that the pulses provide is the timing of events taking place in the detector.

There are, broadly speaking, two types of timing measurement required. First, it may be necessary to know whether two or more events occur within a given time of one another; this would be a 'coincidence' measurement. Secondly it may be necessary to measure the time interval between two particular events using a time spectrometer. The time differences concerned in nuclear measurements usually fall in the range from nanoseconds to milliseconds.

10.2. Coincidence circuits

10.2.1. *Coincidence resolving time*

Coincidence methods are historically the oldest in the timing field. A coincidence unit with, for example, two input channels (see Fig. 10.1) can

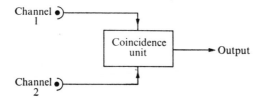

FIG. 10.1. Two-channel coincidence unit.

accept input pulses from two sources and it gives an output pulse only when a pulse in one input channel occurs within a definite time of a pulse in the other channel. The input pulses are first shaped (see Fig. 10.2) into rectangular pulses of a defined duration T, usually the same for each channel, the delay between the leading edges of these pulses and of the input pulses being kept to a minimum. The shaped pulses are fed to a circuit, usually called the mixer, which measures their overlap in time. If one pulse overlaps the other, then the mixer gives a pulse that triggers an output circuit. Clearly an output pulse is produced provided the leading edge of pulse 2 occurs within $\pm T$ seconds of the leading edge of pulse 1. The value T is called the coincidence resolving time.

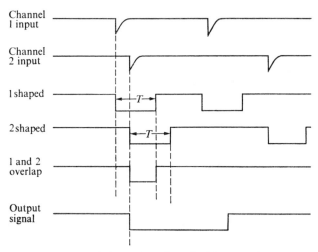

Fig. 10.2. The action of a coincidence unit.

10.2.2. *Chance coincidences*

Suppose that the two input channels of a coincidence unit are supplied with pulses from two completely independent counters detecting particles from two separate radioactive sources. Let the counting rates in the two channels be N_1 and N_2 per second respectively. Then the probability of a pulse in channel 1 being in coincidence, i.e. within $\pm T$ seconds of a pulse in channel 2 is given by $2N_1 T$. Since there are N_2 counts per second in channel 2, the total number of coincidences recorded will be N_c, given by

$$N_c = 2N_1 N_2 T. \tag{10.1}$$

These coincidence counts are due entirely to chance, since the input pulses came from two independent sources. N_c is called variously the chance coincidence rate, the random coincidence rate, or the accidental coincidence rate. If N_c, N_1, and N_2 are measured in the above experiment, a value of T may be deduced.

10.2.3. *Time delays in coincidence experiments*

Fig. 10.3 shows a typical input pulse for a coincidence unit. The pulse must rise above a certain voltage level defined in the coincidence unit before the circuit can respond. The threshold level is often set by an amplitude discriminator. As will be seen from Fig. 10.3, there is an appreciable time delay between the occurrence of an event in the counter and the point at which the coincidence circuit responds. Moreover, as is shown by the dotted lines, this delay varies with the amplitude of the input pulse. This matter will be considered in more detail later (section 10.5.2) in relation to the precision of time measurements. Apart from this delay variation, there is

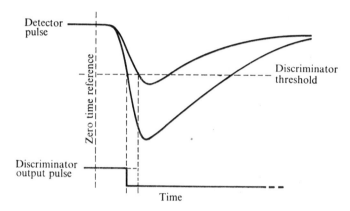

FIG. 10.3. Time shift due to rise time of detector pulse.

also another depending in importance on the type of counter used. Geiger counters, for example, exhibit considerable time-delay variations depending on the position in the counter of the trajectory of the detected nuclear particle. A small Geiger counter may give rise to variations in the region of $\frac{1}{4}$ μs, while for large counters the variation may extend to several microseconds. When using coincidence equipment it is important to arrange compensation for the different mean time-delays in the counters and also to take the variations of delay into account when choosing the coincidence resolving time. These points will be brought out as we consider an actual experiment.

10.2.4. *A coincidence experiment*

In the disintegration of certain radioactive nuclei, two or more particles or radiations may be emitted either simultaneously or in rapid succession. For example, in one process whereby the nucleus ^{60}Co decays to ^{60}Ni, two gamma rays are emitted within 10^{-12} s of one another. Suppose that two

229

gamma-ray detectors are set up near a source of radioactive cobalt (see Fig. 10.4). The detectors are connected to the input channels of a twofold coincidence unit and the output of this unit to a count ratemeter. Of course not every gamma ray absorbed in one detector is associated with a gamma

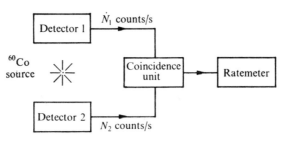

FIG. 10.4. Coincidence experiment.

ray detected in the other, for there are many atoms in the source. Moreover, even if two gamma rays appear to be coincident (i.e. within the resolving time of the coincidence circuit) they are not necessarily associated; the fact that they have given a coincidence count may be due only to chance. Suppose in series with one detector there is introduced a delay that is very long compared to the coincidence resolving time. No genuine coincidences can now be recorded and the output displayed by the ratemeter is due only

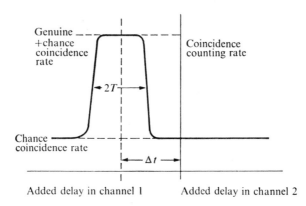

FIG. 10.5. Coincidence delay-curve.

to chance coincidences. As the added delay is reduced, a point is reached where genuine coincidences begin to be recorded as well. If the coincidence counting rate is plotted against added delay, first in one channel and then the other, the curve of Fig. 10.5 is traced out. This curve provides three important pieces of information. First the time Δt is the amount by which

230

the mean time-delays in the two channels differ from one another. Secondly the time $2T$ gives the coincidence resolving time.

Suppose a large added delay is introduced in one channel so that only chance coincidences are recorded, and the coincidence resolving time is then varied. If coincidence counting rate is plotted against coincidence resolving time the result is the straight line in Fig. 10.6 representing the expression $N_c = 2N_1 N_2 T$ derived previously. If the added delay is set so that the operating point is at the middle of the plateau region in Fig. 10.5,

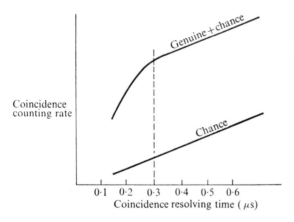

FIG. 10.6. Effect of coincidence resolving time.

both genuine and chance coincidences are recorded and now, as the co-incidence resolving time is varied, quite a different effect appears. If the coincidence resolving time is reduced until it is comparable with the variations of delay in the detectors, some of the input pulses corresponding to genuine coincidences are so delayed that the circuit rejects them. In Fig. 10.6, which might refer to an experiment in which small Geiger counters are used, the fall in genuine coincidences starts at about 0·3 μs. There is therefore a minimum usable value of coincidence resolving time depending on the precision of the input signals.

10.2.5. Choice of resolving time

In performing a practical coincidence experiment it is the genuine coincidence counting rate that has finally to be determined. This may be done by first counting the chance coincidences using a long delay in one channel, and then subtracting them from the genuine-plus-chance rate. The process of subtraction leads, however, to a worsening of the measurement accuracy.

231

Suppose that in equal counting periods $N_{(g+c)}$ genuine-plus-chance co-incidences and N_c coincidences due to chance alone are recorded. Then the number of genuine coincidences

$$N_g = N_{(g+c)} - N_c. \tag{10.2}$$

Now consider the statistics. The probable error in N_g is given by

$$\Delta N_g = 0.67\{N_{(g+c)} + N_c\}^{1/2}. \tag{10.3}$$

If, for example, $N_{(g+c)} = 250$, and $N_c = 150$, then $N_g = 100 \pm 13$ (probable error). If the same number of genuine counts had only ten chance counts associated with them, i.e. $N_{(g+c)} = 110$, $N_c = 10$, N_g would come to 100 ± 7. In a given situation the statistical accuracy can of course be improved in principle without limit merely by counting for a longer period. In the cases mentioned above, however, where the errors differ by a factor of 2, the first count would have to be done over about four times the period of the second to achieve the same accuracy. In this connection it is also necessary to consider the stability of the coincidence resolving time. To a first order this affects only the chance coincidences. Therefore, the smaller the number of these compared with the genuine ones, the smaller is the effect of drift in resolving time during the experiment. The ratio of genuine-to-chance coincidences rate can (at least in principle) be improved in two ways; either by reducing the coincidence resolving time, or by reducing the strength of the source.

The chance coincidence rate N_c can clearly be reduced by decreasing T (see eqn (10.1)) without the genuine coincidence rate being affected (provided suitable counters are used). It is also apparent that $N_c \propto N_0^2$, where N_0 is the number of disintegrations per second in the source, whereas the genuine coincidence rate $N_g \propto N_0$. Thus a reduction in the source strength reduces the chance coincidence rate more than the genuine rate. In practice it is necessary to reach some compromise between coincidence resolving time and source strength in order to achieve worth-while results in a reasonable time.

10.2.6. Anti-coincidence units

Another type of coincidence equipment that is commonly used is the so-called anti-coincidence unit. Dealing again with the simplest two-channel system, the circuit is required to detect all pulses from channel 1 except when a pulse from channel 2 arrives within a given time interval. This interval, called the anti-coincidence resolving time, is defined within the circuit.

10.3. Practical coincidence circuits

10.3.1. *Defining the resolving time*

The first step in designing a coincidence unit is to shape the input pulses so that they have defined duration. Usually it is also necessary to define the amplitude. When the coincidence resolving time required is above 1 μs or so, the duration is usually set by the time constant in RC-coupled monostable circuits. For times below 1 μs it is desirable to use a delay-line circuit since this is a convenient way of producing well-defined short-

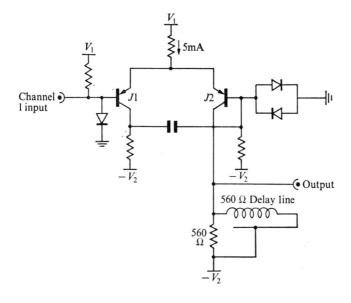

FIG. 10.7. Delay-line controlled pulse-shaping circuit.

duration pulses. This is illustrated by the simplified circuit of Fig. 10.7, where $J1$ and $J2$ are connected as an emitter-coupled monostable circuit. The collector current of $J2$ is quiescently 5 mA, so that when the circuit triggers, a negative voltage pulse of 1·4 V appears across the shorted 560-Ω delay line. The duration of this pulse is twice the delay of the line provided this is less than the natural period of the monostable circuit. When the current is restored in $J2$, a positive pulse is generated and it is arranged that this has no effect. Such a circuit is useful for pulse durations from tens of nanoseconds up to a few microseconds. For the shorter durations in this range a delay line of lower impedance would be chosen because of band-width considerations, outlined in Chapter 2.

233

10.3.2. *Mixer circuits*

One of the simpler circuits for determining coincidence overlap is the diode AND circuit, shown in Fig. 10.8. The diodes D_1 and D_2 are each normally conducting about 1 mA, defined by resistor R_1, so that their common anode potential is roughly 0·7 V. The input channel pulses (see

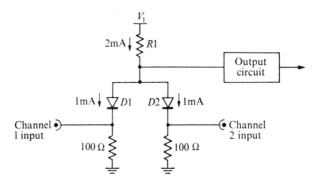

FIG. 10.8. Diode mixer circuit.

Fig. 10.9) are shaped to a defined duration and amplitude (say 1·4 V) so that when a pulse is applied to one diode the current in it is cut off. The common anode potential rises by an amount equal to the input pulse amplitude of 1·4 V for the period of coincidence; this change in potential

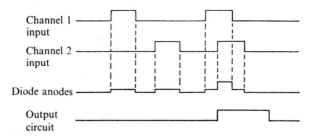

FIG. 10.9. Operation of diode mixer circuit.

is sufficient to trigger a suitable output circuit. This system may be used for a wide range of coincidence resolving times, from nanoseconds upwards, and for multichannel operation.

If it is necessary to define the durations of the individual input pulses to the mixer, a different arrangement may be used, an example of which is illustrated in Fig. 10.10. Two channels with monostable circuits as in Fig. 10.7 are provided, but the collector terminals are joined together and connected to a common shorted delay line. If one input circuit is triggered at a time, pulses of 1·4 V appear across the delay line. If a coincidence occurs the output pulse rises to twice this value. A suitable output circuit will be triggered by coincidence pulses only.

Once again it is necessary that the natural period of the monostable circuit exceeds the required duration of the delay-line pulse. The use of this or similar arrangements for multichannel working is ultimately limited by circuit tolerances. If the contribution from each channel is an amplitude V and there are n channels, the output circuit must respond to a signal nV but not to a signal $(n-1)V$. The situation is further aggravated if the pulse edges overshoot.

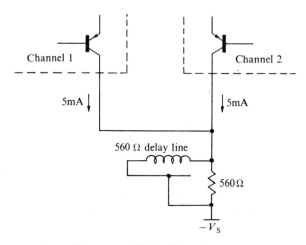

FIG. 10.10. Delay-line mixer.

10.3.3. *Anti-coincidence circuits*

The diode AND circuit of Fig. 10.8 may be modified to give anti-coincidence working merely by the addition of a further diode input channel in which the diode, normally biased off, is turned on by a negative input pulse. This pulse must be arranged to arrive before the coincidence pulses and to be longer in duration so that no output is obtained from the mixer.

10.3.4. *Adjustment of channel delays*

In fast coincidence systems the adjustment of delays in the various input channels is simply achieved by adjusting the lengths of short coaxial cable connections. In slower systems, however, the provision of variable delays presents more difficulty. Variable delay circuits inevitably introduce dead time, which in coincidence input channels is particularly undesirable. Switched lengths of special helically-wound delay-cable may be used, but in practice this tends to be rather bulky (see Chapter 2). It is probably best to provide optional variable delay-circuits of modest range, having recourse to special cable for the longer delays. The restricted bandwidth of delay cable is not usually serious since it is exceptional for short coincidence resolving times to be required at the same time as long delays.

In the usual type of delay circuit, the input signal switches on a fixed current which charges a capacitor and this generates a linear ramp waveform. A variable threshold circuit is triggered by this waveform at a time proportional to its threshold. It may be arranged that when the threshold circuit triggers to give the 'delayed' output pulse, the charging current is switched off and the capacitor discharged.

10.3.5. *A complete fast coincidence unit*

Fig. 10.11 illustrates a practical fast coincidence unit with one anti-coincidence and three coincidence input channels in which the coincidence resolving times are defined by the durations of the input pulses. Extra complication is introduced by the facility to switch off individual channels. The mixer in this case is a tunnel-diode discriminator. In the quiescent condition with no signals present and all channels switched on, the co-incidence channels provide currents I_1, I_2 and I_3 of approximately 5 mA each and the anti-coincidence channel current I_4 of 13 mA in the directions indicated. The 10-mA tunnel diode is therefore conducting a 'reverse' current of 2 mA. Coincidence input signals remove the currents I_1, I_2, and I_3 so that the tunnel diode is switched to the 'high-voltage' state by the current I_4 and the output circuit triggers. For the duration of an anti-coincidence signal I_4 is switched off and the tunnel diode cannot switch to the 'high-voltage' state. Clearly for anti-coincidence operation the input pulse must overlap the related inputs to the coincidence channels. Switching off a coincidence channel has the same effect as applying a continuous input to the channel so that the corresponding 5-mA current into the tunnel diode is permanently removed. When the anti-coincidence channel is switched off, the effect is to maintain the current I_4 in spite of the appearance of any anti-coincidence signals provided these are below a specified amplitude.

10.4. Time spectrometry

10.4.1. *Introduction*

So far this chapter has dealt with the methods for determining whether two or more events occurred within a given time of one another. The second type of measurement referred to at the outset is the determination of the actual time interval between two events. A typical example is time-of-flight work carried out using a pulsed Van de Graaf accelerator (Ferguson,

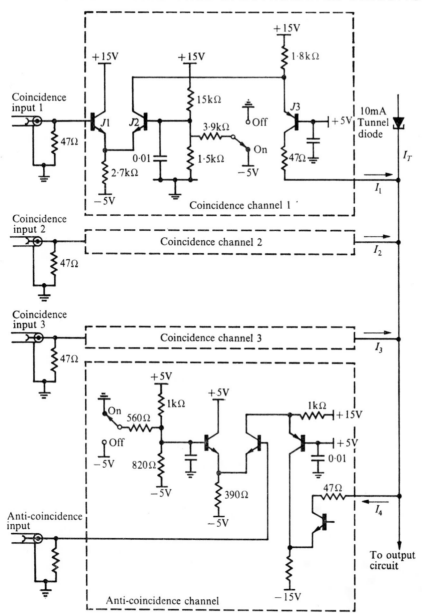

FIG. 10.11. Fast coincidence circuit.

Montague, and Orman 1961). The beginning of the measured interval is defined by a signal synchronized to the time of arrival of the primary beam pulse at the target, and at the end of the flight time another signal is obtained from a radiation detector. Time intervals to be measured in general nuclear work range from milliseconds to nanoseconds, and different methods are used at the extreme ends of this range.

10.4.2. *Measurement of long time intervals*

The equipment used for long time intervals operates in a manner analogous to that of the stop-watch. The output of an oscillator is connected to a register during the time interval to be measured. The number stored in the register after this time is taken as the 'address' of the appropriate channel in a data-storage system and one count is added at that address. The accuracy of the measurement is determined by the period of the oscillator. When the number of channels required is large, and it may be as large as, say, 10^5, the problem becomes one of data handling. It is not usually economically possible to use one core memory location per channel in these cases and one solution is to record the addresses on magnetic tape, event by event, and to analyse the recorded data at the end of the experiment.

10.4.3. *Measurement of short time intervals*

10.4.3.1. *The stop-watch method* of measuring time intervals has been applied down to channel widths of 4 ns, but this is close to the limit set by present techniques and components (Richards 1967). For channel widths of less than 1 ns other methods are used. In essence, most of these methods depend on the expansion of the interval to be measured until it is within the capability of equipment designed for slower speeds.

10.4.3.2. *The chronotron.* The most obvious and historically the first method was to use a large number of coincidence units separated in time by lengths of delay cable. Each unit was provided with its own scaler. A development of this idea was the chronotron, which is still used in a number of forms (Keuffel 1949). Its operation may be understood by inspection of Fig. 10.12.

10.4.3.3. *Vernier systems.* This method was probably first suggested by Chase (1961). The two pulses defining the time interval to be measured are passed separately into two recirculating delay-line stores. The lengths of the lines differ by a definite small amount, t, which determines the channel width. For example, one line might be 100 ns and the other 95 ns long. The channel width is then 5 ns. The recirculation continues until sooner or later the pulses come into coincidence. This is detected by a suitable circuit which stops the process. The number of memory cycles that have occurred then gives the desired time interval in units of t. This system is illustrated in Fig. 10.13. Systems similar to this have been described in which the 'start' and 'stop' input pulses excite oscillatory circuits tuned to slightly different frequencies (Cottini, Gatti, and Giannelli 1956).

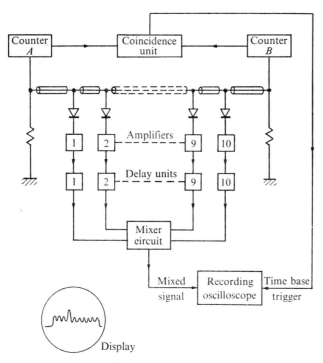

FIG. 10.12. Chronotron principle.

10.4.3.4. *Capacitor timing circuits.* A number of circuits have been devised in which a storage capacitor is charged by a constant current during the time interval to be measured, so that the final voltage across the capacitor is proportional to the time interval. At least three variations on this theme are in common use.

(1) *Time to pulse-height conversion.* In the simplest form of converter in Fig. 10.14 there are effectively two switches between a storage capacitor

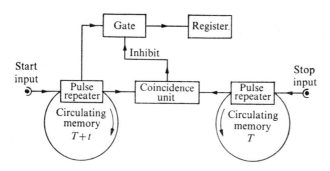

FIG. 10.13. A vernier timing system.

and a source of constant current. One of these, called the 'start' gate, is normally open. The other, the 'stop' gate, is normally closed. The arrival of a start input pulse closes the start gate for a time longer than the interval to be measured and the capacitor is charged by the constant current. On the arrival of the stop pulse, the stop gate becomes open circuit and the

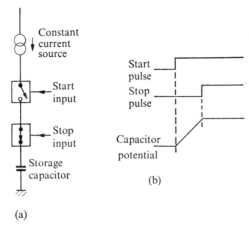

FIG. 10.14. Simple charge-storage timing circuit. (a) Schematic. (b) Waveforms.

capacitor voltage remains stationary at the value proportional to the time interval. In this system nothing happens unless the correct input pulse comes first, but gating has to be provided so as to limit the number of occasions on which maximum amplitude output pulses are recorded simply because no input stop pulse arrives within a given time range.

(2) *'Overlap' converter.* In the 'overlap' type converter of Fig. 10.15 the input pulses are set to a standard duration and amplitude and then applied

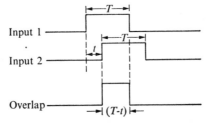

FIG. 10.15. Timing with overlap circuits.

to a coincidence mixer, whose output pulse duration is the overlap time. A constant current I is switched into a capacitor C for the period of overlap. Inspection of Fig. 10.15 shows that if the duration of the pulses after shaping is T, the output voltage changes from a maximum IT/C down to

240

zero as the time interval to be measured increases from zero to T. This circuit has certain practical advantages, among them being that no output appears unless both input pulses are present and within the desired time range. In some applications this advantage may be offset by the fact that no indication is given of which input pulse came first. It is necessary to ensure that the value T is not permitted to drift through any cause, since it is just as important as the constancy of the charging current in its effect on the stability of the circuit.

(3) *Time expander.* In both the time-to-amplitude converters described above the output pulse is passed to a pulse–height analyser. This may be rather paradoxical in that the analyser often operates by converting the amplitude into a time duration. In the time expander of Fig. 10.16 the first

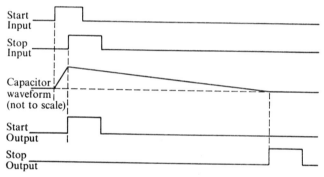

FIG. 10.16. Time-expander principle.

stage is the same as in a simple time-to-amplitude converter, but instead of using the capacitor voltage as the output signal, the capacitor is discharged by a small current until the capacitor voltage is restored to its original value (Ferguson *et al.* 1961). If the charging current were, for example, 20 mA, and the discharging current 100 μA, the time taken for discharge would be 200 times the time taken for charging. An output start pulse is provided at a time synchronized to the input stop pulse, and an output stop pulse at the time when the capacitor voltage has returned to its original value. Thus a new time interval is defined which is 200 times longer than the one we started with. This interval is arranged to fit in with the requirements of the slow timing-devices already described.

10.5. Precision in nuclear-event timing

10.5.1. *Relation between a nuclear event and the corresponding timing signal*

The uncertainties involved in representing nuclear events by electrical timing signals vary considerably between particular cases. As an example,

241

in timing experiments using a particle accelerator, the starting reference is commonly taken from the accelerator itself, whereas the particle to be timed usually arises in a target struck by the primary beam. Because of this there may be a significant timing uncertainty due to variation in the flight time of the primary particles between accelerator and target. In addition, the secondary particles may lose variable amounts of energy within the target, resulting in a further uncertainty. At the other end of the flight path the detector may have a significant sensitive depth and this gives uncertainty in the signal marking the end of the flight time. Practical considerations of this sort apply to all time-of-flight systems and they have to be taken into account in the design and evaluation of experiments.

10.5.2. *Waveform distortion of timing signals*

The definition of a time interval is usually referred to the front edges of electrical pulses. As was pointed out in section 10.2.3 the finite rise-time of these pulses, together with any variation in their amplitude, gives rise to uncertainty in the instant at which the threshold is exceeded at the input of the timer. Usually the amplitude variation arises from the nature of the experiment and nothing can be done about it, but the detector output pulses may sometimes be significantly degraded by the connecting cables.

10.5.2.1. *Transient response of coaxial cables.* The principal cause of attenuation in coaxial cables at frequencies below about 1000 MHz is the skin effect, which gives rise to an attenuation proportional to the square root of frequency. At higher frequencies the dielectric loss becomes appreciable, giving an attenuation contribution directly proportional to frequency, though with modern cables this effect can usually be ignored in practice.

It can be shown that skin effect losses produce a step function response (Orman 1963a)

$$E_{\text{out}} = E_{\text{in}}[1 - \text{erf}\{2^{1/2} bL(t - T)^{1/2}\}. \tag{10.4}$$

In (10.4) E_{out} is the voltage at distance L (feet) from the input end of a semi-infinite uniform cable. E_{in} is the amplitude of the input voltage step function at $t = 0$ and b equals $1 \cdot 45 \times 10^{-8} A$ where A is the attenuation in dB/100 ft at 1000 MHz. Finally, T is the transit time in seconds. It is convenient to define T_0 as the value of $(t - T)$ at which $E_{\text{out}} = 0 \cdot 5 E_{\text{in}}$, i.e. T_0 is the 50 per cent rise-time. Then T_0 equals $4 \cdot 56 \times 10^{-16} A^2 L^2$ seconds. Fig. 10.17 shows the calculated step-function response for some commonly used commercial cables (Orman 1963a). It may at first sight seem surprising that the rise time of the step-function response should be proportional to the square of the cable length. The following simple argument may make

242

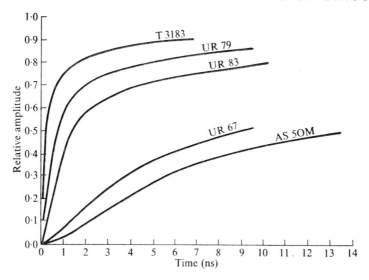

FIG. 10.17. Calculated step-function response for various coaxial cables, length 600 feet.

this clearer. By analogy with the case of a pulse amplifier for which we define a 'cut-off' frequency at which the gain has fallen by 3 dB, we can also think of a 'cut-off' frequency for a length of cable. This will be the frequency f_c at which the attenuation has risen to some value p, i.e. at f_c,

$$A_c = p. \tag{10.5}$$

We have seen that A is proportional to $f^{1/2}$ so that

$$A_c = A_0(f_c/f_0)^{1/2}, \tag{10.6}$$

hence

$$f_c = p^2 f_0/A_0^2 L^2. \tag{10.7}$$

Now the rise time is inversely proportional to f_c and is therefore directly proportional to L^2.

10.5.2.2. *The synchronizing discriminator or 'zero crosser'*. The combination of finite rise time of timing pulses and variation in their amplitude gives rise, as has been said above, to a variation in the instant at which the threshold is exceeded at the input of a timing circuit. This variation is particularly serious for pulses whose amplitude only just exceeds the threshold of a conventional trigger circuit (see Fig. 10.18). This effect is related to the gain-bandwidth of the circuit. Time variations due to these effects can be greatly reduced by the use of special circuits, provided the shape of the input pulses remains substantially constant. This shape often

243

takes the form shown by the full line of Fig. 10.19 which can readily be changed to the bipolar form shown (dash line) by the use, for example, of a shorted delay-line. The start of this pulse is likely to be ill-defined because of bandwidth limitations in the system, but the point X where the signal crosses zero is invariant with amplitude (see Chapter 2). Although it would

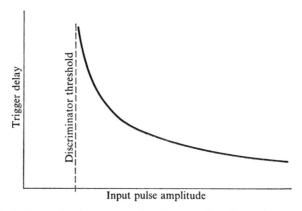

FIG. 10.18. Trigger circuit output-pulse delay as a function of input amplitude.

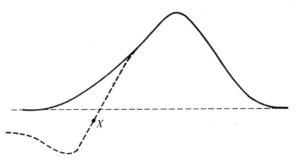

FIG. 10.19. Derivation of 'zero-crossing' waveform.

be difficult to design a circuit that triggers at precisely this point, nevertheless it is possible for a discriminator to trigger at a point close to it, and the difference in timing may be made small by ensuring that the slope of the waveform at X is as steep as possible. Such a circuit is the synchronizing discriminator or 'zero crosser'. Its operation may be briefly explained as follows (Orman 1963b).

A tunnel diode used as a current amplitude discriminator is operated quiescently at point A in Fig. 10.20. The applied bipolar current waveform of Fig. 10.19 reduces the current in the diode and if sufficiently large, drives it into the 'low-voltage' state. As the applied current falls, however, the diode does not return to the 'high-voltage' state until the diode current again

244

exceeds I_{peak}. Thus the voltage across the tunnel diode returns to its initial state at a time very close to the 'zero-crossing' point X in Fig. 10.19. The reason for choosing the high-voltage state as the initial condition is that the transit from the low- to the high-voltage state is faster than the reverse one. When such an arrangement is tested with pulses from a suitable pulse-generator circuit, it is found that the triggering delay changes by as little of 0.7 ns as the input-pulse amplitude is changed from threshold to twenty times threshold. When tested with pulses from an actual photomultiplier, however, the result is quite different. Consider the arrangement shown in

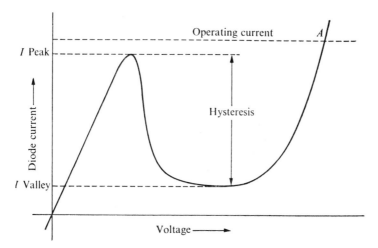

FIG. 10.20. Tunnel-diode characteristic.

Fig. 10.21. A fixed-amplitude electrical signal triggers a fast light pulse generator and also one input of a time-to-amplitude converter. The light pulse illuminates the photo-cathode of the multiplier and a signal from the collector triggers a synchronizing discriminator so as to provide the other input signal for the time-to-amplitude converter. The output of the converter is displayed on a pulse–height analyser. It is arranged that various optical attenuators may be interposed between the light source and the photo-cathode. Fig. 10.22 illustrates the results obtained with this arrangement with three levels of optical attenuation. Care is taken to adjust the photomultiplier EHT supply so that its average output-pulse amplitude remains roughly constant and well above the threshold of the discriminator, but nevertheless the resolution becomes steadily poorer as the light level is reduced. To explain this effect we recall that the output waveform shown in Fig. 10.19 represented not the photomultiplier output-pulse shape, but rather the probability distribution as a function of time for the emitted

245

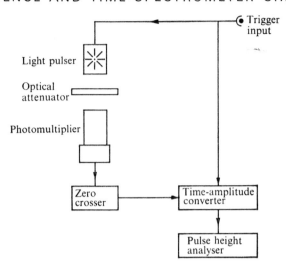

FIG. 10.21. Resolution test set.

electrons. The values $n_e = 500$, 140, and 60 in Fig. 10.22 refer to estimates of the total number of electrons emitted from the photo-cathode for every incident light pulse. When the light level is low and therefore the number of photoelectrons small, the statistical nature of the waveform becomes more pronounced. In other words, at low light levels the shape of the waveform applied to the synchronizing discriminator is not constant and hence the reason for its apparent failure to operate correctly.

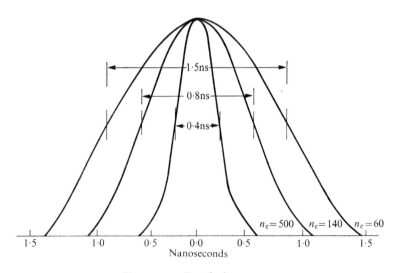

FIG. 10.22. Resolution curves.

In experiments where scintillation light pulses are large but variable, a time resolution of 0·5 ns has been obtained even though the rise time of the pulses applied to the synchronizing discriminator is as large as 10 ns. Resolution tests of the sort described above demonstrate the importance of improving the overall efficiency of scintillator and photo-cathode in scintillation detectors used for timing when the energy dissipated in the detector is relatively low.

References

CHASE, R. L. (1961) *Nuclear pulse spectrometry*, p. 171. McGraw-Hill, New York.

COTTINI, C., GATTI, E., and GIANNELLI, G. (1956) *Nuovo Cim.* **4**, 156.

FERGUSON, A. T. G., MONTAGUE, J. H., and ORMAN, P. R. (1961) *Proceedings of the nucleonic electronics conference*, Belgrade, 1961. I.A.E.A., Vienna.

KEUFFEL, J. W. (1949) *Rev. scient. Instrum.* **20**, 107.

ORMAN, P. R. (1963a) Pulse response of coaxial cables. *AERE Memorandum M1133*.

—— (1963b) *Nucl. Instrum. Meth.* **21**, 121.

RICHARDS, J. M., HINTON, H., and BOWKER, A. J. (1967) Measurement of time of flight using 300 Mc/s scalers. *AERE Report R5422*.

I I *Multi-channel Pulse Data Analysis*

By F. H. WELLS

11.1. Introduction

THE previous chapters in this book have shown how voltage pulses are derived from nuclear detectors, and how the peak amplitude and time of occurrence gives information on the energy and velocity of the incident particles of radiation. Thus, in a nuclear physics experiment, precise measurements of amplitude and time are needed on each nuclear event of interest. Single-channel pulse-amplitude analysers or single-channel pulse-time analysers (coincidence units) may be used if nuclear events having only a narrow energy range are to be studied. However, in most cases a wide range of energy levels is of interest, so that instruments equivalent to large assemblies of such single-channel analysers are needed, viz. multi-channel analysers.

For example, in a typical pulse-amplitude spectrum the voltage pulses from an amplifier between the amplitude limits 1 and 10 V may be sorted into 256 channels, each channel being about 36 mV wide. Thus the number of pulses occurring with peak amplitudes lying between the limits 1 and 1·036 V are separately recorded from those pulses between 1·036 and 1·072 V, and so on. If now the number of pulses obtained in each channel is plotted against that mean-channel amplitude, the pulse-amplitude spectrum of the nuclear source is obtained.

A typical pulse-amplitude curve for a mixed source of uranium and curium is shown in Fig. 11.1, the pulses being derived from the α-particles emitted by the source enclosed in the ion chamber. This curve gives information as to the constituents of the source and the relative abundance of the radioactive isotopes.

Pulse-time spectra occur in neutron spectrometry or the study of the complex phenomena that exist when an element is bombarded with a stream of neutrons. Neutron energy is measured by the time taken for the neutron to travel a known distance, called the flight path; this measurement gives the velocity and hence the kinetic energy may be calculated. Fig. 11.2 shows a typical neutron-absorption spectrum demonstrating the complexity of the spectra.

In some nuclear reactions two or more particles may be involved, and it may then be necessary to measure the energies of these particles. These

248

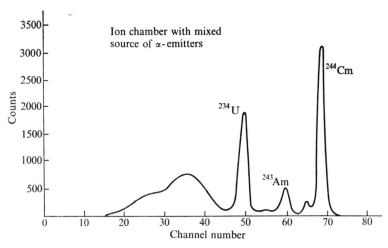

FIG. 11.1. Pulse amplitude spectrum.

measurements give families of spectra and need the use of a very large number of channels in the analyser.

The electronic problem in the design of these analysers resolves itself into three sections:

(a) sorting the pulses into a large number of different categories or 'channels' in terms of peak amplitude or time of arrival.

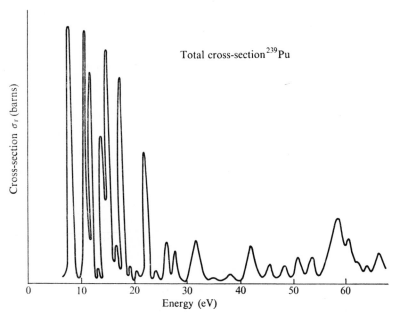

FIG. 11.2. Neutron spectrum.

249

(b) scaling the number of events occurring in each of these categories during the experiment,

(c) recording the numbers given by this scaling operation, these numbers being the data obtained by the experiment.

11.2. Accuracy requirements of pulse analysers

11.2.1. *Stability and accuracy*

The paramount consideration in the design of this equipment is that the channel width and also the mean amplitude or time position of each channel shall be as stable and accurate as possible. If the pulse analyser is receiving input pulses having a completely random distribution of amplitudes or 'times' over the range accepted by the equipment, then each channel should register the same number of pulses, apart from statistical fluctuations of the numbers. If now one channel is a little greater in width than the rest then the number of pulses received in this channel will be larger, giving an artificial peak in the pulse-amplitude spectrum. Thus in the case of a fairly uniform spectrum it is absolutely essential that all channels be of as nearly the same width as possible. A practical compromise is to try to ensure that the difference between individual channel widths should be somewhat less than the possible errors due to statistical fluctuations. If a total count of the order of 10 000 pulses per channel can be obtained then each channel width should be within ±0·5 per cent of the average channel width. Since in a typical pulse-amplitude case the average channel width might be 36 mV then the differences between channels in this example should not exceed ±0·2 mV—a very stringent requirement. It should be noted that a gradual change of channel width from one end of the spectrum to the other is not so important as sudden changes of channel width between adjacent channels. The former error gives a slight tilt to the spectrum that may be allowed for by a careful calibration of the equipment; the latter type of error may give an artificial peak or valley in the spectrum and possibly cause confusion and an erroneous analysis of a complex spectrum, except in the case when a digital computer is being used for analysis; see section 11.8.

The mean amplitude or time position of each channel must also be as stable as possible. In the case of a uniform, level, spectrum this position may be allowed to change a little without distortion of the resultant spectrum. However, in the case of a spectrum having a number of sharp peaks, as shown in Fig. 11.1, any movement of the channel position during the counting time will result in a broadening of the peaks and extra counts in the valleys, giving a considerable loss of resolution. In general, the channel position should not move by more than a fraction of a channel width during the time of an experiment. In the case of a 256-channel amplitude analyser this requirement necessitates a stability of the order of ±0·1 per cent of the mean channel position—again a very stringent requirement for high-amplitude channels.

11.2.2. *Input pulse shapes to amplitude analysers*

The spectrum information given by an analyser will only be useful if the peak amplitudes of the input pulses are a direct function of the electric charge released at the nuclear counter (ion chamber, proportional counter, scintillation counter, or semiconductor counter). If this relation is not accurately maintained then the pulse amplitude spectrum will be in error, and it may not be possible to translate the results back into the energy spectrum of the nuclear particles or radiations being measured.

Assuming that the amplifier system is free from such defects as non-linearity and overloading troubles there are two fundamental considerations that limit the accuracy at *high* counting rates.

(a) *Pulse rate*. If the number of pulses per second is sufficiently great then some pulses may overlap in time and so their waveforms are super-imposed, giving rise to a much larger pulse. Thus the counting rate must be kept low enough for there to be little chance of this happening. As an example, if we take a pulse shape having rise and fall exponential time constants of $1~\mu s$ then the effective width of the pulse, before its amplitude has fallen to less than 0·5 per cent of its peak value, will be about $6~\mu s$. If now we have a number of equal-amplitude pulses arriving at random, then the average number of pulses per second should not exceed 1700 if less than 1 per cent of these pulses must not overlap each other. This is an extreme case, and in practice the effect is complicated by the wide range of amplitudes present in the input pulses.

(b) *Amplifier a.c. coupling*. The amplifier and connections to the pulse amplitude analyser are normally a.c. coupled, since it is impractical to use high-gain d.c. amplifiers for this purpose. Thus the peak amplitude of any pulse measured with respect to earth potential will be affected by the average number of pulses per second passing through the amplifier. If the coupling time constants in the amplifier are long enough this change will affect all pulses equally, giving a shift of the pulse amplitude spectrum. For example, a calculation on the case given above of equal-amplitude pulses with $1~\mu s$ time constant (rise and fall) shows that if the average number of pulses per second is 1000 then this change of amplitude is of the order of 0·2 per cent. Again, this effect is very dependent on the shape of the pulse amplitude spectrum.

Both the above effects (a) and (b) may be present and should be carefully considered when counting at high input rates. In practice the presence or absence of such effects as these may be measured by taking several pulse amplitude spectra, using the same nuclear source, counter, and amplifier, but progressively increasing the counting rate by increasing the source strength or by other means.

The design of amplifiers for pulse amplitude spectrum work will not be further considered, and it will suffice to say that considerable improvement on the above two effects may be obtained by the use of double delay-line pulse-shaping techniques to give pulse shapes as in Fig. 11.3; see Chapter 6.

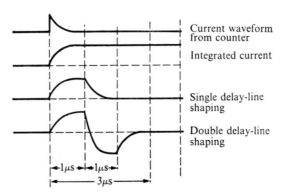

FIG. 11.3. Double delay-line pulse shaping.

11.2.3. *Amplitude resolution and number of channels*

The above two factors are obviously directly related to the shape of spectrum measured. To portray a sharp peak properly the channel width should be kept correspondingly narrow for good definition. The sharpness of a peak is mainly determined by the performance of the nuclear counter, pulse amplifier, and in most experiments with existing counters the channel widths need not be much less than 0·5 per cent of the maximum pulse amplitude. Occasionally rather better resolution is required. For small amplitude pulses derived from low-energy nuclear particles the energy resolution of the counters is usually much worse, and the analyser channel width may be made larger in comparison to its mean amplitude. Instruments with 100- to 512-channels are commonly used, but in some high precision work with solid-state counters at least 1000 channels may be required.

11.2.4. *Time spectra resolution and number of channels*

Neutron spectrometry by 'time of flight' methods normally employs a pulsed source of neutrons, and this pulse duration gives an uncertainty in the determination of 'starting' time. This effect is usually the main source of error, but by using a long flight path a very accurate determination of neutron velocity is possible. The number of timing channels required commonly lies in the range 200–4000 in order to make full use of the accuracy inherent in the method.

11.2.5. *Number of channels for multi-particle analysis*

In the case of a nuclear reaction involving two or more particles the number of analyser channels required is the product of the number of channels needed to define the energy spectra of the individual correlated events. For example, a (n,γ) reaction may need 500 timing channels for neutron energy and 200 amplitude channels for γ-energy, so giving a total of 100 000 channels, and sometimes even higher channel numbers might be involved.

11.3. Analogue design methods in multi-channel amplitude analysers

The preceding sections have shown that the accuracy of some pulse amplitude analysers need not be better than ± 1 per cent, so that an instrument design based on analogue rather than digital techniques can sometimes be used successfully. This is not the case for pulse time analysers where analogue techniques cannot be used because the accuracy demanded is too high. Two analogue instruments for pulse amplitude analysis have been commonly used and these will now be briefly described.

11.3.1. *Chart-recording single-channel analysers*

One of the simplest designs of spectrometer is to use a single-channel pulse amplitude analyser that counts all the input pulses occurring between two defined amplitude limits. The mean amplitude of this channel is then moved slowly over the whole range of input amplitudes to be measured and the rate of arrival of pulses received in the channel is continuously recorded. The channel width is kept constant during this process so that the whole spectrum is gradually recorded. Fig. 11.4 shows a block diagram of the equipment involved.

The equipment comprises two amplitude discriminators set at levels of V_2 and V_3 volts accepting the pulses from a back-biased amplifier of voltage gain A. This amplifier only gives an output when the input signal exceeds a peak amplitude V_1. Two discriminators define the channel boundaries and their outputs are mixed in an anti-coincidence circuit. An output pulse is obtained only when the lower discriminator is triggered without the upper discriminator being triggered. This means that the input pulse amplitude to the discriminators must lie between the limits V_2 and

18

V_3. The corresponding input to the amplifier must then be between the limits $(V_1 + V_2/A)$ and $(V_1 + V_3/A)$. The channel width is thus $(V_2 - V_3)/A$ set at a mean amplitude of $\{V_1 + (V_2 + V_3)/2A\}$.

The reason for using an input amplifier is to improve the stability of the channel width. If the amplifier were not used the two discriminator voltage levels would have to be spaced one channel-width apart, only 36 mV in a practical case. Any unsteadiness in the level of either discriminator would then have a large effect on the channel width. A variation of 3 mV in one

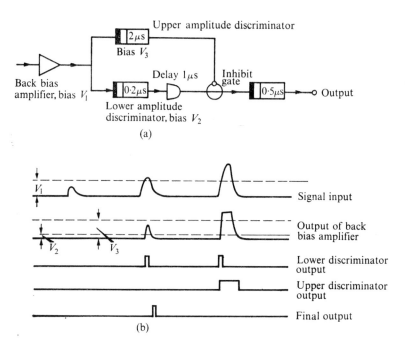

(a)

(b)

FIG. 11.4. Single-channel analyser. (a) Schematic. (b) Waveforms.

discriminator with respect to the other might easily occur giving a 10 per cent change of channel width. Hence by using the amplifier the difference between the two discriminator levels may be correspondingly increased, thus improving the stability of channel width, provided the amplifier gain is stable; a gain of 50 could be used for this purpose. The maximum input to this amplifier might be about 10 V, so that the amplifier would have to give an output up to 500 V. This is impracticable. Use is made of the fact that only pulses whose amplitude lie in the channel width need be amplified; the amplifier is operated in Class C with the input biased off by a controllable voltage V_1. Again, if the input amplitude is greater than the upper channel width then the pulses are of no interest and the amplifier can be allowed

254

to overload on these pulses. A typical circuit for this type of amplifier is shown in Fig. 11.5. This method then gives the complete spectrum with a minimum of equipment making it attractive for many experiments. The major disadvantage is that it relies on the source of pulses remaining constant during the measurement, since the amplitude channels are

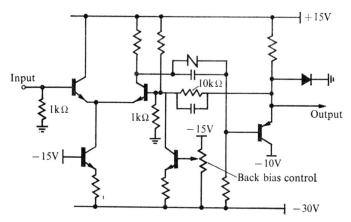

FIG. 11.5. Back bias amplifier.

measured in turn rather than simultaneously. For the same reason the method is extremely wasteful of experiment time and sufficient time per channel must always be allowed to obtain a statistically significant result. These two facts make the use of this method impossible for a large number of applications where it becomes essential to measure the outputs of all channels simultaneously rather than in sequence (Breitenberger 1953, Farley 1954).

11.3.2. *Gray wedge analysers*

An analyser design using a photographic recording technique is shown in Fig. 11.6 (Bernstein, Chase, and Schardt 1953). The input pulse is fed through a back-biased amplifier as in Fig. 11.5, and the output pulse from

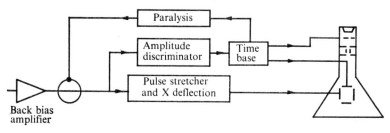

FIG. 11.6. Gray wedge amplitude analyser.

this amplifier charges a storage capacitor to the peak value of the pulse, Fig. 11.6. The output pulse from this capacitor produces the Y deflexion on a cathode-ray tube. This pulse also triggers an amplitude discriminator set at a low level which, after a short delay, triggers a time base giving a short duration X sweep. The electron beam is normally biased off and is switched on by a brightening pulse for the duration of the time base. Thus a horizontal line is displayed with the Y position proportional to the input

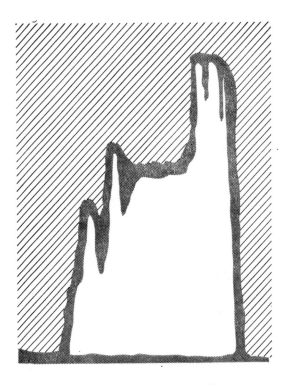

FIG. 11.7. Gray wedge photograph (Chase 1961).

signal amplitude less the back-bias voltage. The storage capacitor is discharged at the end of the time base, and all circuits return to their quiescent condition so that another input pulse can be received.

A recording camera is used and the film is left exposed to the face of the cathode-ray tube for the whole counting period so that an integrated record is obtained, the denseness of the photographic record indicating the number of input pulses of any given amplitude. The denseness of this record may then be examined by a recording densitometer equipment to give a graph of the pulse amplitude spectrum. Alternatively a grey optical wedge may be inserted between the camera and cathode-ray tube screen. This wedge is positioned so that the light from one end of each time base is attenuated with respect to the other end. In this case the final record looks like Fig. 11.7 (Chase 1961), and the line of equal photographic denseness will portray the pulse amplitude spectrum.

The equipment is relatively simple, but the result is qualitative rather than quantitative, and experience is required in determining the optimum settings of beam brightness or camera aperture. However, the total dead time of this apparatus need not be longer than 10 μs, and much shorter times could be achieved with circuit development (Flynn and Johnston 1957), so that the method is applicable to the measurement of high counting rates with short duration input pulses. This is the great advantage of this technique over the digital methods described below. The inconvenience of the photographic technique can be partially overcome by using a polaroid land camera system.

11.4. Combined analogue and digital design methods in multi-channel amplitude analysers

The analogue part of the analyser is the sorting process (see section 11.4.2) and needs precise and stable circuit design, while the digital section provides the scaling and recording functions (see sections 11.6 and 11.7). We will first consider the sorting process, usually termed *analogue to digital conversion*.

The input information occurs as pulses, but the relevant parameter, whether it be amplitude or time, may have any value over a given range. However, the number of channels in the analyser is limited to a relatively small number, so that the process of sorting into channels will be to 'label' each pulse on arrival with the channel number nearest to its actual value. This channel number forms the 'digital' code for each pulse, but is, of course, only an approximation to the true value. It is in these circuits that all the accuracy requirements of section 11.2 must be met.

11.4.1. *Multi-channel amplitude discriminators*

The earliest type of analogue-to-digital converter used a number of separate amplitude discriminators individually set to define the limits of each channel. The outputs of these circuits were then sorted and routed to separate scalers, one for each amplitude channel, such that each scaler reading represents the number of input pulses whose amplitude fell in the corresponding amplitude channel (Cooke-Yarborough *et al.* 1950, Moody *et al.* 1951). The amplitude selection circuit thus acts as a multi-position switch.

The disadvantage of this method is that the channel widths cannot be kept sufficiently uniform without constant resetting of the amplitude discriminator levels, and, for this reason, the method has now become obsolete except in some special cases. In addition, it became uneconomical where channel numbers exceeded about 30. One advantage, however, is that the circuits can be designed to have quite short paralysis times after each pulse input (a few microseconds), so that this technique is still useful for some applications demanding very high-speed amplitude analysis.

11.4.2. *Analogue-to-digital converter designs for ensuring equality of channel widths*

The basis of the circuit design is to try and define each of the channel limits by using the same circuit operations performed by one common circuit element. Two methods have been developed as follows:

(i) The peak pulse amplitude is converted into a time interval directly proportional to amplitude. This time interval is then subdivided into a number of equal parts by using a timing oscillator technique.

(ii) The peak pulse amplitude is measured by counting the number of defined and constant amplitude increments needed to give approximately the same amplitude as the analogue input.

The application of these principles in the design of some types of amplitude-to-digital converters will now be considered.

(a) *Hutchinson–Scarrott amplitude-to-digital converter.* This circuit principle is shown in Fig. 11.8. The circuit still uses valves since no instruments of this type have been designed recently (Hutchinson and Scarrott 1951), but a transistor design could easily be derived if needed. The input, after amplification, charges a storage capacitor C to the peak value of the pulse through diode D_1. This amplitude is then translated into time information in the storage part of the instrument. A Miller-type linear saw-tooth generator is used, giving $200V$ swing with quick fly-back. A 'channel' oscillator generates 100 equally spaced pulses in the total period of run-down. The oscillator is quenched during the fly-back and then allowed to start again when the run-down commences, as shown in Fig. 11.8(b). This saw-tooth generator is then compared in voltage with the input signal as given by the storage capacitor. When the two waveforms are of equal voltage during the run-down, the circuit following the comparator is triggered, and this in turn selects the next 'channel' pulse from the oscillator. Thus the number of channel pulses between the start of the run-down and this selected channel pulse indicates the amplitude of the input pulse in digital code. The reference saw-tooth and digital section of this design does not have to be locked in time to the input pulse signal and this fact makes it suitable for use with circulating-type digital memories (section 11.6.1). The capacitor waveform can be restored to zero as soon as the digital selection process is completed. One disadvantage of the circuit is that small capacitor leakages can give a slight tilt to the waveform, giving an erroneous amplitude measurement.

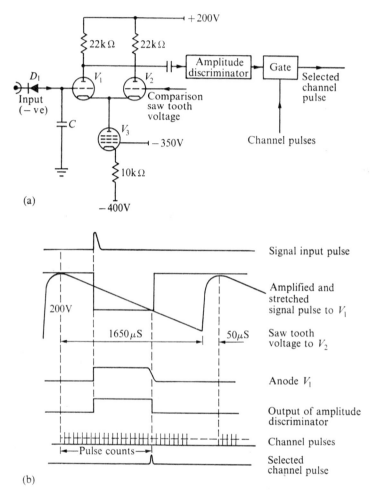

FIG. 11.8. Hutchinson–Scarrott amplitude to digital converter. (a) Circuit. (b) Waveforms.

(b) *Wilkinson-type amplitude-to-digital converter*. This method (Wilkinson 1950) uses pulse amplitude-to-time conversion and then measures the time intervals by oscillator and scaling methods. Fig. 11.9 shows the block diagram and a simple transistor circuit arrangement; a more refined circuit (Bonsignori, Malosti, and Pelligrini 1963) employs amplifiers and negative

259

feedback to reduce the non-linearities caused by the imperfect diode characteristics.

Transistors J_1, J_2, J_3 in Fig. 11.9(b) form a back-biased amplifier giving a negative output pulse that is applied through J_4 to charge the capacitor

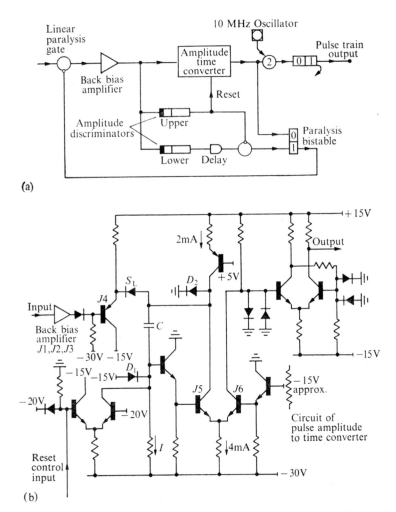

(a)

(b)

Fig. 11.9. Amplitude-digital converter. Wilkinson method. (a) Schematic. (b) Circuit.

C to the peak value of the pulse; transistor J_5 is just cut off during this period. After the peak of the input pulse the voltage across C charges in a linear fashion at the rate determined by the current I (40 μA) through C, until the rise of potential at the collector of J_5 is stopped by D_2. The base potential of J_4 then falls about 0·5 V until caught by diode D_1, when the circuit is back to its quiescent condition. The duration of the linear voltage change across C is proportional to the peak input pulse amplitude and is indicated by the time $J5$ is conducting. Hence the collector waveform from $J6$ is a suitable output for driving a bistable to give a good waveform for opening the gate, allowing the pulses from 10 MHz oscillator through to the output. The number of these output pulses is thus proportional to the peak input amplitude, and is scaled to obtain a digital number proportional to the pulse amplitude.

The circuit must be designed to be temperature independent. If a sufficiently good performance in this respect cannot be easily achieved then the temperature should be stabilized by a suitable oven enclosure. Likewise it is necessary for the oscillator to be stable in frequency to at least 0·1 per cent (for a 256-channel instrument) and this is conveniently achieved by a crystal oscillator. However, this means that the timing pulses are not synchronized to the capacitor timing waveform so that the number of pulses in the output will be uncertain to one pulse, or the channel position of the pulse will be uncertain to $\pm\frac{1}{2}$ clock period or $\pm\frac{1}{2}$ channel. This effect may be reduced to $\pm\frac{1}{4}$ or $\pm\frac{1}{8}$ channel using a higher frequency oscillator and then following the oscillator gate by a scale of two or scale of four. The effect is, however, random in character so that the channel position will be correct when averaged during the course of the experiment. The channel width is always correct.

(c) *Kandiah-type amplitude-to-digital converter.* This method (Kandiah 1962) uses the technique of section 11.4.2(ii) and a circuit is given in Fig. 11.10. The input pulse first charges a capacitor $C1$ in series with much larger $C2$ to the peak amplitude through an amplifier system consisting of $J1$, $J2$, and $J3$ stabilized by negative feedback. After the peak of the pulse has passed, $J2$ and $J3$ are cut off by $J1$ being in full current and this condition is detected by the balance detector. The square wave generator is then started and the capacitor $C2$ is discharged in small equal steps at a rate determined by a pulsed oscillator until the voltage across $C1$ and $C2$ passes through its original starting potential. At that point $J2$ begins to conduct and the balance detector stops the square wave generator. The capacitor discharge is then stopped and the number of pulses from the oscillator during this action represents the pulse train. $C1$ and $C2$ are subsequently restored to their initial potentials. Stabilities of the order of millivolts have been achieved with this method.

11.4.3. *Paralysis time and auxiliary logic*

The paralysis time of each of methods (a)–(c) in the previous section consists of the time to charge a capacitor followed by the discharge time with the pulse train; values between 50 and 200 μs are common. Further development may be expected towards reduction of this dead time by the use of higher speed pulse trains of 50 MHz rate which would reduce the dead time of a 256-channel converter to perhaps 10 μs (Gere and Miller 1967).

This paralysis time may be reduced further still by restricting the analysis to only those pulses of interest. The upper discriminator in Fig. 11.9 is typical of all systems and is used to provide a rapid reset of the amplitude to digital converter; the reset occurs after those input pulses which have

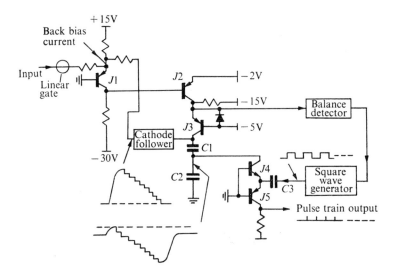

FIG. 11.10. Pulse amplitude-digital converter. Kandiah method.

amplitudes greater than the range of interest. This re-set action reduces the dead time of the circuit, especially in those cases where the spectrum of pulses contains a majority of such 'unwanted' high amplitude pulses.

Likewise a 'lower' amplitude discriminator is included so that no output is obtained until the input pulse amplitude has reached this lower amplitude limit. The paralysis bistable will only be set if the input pulses are within the channel defined by the two discriminators and so may be used to inhibit further input pulses until the completion of the analogue-to-ditigal circuit waveform. This paralysis bistable waveform can also be used as a measure of the total paralysis time during the analogue-to-digital conversion. By integrating this time during the total experimental period a mean value of paralysis time is obtained. This figure is sometimes needed for the interpretation of the resulting amplitude spectrum.

11.4.4. 4000-*channel short dead time multi-channel amplitude analyser technique*

The techniques described in (b) and (c) of section 11.4.2 can both be used for 1000- to 4000-channel analysers provided great care is taken in the circuit design to achieve the necessary stability and channel width uniformity. However, the dead time may become prohibitively long (perhaps 1000 μs) for the 4000-channel analyser. If a shorter dead time is needed then either very high-speed pulse trains are needed (between 100 and 300 MHz) or analogue-to-digital conversion must be by a 'successive approximation' method.

The former technique is becoming possible (Gere and Miller 1967) but involves very high-speed circuit design while the latter has always been rejected for multi-channel analyser work because of the very poor channel-

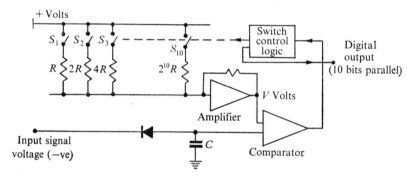

FIG. 11.11. Principle of successive approximation analogue-to-digital conversion.

width uniformity. However, a technique has been described for using such a successive approximation converter design and still obtaining uniform channel width while preserving the advantages of high speed. The design achieves 10 μs dead time for a 1000-channel instrument (Cottini, Gatti, and Svelto 1963).

We will first consider the basic design principles of the successive approximation analogue-to-digital converter and then how to use this design successfully for multi-channel analysers.

(a) *Successive approximation analogue-to-digital conversion.* Fig. 11.11 shows the simplest concept of these systems. The input pulse charges a capacitor C to its peak value and the capacitor holds this voltage undisturbed during the measurement. Switch S_1 is first closed and the output voltage V is compared with the capacitor voltage; if V is too large S_1 is then opened, but if V is too small S_1 is kept closed. Switch S_2 is then closed and a similar

263

comparison is made with S_2 being opened if V is too large but left closed if V is too small. Switches S_3 to S_{10} are similarly operated in succession. The component values of the resistors are so arranged that the amplifier output V due to switch S_1 being closed on its own (S_2 to S_{10} open) will give a voltage exactly equal to half the maximum amplitude voltage scale. Likewise S_2 on its own gives V equal to a quarter of the amplitude scale, S_3 gives one-eighth, and so on for S_4 to S_{10}. The switches take the form of transistors that can be saturated when closed and are controlled from bistables so that the time between switch actions can be reduced to about 1 μs.

At the end of this process the amplifier output V is the sum of that derived by the addition of all the switch settings and for a ten-switch system can be within 0·1 per cent ($\pm$0·05 per cent) of the correct value in 10 μs, with the switch control bistables indicating the amplitude in binary code. The channel-width accuracy in a multi-channel system will depend on the stability and accuracy of the values of the graded resistors (R, $2R$, etc. in Fig. 11.11) and also on the matching of the switch characteristics S_1–S_{10}. Probably the worst variation of channel width can occur between the channel 0·499 to 0·500 of full scale (switch S_1 open and switches S_2 to S_{10} closed) and the channel 0·500 to 0·501 of full scale (switch S_1 closed and switches S_2 to S_{10} open). If now the accuracy of R in Fig. 11.11 is $\pm$0·05 per cent then the ratio between the widths of these two adjacent channels may be as large at 1·7:1, an intolerable condition for a multi-channel analyser (see section 11.2.1). Thus the accuracy of R should be improved as far as possible to perhaps $\pm$0·01 per cent, giving a channel-width ratio of not worse than 1·1:1, or a possible $\pm$10 per cent variation of channel width. This variation is much improved but is still too great for the majority of multi-channel applications. The effect of this variation may be greatly reduced by using the technique described below, which makes the 'effective' channel-width uniformity good enough for use on a 1000- to 4000-channel analyser. The 'successive approximation' analogue-to-digital converter should be made as good as possible before applying the following 'Gatti' technique.

(b) *Reduction of the effect of channel-width non-uniformities in successive approximation analogue-to-digital converter by random-base-line additions.* This method by Gatti (Cottini, Gatti, and Svelto 1963) is shown in Fig. 11.12. The principle is to add a known voltage to each input pulse amplitude and then to subtract from the digital output (channel) of the converter a number equal to the channel number of this added voltage; this added voltage is varied after each input pulse. For example, if the input pulse corresponds to channel 500 (between 0·499 and 0·500 of full scale) and if the added voltage is 0·100 of full scale then the addition will fall into channel 600 (between 0·599 and 0·600 of full scale). The analogue-to-digital converter then operates on the combined voltage (600) and finally the number 100 is subtracted from the digital output; the next time that an input pulse corresponding to channel 500 occurs the added voltage will most likely be different to 0·100 of full scale. Thus, during the course of the experimental counting period, all the pulses appropriate to channel 500 will be using a large assortment of channels in the analogue to digital converter circuit, since the added voltage may vary widely. The nett effect is that the width of channel 500 will appear to be the average of all channels in the converter

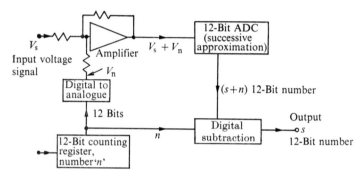

Fig. 11.12. Random base-line addition to successive approximation analogue-to-digital conversion.

circuit and similarly with all the other channel numbers. By this means the widths of all channels are equalized, provided sufficient input pulses are received to give a time-averaging effect.

The number of input pulses required before this average is achieved will vary with the spectrum shape and this calculation is beyond the scope of this paper. The original reference cited (Cottini, Gatti, and Svelto 1963) may be consulted for a more detailed study of this complex system.

11.5. Multi-channel time discriminators

In this case the analogue quantity is the time of occurrence of a pulse with respect to another pulse. The commonest application in nuclear physics is the measurement of neutron energy by timing the neutron over a given flight path, thus accurately determining the velocity and hence the kinetic energy. This necessitates a timing accuracy between 0·1 to 0·01 per cent of the time of flight. The electronic method is simple as in Fig. 11.13

where a gate is opened by the one pulse (start) and closed by the other pulse (stop). This gate switches pulses from a time reference crystal oscillator to the output for the duration between start and stop pulses giving a pulse train output, the number of pulses being a digital representation of the time to be measured.

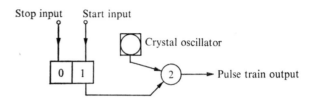

FIG. 11.13. Pulse time-digital converter.

The similarity of the pulse train outputs from sections 11.4.2(b) and (c) (pulse amplitude) and 11.5 (pulse time) should be noted, and the same remarks about random jitter of the channel position (time delay) apply.

11.6. Digital section of multi-channel analyser

The work of this section comprises the two stages in sections 11.4.2 and 11.4.3, and for this purpose a digital memory is required plus facilities for performing the simple arithmetic required for scaling. Three types of memory have been found suitable: (a) circulating delay-line memory, (b) ferrite core store, and (c) mechanical counting registers. Of these three the ferrite core store is now the most commonly used, although the circulating delay line usually gives a cheaper but slower speed memory.

11.6.1. Delay-line memory analysers

The first design of this type (Hutchinson and Scarrott 1951) uses the converter of section 11.4.2(a) combined with a circulating delay line which in one particular instrument takes the form of a nickel magnetostriction element of 1700 μs delay. Other designs use quartz or mercury delay lines. At the start of the run-down in Fig. 11.8 a second oscillator called a 'digit' oscillator is started, the frequency of this being set at fourteen times the frequency of the channel oscillator. These two oscillators are kept exactly in synchronism by a suitable cross-connection. Thus this digit oscillator subdivides each channel period into fourteen time intervals. When the run-down starts a group of three or four digit pulses are injected into the delay line. This gives current pulses about 0.3 μs wide into the driving coil on the line, sending a pressure wave down the nickel wire by the magnetostriction effect. These pulses arrive at the end of the wire after a delay of

266

1700 μs and the pressure wave changes the magnetic flux through a receiver coil on the wire, so producing a voltage pulse. The run-down waveform is adjusted so that the total duration is about 1650 μs when the fly-back occurs; hence it is back at the starting point when the pulses appear at the receiver. These pulses then restart the run-down; consequently this circuit is always locked to the time delay of the delay line.

When an input is received a channel pulse appropriate to the input amplitude will have been selected as described in section 11.4.2(a). This channel pulse then gates the next digit pulse and this in turn gives a pulse into the delay line, thus storing one 'count' at a particular time interval relative to the start pulses in the delay line. When this stored pulse reaches the line receiver it is shaped, gates the next digit pulse and then is again injected into the line through the transmitter. Thus once the pulse is injected it is continuously circulated. It should be noted that the delays are such that this pulse is always the *first* digit pulse following the channel pulse appropriate to the original input pulse.

Now consider a second input pulse of the same amplitude as the first pulse. Again the same channel pulse will be selected in the comparison circuit so that the circuit will try to inject into the delay line the first digit pulse following this channel pulse. This event occurs at the same time as the stored pulse (due to the first input signal) arrives at the receiver ready to be recirculated. When this happens a coincidence circuit operates which removes the pulse from the transmitter and instead selects the *second* digit pulse after the channel pulse. This second pulse is then fed into the delay line and is continuously recirculated. When the third input pulse of the same amplitude arrives it is stored by injecting a pulse into the *first* digit position of the channel in the delay line. The fourth input pulse of this amplitude to arrive cannot be stored in either the first or second digit places, once these are full. The coincidence circuit in the feedback loop then operates to cancel the first and second digit pulses, but instead injects a pulse into the delay line in the position of the *third* digit pulse.

The result of this process is that the input signals whose amplitudes fall within one channel are stored as a number of digit pulses up to a total of fourteen, following the appropriate channel pulse time. These digits form a binary scaler and so may store a total of about 16 000 input pulses. Since there are 100 channel pulses in this instrument the delay line is equivalent to 100 scalers, each having a total capacity of about 16 000 counts stored as a pattern of digit pulses.

Existing equipments display the pulses in the line on a cathode-ray tube raster as shown in Fig. 11.14. In this display the linear run-down forms the X time base, thus being equivalent to an amplitude scale. This is sub-

267

divided into channels by the Y time bases which are driven by the channel pulses. Each digit stored in the line is then used for intensity modulation of the cathode-ray tube resulting in a direct display of the pulse amplitude spectrum.

When a pulse has been finally accepted and stored in the delay line a circuit is triggered which returns the storage capacitor to its quiescent condition and the analyser can accept another input pulse. The input circuits prior to the storage capacitor have also been paralysed after each

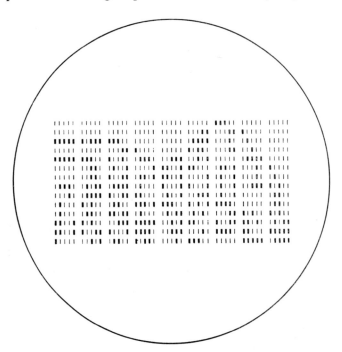

FIG. 11.14. Hutchinson–Scarrott analyser display.

input pulse into the storage capacitor to prevent erroneous results arising from a second input pulse arriving before the first pulse has been stored in the delay line. This paralysis time may be as long as 1700 μs or very short, depending on the time between the input pulse arriving and the instant the linear run-down is at the same amplitude. Thus on the average the dead time is 850 μs.

Developments of these instruments using quartz delay lines have been able to reduce the delay to 1200 μs, giving a mean dead time of 600 μs plus the possibility of storing up to one million pulses in each of 100 channels.

The outstanding advantage of this instrument is the low cost coupled with the display, which shows directly the shape of the pulse spectrum.

The user can observe the pattern building up as pulses are counted. Thus it becomes an 'oscilloscope' for nuclear physics research.

For a permanent record of the result the cathode-ray tube may be photographed, or the digital information written down and possibly translated into decimal form by the use of mathematical tables or more complicated automatic calculating machinery. The disadvantages of this instrument are:

(a) Since the delay-line storage is a transient memory device any electrical interference picked up by the delay-line receiver may disturb the pattern of pulses circulating and hence lose completely the results of an experiment.

(b) The reading out of the final answer is laborious, although it is possible to device automatic printing-type machinery for this purpose; this latter facility greatly increases the total cost of the equipment.

Although the circuits are rather complex the total quantity of equipment is reasonably small, and the cost is rather less than the alternative design described in the next section.

11.6.2. Wilkinson mechanical-register storage analyser

The pulse train obtained from the converter in section 11.4.2(b), is scaled as shown in Fig. 11.15. The binary stages in this scaler actuate relays so

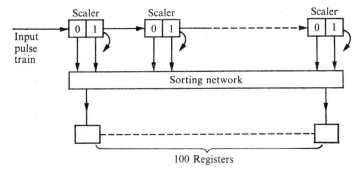

FIG. 11.15. Wilkinson-type amplitude analyser. Storage and display system.

that the number in the scaler is translated into the setting of an appropriate number of relay contacts. These contacts are then wired to give 100 outputs, corresponding to the 100 different possible scaler numbers, and these outputs are then connected to 100 mechanical registers. Thus when the signal input pulse has been translated into a scaler setting, power is supplied through the relays to one out of the 100 registers and one pulse is recorded on the register corresponding to the number in the scaler. After this event

19

the scaler is reset to zero and the signal amplifier paralysis circuits returned to their quiescent condition, allowing the instrument to accept another input pulse. This method is slow because of the slow speed of operation of relays and registers, so that it is only suitable for experiments where the input pulse rate is very low. The dead time after each input pulse recorded is about 200 μs.

11.6.3. *Ferrite-core storage analysis*

(a) *Scaling*. This method (Byington and Johnstone 1955) applies the computer memory technique (Rajchman 1953) to the problem of storing the numbers of pulses in each of the channels of the Wilkinson-type analyser.

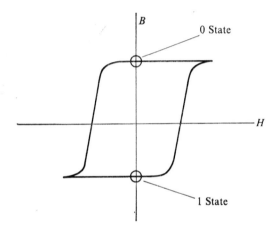

FIG. 11.16. Ferrite square-loop magnetic store.

The memory is based on the use of small 2 mm-diameter iron toroidal cores made of a ferrite material having a square-loop hysteresis curve as shown in Fig. 11.16. The core may be permanently magnetized in one direction or the other as indicated at 'o' and '1'. It may thus be considered as a binary store recording the figures o or 1. If now 16 of these cores are arranged to record the setting of a 16-stage binary scaler any figure up to about 64 000 may be recorded by the combination of these cores: alternatively 20 cores may be used to record in decimal form the setting of a 5-decade scaler based on the use of 20 (binary coded decimal) scaling stages allowing up to 100 000 pulses to be stored. These methods are used for the analyser.

The mode of operation of a typical store will be described using Fig. 11.17. The input pulse amplitude is first converted into a pulse train (section 11.4.2(b)) and then scaled in the 'address' register as in Fig. 11.15. The setting of this scaler then selects the appropriate one out of 512 columns

270

of the store and a pulse of current $+I$ (typically 450 mA in one turn) is fed into this column. The cores in this column will then all be returned to the 'o' state and the horizontal 'read' wires will have voltage pulses induced on them if any core has been changed from '1' to 'o'. These 'read' pulses will then set the corresponding stages of the 'contents' register to the '1' state, whilst other stages not receiving read pulses will be left in the initial 'o' state. Thus the binary number originally stored in the selected column of

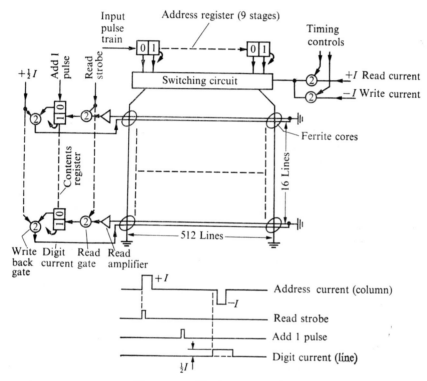

FIG. 11.17. Ferrite-core storage.

cores has now been transferred to the 'contents' register and the cores will all be at 'o'. The next stage is to add 1 to the contents register, this acting as a scaler, so that a new binary number is now in this register. The final step is to write this new number back into the selected column of cores, which is done by pulsing the column again with a current of $-I$ as shown in Fig. 11.17. At the same time a pulse of $+\frac{1}{2}I$ is fed through each of the horizontal digit wires driven from any stage of the contents register that is at 'o'.

Of the 16 cores in the selected column those cores controlled by the contents register stages at 'o' will have a current of $-\frac{1}{2}I$ and be left in the 'o' state, whilst the others will have the full current $-I$ and be set to the '1' state. This procedure has therefore set the selected column to the same number as in the 'contents' register. The address and contents registers are then reset to zero and another input pulse can be accepted. This process has provided scaling facilities on each analyser channel, but instead of 512 scalers being needed only one scaler (contents register) has been used, together with a large ferrite core memory. This procedure is possible because only one channel is in operation at any one time.

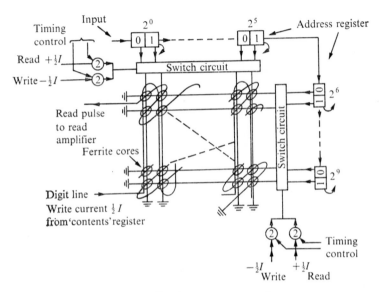

FIG. 11.18. Ferrite-core storage plane.

There remains the problem of selecting one column out of 512 as determined by the address register setting. To simplify this switching problem each horizontal line of cores in Fig. 11.17 is replaced by a plane of cores as in Fig. 11.18, having X and Y drive wires controlled by the two halves of the address register. The switching of current into one X and one Y wire may then be reasonably done through suitable diode arrangements controlled by the 'address' register. The action of the whole store is the same as in Fig. 11.17, but the column current pulse $+I$ is now replaced by $+\frac{1}{2}I$ in a selected X wire and $+\frac{1}{2}I$ in a selected Y wire. The core at the junction of X and Y will thus be the selected and only one to receive a total current of I. The total store will consist of 16 such planes and is usually wired as a compact three-dimensional assembly.

The total dead time of this analyser is the time for the converter, section 11.4.2(b), plus about 12 μs for the operation of the store.

272

(b) *Display*. This form of ferrite storage does not show any visual indication of numbers so that reading out methods are needed. This is simply done by selecting each column in turn and reading its value into the 'contents' register as described above. This register value can then be taken out on to paper tape or punched cards for use by a digital computer, or give a printed record on a typewriter, or be converted to an analogue voltage for the Y deflexion on a display cathode-ray tube.

11.6.4. *Multi-channel time interval analysers*

The time-interval-to-pulse-train converter has been described in section 11.5 and the ferrite core storage of section 11.6.3 can then be used to count the number of events occurring in each time channel to give the time spectrum information. The channel storage capacity needed will vary from 500 to 4000 channels for this class of work (Schumann 1956) so that these instruments can be quite complex and costly.

11.7. Use of ferrite store as multiscaler

The store described in section 11.6.3 is performing the same function as 512 scalers and this may be applied to any experiment that requires large numbers of scalers together with automatic display and output facilities. There is one all-important restriction, namely only one scaler can be used at any one time. However, there are a large number of experiments, such as measurements of the decay of radioactivity of short-lived isotopes, where this restriction can be accepted and multi-channel analyser stores can be effectively used in these cases. The action of the store in such a decay time experiment is that the first channel is first selected by setting the address register to 'one' and then reading the corresponding first column of cores into the contents register. This register is then used as a scaler on the experiment for a pre-set time. After this time the number in the contents register is written back into the store column and then this contents register is re-set to zero. The address register is then advanced one channel to select the second column of cores and the cycle of events is repeated. Thus the store has been used as many scalers, each counting the radioactivity at different times, and the display will show the decay curve of the short-lived isotope. This is one example of many applications.

11.8. Small digital computers for digital storage in analysers

A small digital computer will commonly contain a ferrite core memory of 4000 words (or channels), each word containing 12–24 bits (or ferrite cores). Fig. 11.19 shows the arrangement and should be compared with Fig. 11.17. It will be seen that the diagrams are similar except for the inclusion of an arithmetic unit in the digital computer. It follows that the small computer is well suited to undertake all the digital storage section of the multi-channel analyser provided that the simple arithmetic of advancing the register by 'one' during the storage process can be accomplished in a few microseconds.

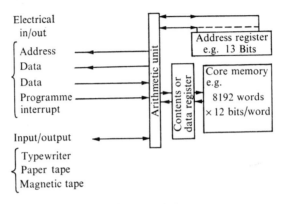

FIG. 11.19. Small digital computer.

There are several advantages of the small computer over the special purpose analyser store:

(a) The computer may be used for simple spectrum analysis after the data has been collected, such as subtracting one set of spectrum data from another.

(b) One computer store can accept data from several analogue-to-digital converters and so fulfil the role of several multi-channel analysers, for example six 512-channel analysers.

(c) Small digital computers are being widely used in many process control applications, so that large numbers are being manufactured, thus reducing production costs and giving reliable instruments.

The disadvantage of the computer is the relatively high cost if only one multi-channel analyser of low-channel capacity is required. In this case the cost of the special purpose analyser will be substantially lower than the computer.

11.9. Multi-spectrum amplitude and time recorders using magnetic tape

So far we have considered the energy analysis of a series of discrete nuclear particles, but in some experiments one or two other parameters may have to be recorded for each particle. One such case occurs in neutron scattering measurements where scattering angles as well as neutron energy must be recorded. Again, two particles or quanta may be correlated and so two sets of related 100-channel amplitude spectra must be recorded, giving 10 000 channels needed in the storage scalers. Thus the ferrite core storage of the analyser becomes very large and expensive for this class of work.

The total cost of the instrument may be reduced, particularly when several experiments are being done at the same time, by the introduction of magnetic tape recorders to store the primary information, as expressed by the train of pulses from an amplitude or time-of-flight converter during the course of the experiments. In the instruments at A.E.R.E. (Wells, Hooton, and Page 1960) the recording is in digital form as a binary code across the width of a 16-track 1-in. tape. The recording equipment in use during the experiment will then be kept to the absolute minimum, consisting of the pulse amplitude or time-of-flight to pulse train encoders, auxiliary information coding, and the magnetic tape-recording drive circuits.

After the experiment the reel of tape is a permanent record of the data and this is analysed by running it very fast through a separate reading-tape transport equipped with a multi-channel ferrite core store, ideally a small computer with 16 000-channel storage. This instrument is set to select any 16 000 adjacent channels and the information is read from the tape into the store, giving the total number of events that occurred in each of these channels during the experiment. The ferrite store information is then taken out on to punched cards or computer magnetic tape and the input data magnetic tape is run through again, selecting a different batch of 16 000 channels, and so on, until all the information of interest on the magnetic tape reel has been read off. Only the one analyser is needed for many recorders since the recording process takes place at low tape speeds while the analyser runs the tape at 120 in./s.

There are two points of interest in an experiment that controls the design of the tape-recording system:

(a) The experimental data occurs at approximately random time intervals so that sufficient paralysis time must be included in the tape-recording circuits to avoid one event being recorded too soon after the previous event. This time is chosen to give a minimum bit spacing along the tape of 0·005 in. for the 1-in. tape system, and the tape speed must be such that this paralysis time does not cause too much loss of information; for example in one experiment the paralysis time was 5 mS (tape speed 1 in./s), causing an average loss of 2·5 per cent of the data. This requirement also has the result that only a small proportion of the tape length carries a record.

(b) If a magnetic tape error in recording or reading occurs then such error will only affect the data of *one* nuclear event. The electronic design must be such that in the event of a 'drop-out' occurring this error is detected and the whole data on that particular nuclear particle is rejected, so that 'drop-outs' cause only a small loss of information; the tape-recording method is chosen to facilitate the error detection and reject circuit action (Wells 1960).

The wastage of tape due to effect (a) above can be greatly reduced if a 'buffer' store is introduced in the recorder. Fig. 11.20 shows a block diagram

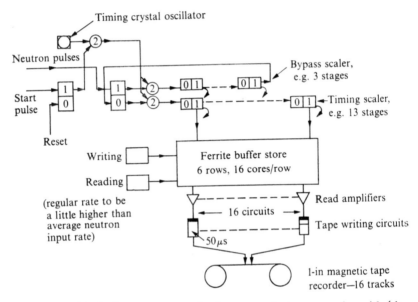

FIG. 11.20. Typical arrangement of 1-in. magnetic tape recorder with 'de-randomizer' store.

of such a recorder arranged for neutron energy measurement by 'time-of-flight' methods. The timing gate is opened by the 'start' pulse and allows timing pulses from a crystal oscillator to be passed to a binary scaler. This scaler continues counting till a neutron pulse arrives from the scintillation counter at the end of the flight path. The binary scaler then stops counting and its binary number is written into the temporary or buffer store, which uses small square-loop ferrite cores. The scaler is then allowed to continue counting, after a suitable correction for lost time, until the next neutron pulse arrives, and so on. The buffer store information is read out in a regular manner to the magnetic tape so that the store is being 'filled' with neutron binary information at random time intervals and is being 'emptied' at regular time intervals. The rate of reading out and the capacity of the store must be chosen to give a negligible chance of too many neutron events occurring in a short time and 'overfilling' the store. On the other hand, the store will often be empty when the tape is ready to receive information so some of the tape is still not used. Thus the introduction of this buffer store increases the recorded fraction of each reel of tape from perhaps 2 to 50 per cent at the expense of increased complexity.

References

BERNSTEIN, N., CHASE, R. L., and SCHARDT, W. (1953) *Rev. scient. Instrum.* **24**, 437.

BONSIGNORI, C., MALOSTI, D., and PELLIGRINI, U. (1963) *Nucl. Instrum. Meth.* **20**, 362.

BREITENBERGER, E. (1953) *Phil. Mag.* **44**, Ser. 7, 987.

BYINGTON, P. W. and JOHNSTONE, C. W. (1955) *I.R.E. Convention Record*, Vol. 3, part 10, p. 204.

CHASE, R. L. (1961) *Nuclear pulse spectrometry*. McGraw-Hill, New York.

COOKE-YARBOROUGH, E. H., BRADWELL, J., FLORIDA, C. D., and HOWELLS, G. A. (1950) *Proc. Instn elect. Engrs*, Part III, Vol. 97, No. 46.

COTTINI, C., GATTI, E., and SVELTO, V. (1963) *Nucl. Instrum. Meth.* **24**, 241.

FARLEY, F. J. M. (1954) *J. scient. Instrum.* **31**, 241.

FLYNN, J. T., JOHNSTON, F. A. (1957) *Rev. scient. Instrum.* **28**, 867.

GERE, E. A. and MILLER, G. L. (1967) *I.E.E.E. Trans. Nucl. Sci. NS*-**13** (3), 508.

HUTCHINSON, G. W. and SCARROTT, G. G. (1951) *Phil. Mag.* **42**, 792.

KANDIAH, K. (1962) *Nuclear electronics*, Vol. 2, p. 11. I.A.E.A. Vienna.

MOODY, N. F., BATTELL, W. J., HOWELL, W. D., and TAPLIN, R. H. (1951) *Rev. scient. Instrum.* **22**, 555.

RAJCHMAN, J. A. (1953) *Proc. Instn Radio Engrs* **41**, 1407.

SCHUMANN, R. W. (1956) *Rev. scient. Instrum.* **27**, 686.

WELLS, F. H., HOOTON, I. N., and PAGE, J. G. (1960) *J. Instn Radio Engrs* **20**, 749.

WILKINSON, D. H. (1950) *Proc. phil. Soc.* **46**, 508.

I2 *Logic and Logic Circuitry*

By J. M. RICHARDS

12.1. Introduction

THE design of a system for control and data handling in experiments does not usually require electronic circuit development or the use of circuits with close tolerances. It does, however, need careful thought in system planning and logical thought in the design of the individual units. This chapter describes some techniques of logical thought and several of the devices available to the designer of control and recording apparatus.

12.2. Logic

Over the last 3000 years a rigorous method of thought has grown up which now forms the basis of mathematics, science, and some philosophy. A set of postulates is propounded to be accepted without argument; these postulates are then taken by means of careful logical argument to various logical conclusions. The conclusions can be compared with observation in any sector of the universe and the extent of agreement is a measure of the usefulness of the postulates in that sector.

In *The laws of thought* George Boole (1815–64) treated the basic arguments required for this process of logical deduction in the same way and founded Boolean algebra, a systematic treatment of logical argument.

12.2.1. *Boolean algebra*

A number of different sets of basic postulates for Boolean algebra can be chosen. One set with the property that no postulate can be derived from the remaining postulates is given below.

A class of elements B together with two binary operations $+$ and $.$ is a Boolean algebra, if and only if, the following postulates hold:

p_1. The operations $+$ and $.$ are commutative.

p_2. The operations $+$ and $.$ are associative.

p_3. There exist in B distinct identity elements o and 1 relative to the operations $+$ and $.$ respectively.

p_4. Each operation is distributive over the other.

p_5. For every element a in B there exists an element $\bar{a}$ in B such that $a + \bar{a} = 1$ and $a.\bar{a} = 0$.

The operations + and . do not have the same meanings that they have in conventional algebra, and for this reason some texts use special symbols for the operations of Boolean algebra. The use of + and . in Boolean algebra is so widespread and convenient, however, that the reader is advised to become accustomed to this notation in Boolean algebra.

The function $a + b$ is conventionally known as the *union* of a and b and the function $a.b$ as the *meet* of a and b. As with multiplication in normal algebra the meet symbol is often omitted and understood to exist, i.e. $ab = a.b$.

An algebra with these postulates has application not only in verbal logic but also in probability theory and, more important for this chapter, in switching circuits.

12.2.2. *The meaning of the postulates*

The meaning of these postulates in verbal logic is illustrated in the following example:

Class and sets. The *class of men* is represented in Fig. 12.1 by the area

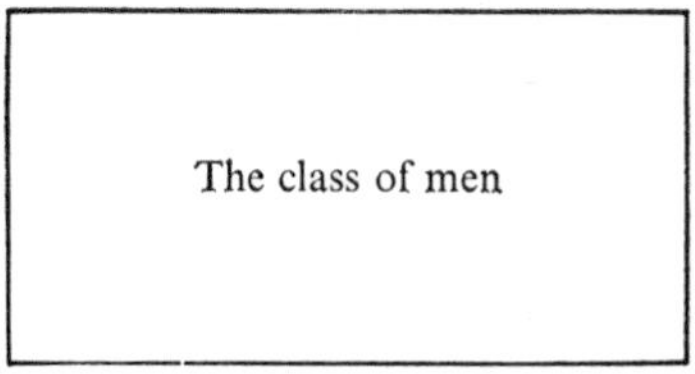

```
┌──────────────────────┐
│                      │
│                      │
│    The class of men  │
│                      │
│                      │
└──────────────────────┘
```

FIG. 12.1. The representation of a class.

within the rectangle. Any set of men can then be represented by a smaller area within the rectangle. Fig. 12.2 shows a circle representing the *set of men who are married*. This set can also be represented by the symbol *m*.

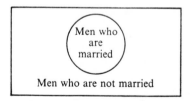

Men who are not married

FIG. 12.2. Sets in a class.

The Venn diagram constructed in this way enables logical relationships between different sets of men to be indicated pictorially. Fig. 12.3, for example, shows the relationship between the *set of men who are married* (*m*) and the *set of men who are happy* (*h*). The relative sizes of the different areas

279

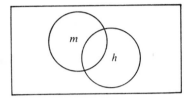

FIG. 12.3. The relationship between two sets.

have no significance, but the diagram indicates that there is a *set of men who are happy and married*, a *set of men who are unhappy and married*, a *set of men who are happy and not married*, and a *set of men who are not happy and not married*.

Binary operations. The two binary operations $+$ and . are illustrated in Figs. 12.4 and 12.5. The union $+$ is an 'OR' operation, thus $m + h$ is the

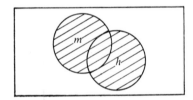

FIG. 12.4. $m + h$ of Fig. 12.3 shaded.

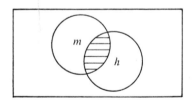

FIG. 12.5. $m.h$ of Fig. 12.3 shaded.

set of men who are happy or married or both. On the other hand, the meet is an AND operation so that $m.h$ is the *set of men who are both happy and married*.

12.2.3. *The postulates*

P1. The commutative postulate states that

$$m + h = h + m \quad \text{and} \quad m.h = h.m.$$

Thus the *set of men who are both married and happy* is identical to the *set of men who are both happy and married*.

280

P2. Let *d* represent the *set of men who own dogs*. The associative postulate states that

$$m + (h + d) = (m + h) + d = m + h + d$$

and

$$m.(h.d) = (m.h)d = m.h.d.$$

Thus the *set of men who are married and both happy and own dogs* is identical to the *set of men who are both married and happy and also own dogs*.

P3. There is a *set of no men* 'o' in the class of men such that the *set of men who are happy or no men* is identical to the *set of men who are happy*

$$h + o = h.$$

There is also a *set of all men* 'I' in the class of men such that the *set of men who are happy and also men* is identical with the *set of men who are happy*.

$$h.I = h.$$

P4. This postulate states that

$$m.(h + d) = m.h + m.d,$$

i.e. the *set of men who are married and either happy or own dogs or both*, is identical with the *set of men who are married and happy or married and own dogs or both*.

The postulate also requires that

$$m + h.d = (m + h).(m + d),$$

i.e. the *set of men who are married or both happy and own dogs* is identical with the set of men who belong not only to the *set of men who are married or happy* but also to the *set of men who are married or own dogs*.

This is illustrated in the Venn diagram, Fig. 12.6

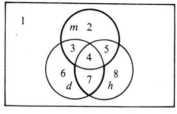

FIG. 12.6. Venn diagram showing $m + hd = (m + h)(m + d)$.

m is represented by the areas	2 3 4 5,	
$h.d$ is represented by the areas	4	7,
so that $m + hd$ is represented by the areas	2 3 4 5	7,
which is the area within the thick line.		
$m + h$ is represented by the areas	2 3 4 5	7 8,
$m + d$ is represented by the areas	2 3 4 5 6 7,	
so $(m+h)(m+d)$ is represented by the areas	2 3 4 5	7,
which is again the area within the thick line.		

P5. As well as the *set of men who are happy* (h) there is a *set of men who are not happy* ($\bar{h}$), such that all men are either happy or not happy, $h + \bar{h} = 1$ and no man is both happy and not happy, $h.\bar{h} = 0$.

The examples above show the greater conciseness and precision of Boolean algebra in comparison to verbal logic.

The principal limitation of Boolean logic is contained in the important fifth postulate.

12.2.4. *The theorem of duality*

Every statement or algebraic identity deduceable from the postulates of Boolean algebra remains valid if the operations $+$ and . and the identity elements 0 and 1 are interchanged throughout. This theorem follows at once from the symmetry of the postulates with respect to the two operations and the two identities.

12.3. Application of Boolean algebra to switching circuits

Logic switching circuits use switching devices such as relays, diodes, and transistors. As a consequence, logic signals have one of two states, i.e. a signal line will be in one of two voltage ranges, or be carrying one of two ranges of current.

One of these states is the '0' and the other the '1' state of the signal. The selection is arbitrary but usually the same convention is used throughout a unit. An example of such a convention is the Harwell 2000 Series type 1 binary level where

'1' state signifies signal line between -5 and -10 V,

'0' state signifies signal line between $+1$ and -1 V,

There is a regrettable tendency for the '0' state to be known as the 'signal absent' and the '1' state to be known as the 'signal present' state, but this nomenclature can lead to confusion and should be avoided.

Logic signals are ideal for the application of Boolean logic as the state of a signal can be described as *in the '1' state* or *not in the '1' state* and these two states are separated by a clear division.

In Fig. 12.7 the rectangle represents all possible states of a system. A circle represents all these states in which the signal labelled x is in the '1' state. The rest of the rectangle, therefore, represents all the states in which the signal labelled x is in the '0' state.

The equation $z = x.y$ signifies that signal z is only in its '1' state when signal x and signal y are both in their '1' states.

The equation $z = x + y$ signifies that the signal z is in its '1' state when signal x is in its '1' state or signal y is in its '1' state or both; clearly z is in its '0' state when both signals x and y are in their '0' states.

The signal $\bar{x}$, (pronounced 'x bar'), will be in its '1' state when x is in its '0' state, and vice versa, hence $\bar{x}$ is known as the inverse of x.

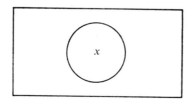

FIG. 12.7. Venn diagram of a logic circuit.

12.3.1. *Aids in the use of Boolean algebra*

Three aids in the use of Boolean algebra are illustrated below in the proof of De Morgan's law, one of the most useful relations in Boolean algebra:

$$\overline{x + y} = \bar{x}.\bar{y}.$$

This will first be proved algebraically and then by Venn diagram, by truth table, and by Karnaugh plot.

Proof by Boolean algebra. If De Morgan's law is correct $\bar{x}.\bar{y}$ and $x + y$ are inverses as defined by the fifth postulate. This provides two tests for the inverse

$$a.\bar{a} = 0 \quad \text{and} \quad a + \bar{a} = 1.$$

In this case

$$\bar{x}.\bar{y}.(x + y) = \bar{x}.\bar{y}.x. + \bar{x}.\bar{y}.y \qquad \text{(by Postulate 4)}$$
$$= \bar{x}.x.\bar{y}. + \bar{x}.\bar{y}.y; \qquad \text{(by Postulate 1)}$$

but

$$\bar{x}.x = 0 \quad \text{and} \quad \bar{y}.y = 0,$$

therefore

$$\bar{x}.\bar{y}.(x + y) = 0.\bar{y} + \bar{x}.0$$

$$= 0.$$

283

Further:

$$\bar{x}.\bar{y}. + x + y = x + y + \bar{x}.\bar{y} \qquad \text{(by Postulate 1)}$$
$$= (x + y + \bar{x})(x + y + \bar{y}) \quad \text{(by Postulate 4)}$$
$$= (x + \bar{x} + y)(y + \bar{y} + x) \quad \text{(by Postulate 1)}$$
$$= (1 + y)(1 + x); \qquad \text{(by Postulate 5)}$$

but

$$1 + y = 1 \qquad \text{and} \qquad 1 + x = 1,$$

therefore

$$\bar{x}.\bar{y} + x + y = 1,$$

therefore

$$x + y \text{ is the inverse of } \bar{x}.\bar{y},$$

therefore

$$\overline{x + y} = \bar{x}.\bar{y}.$$

Proof by Venn diagram. A study of Figs. 12.8–12.11 will show that $\overline{x + y} = \bar{x}.\bar{y}$.

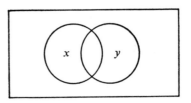

FIG. 12.8. Venn diagram of x and y.

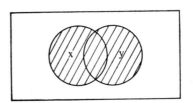

FIG. 12.9. Venn diagram with $x + y$ shaded.

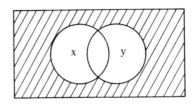

FIG. 12.10. Venn diagram with $\overline{x + y}$ shaded.

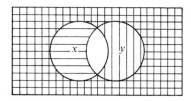

FIG. 12.11. Venn diagram showing $\bar{x}.\bar{y}$.
$\bar{x}$ has vertical shading. $\bar{y}$ has horizontal shading. Thus $\bar{x}.\bar{y}$ is
shaded in both directions.

Proof by truth table. A truth table is a tabulation of every possible condition
of a logical expression. In this example there are four possible conditions
and these are shown in Table 12.1.

TABLE 12.1

Truth table for $\overline{x+y}$ and $\bar{x}.\bar{y}$

Column	I	2	3	4	5	6	7	8
Condition	x	y	$x+y$	$\overline{x+y}$	$\bar{x}$	$\bar{y}$	$\bar{x}.\bar{y}$
I	0	0	0	I	I	I	I
2	0	I	I	0	I	0	0
3	I	0	I	0	0	I	0
4	I	I	I	0	0	0	0

If we compare column 5 and column 8 it will be seen that they are equal
for every possible condition of x and y, therefore $\overline{x+y} = \bar{x}.\bar{y}$.

Proof by Karnaugh plot. A Karnaugh plot is a systematic form of Venn
diagram. Fig. 12.12 shows a Karnaugh plot for one variable. The rectangle

FIG. 12.12. Karnaugh plot of x.

represents all possible states of the system, while the right-hand half of
the rectangle represents those states of the system in which the signal x
is in the '1' state. The left-hand half of the rectangle therefore represents
those states in which the signal x is in the '0' state.

	$\bar{x}$	x
$\bar{y}$	$\bar{x}.\bar{y}$	$x.\bar{y}$
y	$\bar{x}.y$	$x.y$

FIG. 12.13. Karnaugh plot of two elements x and y.

Fig. 12.13 shows a Karnaugh plot for two elements x and y. The right-hand half of the plot still represents all those states of the system for which x is in the '1' state, and the bottom half represents all those states of the system for which y is in the '1' state. Thus the plot is divided into four equal squares, corresponding to the four possible combinations of x and y.

	$\bar{x}$	x
$\bar{y}$	0	1
y	1	1

FIG. 12.14. Karnaugh plot of $x + y$.

Fig. 12.14 shows the state of the expression $x + y$ for each combination. The plot is obtained by setting $x + y$ in the '1' state whenever x is in the '1' state and whenever y is in the '1' state. Fig. 12.15 shows a plot of $\overline{x + y}$.

	$\bar{x}$	x
$\bar{y}$	1	0
y	0	0

FIG. 12.15. Karnaugh plot of $\overline{x + y}$.

Since $\overline{x + y}$ is the inverse of $x + y$, Fig. 12.15 can be derived from Fig. 12.14 by converting all '1' states to '0' and vice versa. Comparison of Figs. 12.15 and 12.13 shows that $\overline{x + y} = \bar{x}.\bar{y}$.

These examples give a good indication of the usefulness of the different aids to logic design. The algebraic approach is the most rigorous and versatile, but for many problems a Karnaugh plot is the most useful.

12.4. The use of Boolean algebra in logic design

Boolean algebra has two main uses in designing switching circuitry:

(a) the specification of the logical effect required,
(b) a specification of the circuitry required to produce this effect.

286

Thus a Boolean expression can be treated merely as a statement of the aim of a particular logic expression or as an instruction to implement the logic function with a number of logic elements. The normal procedure is to generate an initial expression for the aim of the design and then to simplify it to a form requiring the minimum number of logic elements.

The simplest Boolean expression will usually depend on the application: $a+b$ and $\bar{a}.\bar{b}$ are logically equivalent but, as formulae for circuits, the former would be simpler in NOR logic, while the latter might be the best form in relay circuitry. Nevertheless, the circuit required to perform a given logic function can sometimes be simplified considerably by algebraic manipulation. The following example illustrates this.

Example. Suppose a signal e is required and has been specified by

$$e = \bar{a}\,\bar{b}\,\bar{c}\,\bar{d} + \bar{a}\,\bar{b}\,c\,d + \bar{a}\,\bar{b}\,\bar{c}\,d + \bar{a}\,\bar{b}\,c\,\bar{d} + a\,b\,\bar{c}\,d + a\,\bar{b}\,c\,d + \bar{a}\,b\,c\,d +$$
$$+ \bar{a}\,b\,\bar{c}\,\bar{d} + \bar{a}\,b\,\bar{c}\,d.$$

To simplify in Boolean algebra, first look for eight terms containing a single element. There are none. Then look for groups of four terms containing a common pair of elements. There are three: $\bar{a}\,\bar{b}$, $\bar{a}\,\bar{c}$, and $\bar{a}\,d$. Hence we can write e as follows, remembering that a term can be repeated as often as is convenient since $x + x = x$.

$$e = \bar{a}\,\bar{b}\,\bar{c}\,\bar{d} + \bar{a}\,\bar{b}\,c\,d + \bar{a}\,\bar{b}\,\bar{c}\,d + \bar{a}\,\bar{b}\,c\,\bar{d} +$$
$$+ \bar{a}\,\bar{b}\,\bar{c}\,\bar{d} + \bar{a}\,\bar{b}\,\bar{c}\,d + \bar{a}\,b\,\bar{c}\,\bar{d} + \bar{a}\,b\,\bar{c}\,d +$$
$$+ \bar{a}\,\bar{b}\,c\,d + \bar{a}\,\bar{b}\,\bar{c}\,d + \bar{a}\,b\,c\,d + \bar{a}\,b\,\bar{c}\,d +$$
$$+ a\,b\,\bar{c}\,d +$$
$$+ a\,\bar{b}\,c\,d.$$

Next we try to pair the last two terms, if possible, with terms having three common elements. This involves repeating two more terms. Therefore

$$e = \bar{a}\,\bar{b}\,\bar{c}\,\bar{d} + \bar{a}\,\bar{b}\,c\,d + \bar{a}\,\bar{b}\,\bar{c}\,d + \bar{a}\,\bar{b}\,c\,\bar{d} +$$
$$+ \bar{a}\,\bar{b}\,\bar{c}\,\bar{d} + \bar{a}\,\bar{b}\,\bar{c}\,d + \bar{a}\,b\,\bar{c}\,\bar{d} + \bar{a}\,b\,\bar{c}\,d +$$
$$+ \bar{a}\,\bar{b}\,c\,d + \bar{a}\,\bar{b}\,\bar{c}\,d + \bar{a}\,b\,c\,d + \bar{a}\,b\,\bar{c}\,d +$$
$$+ a\,b\,\bar{c}\,d + \bar{a}\,b\,\bar{c}\,d +$$
$$+ a\,\bar{b}\,c\,d + \bar{a}\,\bar{b}\,c\,d.$$

This expression can now be factorized and simplified. Therefore

$$e = \bar{a}\,\bar{b}(\bar{c}\,\bar{d} + c\,d + \bar{c}\,d + c\,\bar{d}) +$$
$$+ \bar{a}\,\bar{c}(\bar{b}\,\bar{d} + \bar{b}\,d + b\,\bar{d} + b\,d) +$$
$$+ \bar{a}\,d(\bar{b}\,c + \bar{b}\,\bar{c} + b\,c + b\,\bar{c}) +$$
$$+ b\,\bar{c}\,d(a + \bar{a}) +$$
$$+ \bar{b}\,c\,d(a + \bar{a}).$$

now

$$\bar{c}\,\bar{d} + c\,d + \bar{c}\,d + c\,\bar{d} = 1, \quad \text{and} \quad a + \bar{a} = 1,$$

therefore

$$e = \bar{a}\,\bar{b} + \bar{a}\,\bar{c} + \bar{a}\,d + b\,\bar{c}\,d + \bar{b}\,c\,d$$
$$= \bar{a}(\bar{b} + \bar{c} + d) + d(b\,\bar{c} + \bar{b}\,c).$$

The steps are much easier to follow if the formula is represented by a Karnaugh plot as in Fig. 12.16. It can be seen that $e = 1$ throughout large areas of the plot, including the top left-hand quarter. Thus the four terms that describe this quarter, $\bar{a}\,\bar{b}\,\bar{c}\,\bar{d} + \bar{a}\,\bar{b}\,c\,d + \bar{a}\,\bar{b}\,\bar{c}\,d + \bar{a}\,\bar{b}\,c\,\bar{d}$, can all be covered by the composite term for the whole quarter, $\bar{a}\,\bar{b}$.

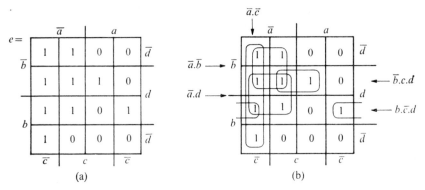

FIG. 12.16. Karnaugh plots of e. (a) Plot of e. (b) Plot of e, showing related terms.

In a similar way the left-hand column of the plot can be represented by $\bar{a}\,\bar{c}$, and this term can replace the four terms, $\bar{a}\,\bar{b}\,\bar{c}\,\bar{d} + \bar{a}\,\bar{b}\,\bar{c}\,d + \bar{a}\,b\,\bar{c}\,d + \bar{a}\,b\,\bar{c}\,\bar{d}$. Thus all the terms in the $\bar{a}$ side of the plot can be described by $\bar{a}\,\bar{b} + \bar{a}\,\bar{c} + \bar{a}\,b\,c\,d$. We can achieve further simplification by recognizing that $\bar{a}\,b\,c\,d$ is part of the block $\bar{a}\,d$, and so the seven original terms incorporating $\bar{a}$ can be represented by three terms $\bar{a}\,\bar{b} + \bar{a}\,\bar{c} + \bar{a}\,d$. The two terms on the right-hand side of the plot cannot be included in groups of four terms. They can, however, be paired with terms on the left-hand side of the plot. $a\,\bar{b}\,c\,d$ can be paired with its neighbour $\bar{a}\,\bar{b}\,c\,d$, and these can be described together by the term $\bar{b}\,c\,d$.

Finally, $a\,b\,\bar{c}\,d$ can be paired with $\bar{a}\,b\,\bar{c}\,d$ and described by $b\,\bar{c}\,d$. This last pair shows that the plot should be considered to be linked from side to side and from top to bottom. Thus, using the Karnaugh plot as a guide, we have reduced the original formula to

$$e = \bar{a}\,\bar{b} + \bar{a}\,d + \bar{a}\,\bar{c} + \bar{b}\,c\,d + b\,\bar{c}\,d,$$

which is the expression obtained previously by purely algebraic methods.

288

In many cases some of the possible combinations of signals will never occur in practice; the value of e in these cases is immaterial. Suppose that in the above example the terms $a\,b\,\bar{c}\,\bar{d}$, $a\,\bar{b}\,\bar{c}\,d$, $a\,b\,c\,d$, and $\bar{a}\,b\,c\,\bar{d}$ will never occur. The revised Karnaugh plot is shown in Fig. 12.17 with these terms represented by X. Brief inspection of Fig. 12.17 shows that $e = \bar{a} + d$ would be a satisfactory coding.

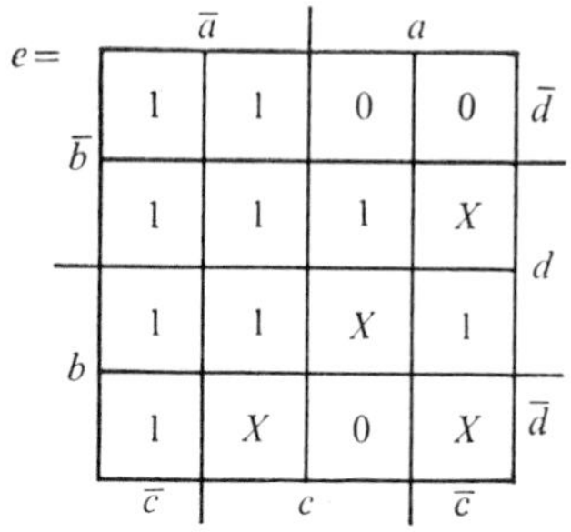

FIG. 12.17. Karnaugh plot of e with terms that will never occur represented by X.

An alternative switching network is obtained by analysing the 'o' states in the plot. In the case above, $e = \overline{a\,d}$ is an alternative solution.

Use of the Karnaugh plot for more than four variables requires several of the above diagrams and the extension beyond six variables is cumbersome.

12.5. Logic elements

12.5.1. *Practical limitations of logic elements*

The logical design of a unit must take into account the practical limitations of the logic elements used. Common limitations are:

(i) *Fan-out, or fan-in.* A signal derived from a logic element is only able to drive a limited number of elements, the fan-out number. There is also a limitation on the maximum number of inputs allowed to a single element, the fan-in number.

(ii) *Signal degradation.* The output of some logic elements, for example a diode, is a degraded form of the input signal, so that, in a chain of logic elements, signals require periodic regeneration.

(iii) *Logic noise.* While switching elements are operating surge currents may flow in earth or signal lines. These surges may trip other circuits in the same system. In extreme cases the whole logic system will oscillate.

(iv) *Delays.* In high-speed logic circuits the physical time taken for a signal to travel to its destination may make it too late to implement

289

the required decision. At lower speeds similar effects can be observed but the delays occur within the logic elements.

(v) *Racing*. If correct operation depends on one signal arriving before another, care must be taken to ensure that the chain of events leading to the arrival of the second signal can never be completed before the first signal is established.

12.5.2. *Logic series*

The design of a logic system usually requires a range of mutually compatible logic circuits. The number of different circuits needed is, however, limited, so that there is scope for mass production of a small range of perhaps ten to twenty different configurations of each type of logic element. A typical range would include

several versions of the basic gate,
several bistables,
half adders and full adders for addition circuits,
a power driver to drive a large fan-out, and
converters to other logic systems.

These series are nearly always in one of the following forms:

(a) plug-in boards,
(b) packaged circuits, in which conventional components are encapsulated in a standard cover,
(c) integrated circuits, in which the circuitry is manufactured on a slice of silicon and encapsulated in a number of standard packages.

The task of the logic designer is then to specify the various circuits and their interconnections.

12.5.3. *Large-scale integration*

The task of the logic designer is likely to be affected considerably by the advent of large-scale integration. Integrated-circuit manufacturers are developing the ability to construct a very large number of interconnected gates on one slice of silicon, so that it is possible to construct complete sub-units in one encapsulation. When sub-units are required in reasonable quantities this technique can show drastic cost reduction. It also leads to considerable reduction in size and to an increase in operating speed because of the reduced parasitic capacitances.

The term 'medium-scale integration' is used to describe elements that provide complex functions but do not contain sufficient circuitry to merit the title 'large-scale integration'.

290

12.6. Logic symbols

Frequently the details of the circuit elements used need hardly be considered in the logic design of a unit. A useful diagram can then be drawn using symbols for the logical functions without showing further circuit detail.

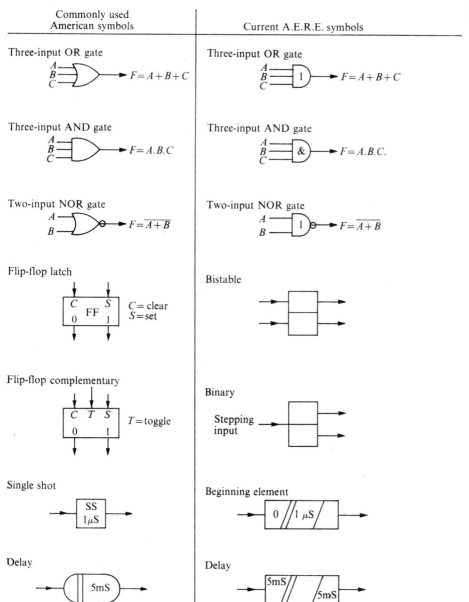

Commonly used American symbols	Current A.E.R.E. symbols

Three-input OR gate $\quad F = A + B + C$

Three-input OR gate $\quad F = A + B + C$

Three-input AND gate $\quad F = A.B.C$

Three-input AND gate $\quad F = A.B.C.$

Two-input NOR gate $\quad F = \overline{A + B}$

Two-input NOR gate $\quad F = \overline{A + B}$

Flip-flop latch $\quad$ C S FF 0 1 $\quad$ C = clear S = set

Bistable

Flip-flop complementary $\quad$ C T S 0 1 $\quad T$ = toggle

Binary $\quad$ Stepping input

Single shot $\quad$ SS 1μS

Beginning element $\quad$ $0 \; / 1\,\mu$S

Delay $\quad$ 5mS

Delay $\quad$ 5mS $\; / \; 5$mS

FIG. 12.18. Comparison of logic symbols.

291

National standard logic symbols for this purpose are promulgated both in Britain (BS 530) and in the United States (*American standard for graphic symbols for logic diagrams* Y 32.14). Various other conventions are also used.

At the time of writing the International Electrotechnical Commission is preparing a standard, and a new British Standard (BS 3939, Section 24) is being developed in the light of the international discussions.

Fig. 12.18 compares the most widely used American symbols with the logic symbols currently used at A.E.R.E. Harwell. The latter symbols are based on a proposed I.E.C. Standard.

12.7. Limitations in the use of Boolean algebra

The major limitation in the use of Boolean algebra in logic circuit design is that some fairly simple switching circuits are described by rather complex Boolean expressions, so that it is not always easy to use Boolean algebra

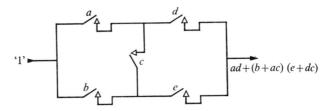

FIG. 12.19. Simplest realization of the expression $ad + (b + ac)(e + dc)$.

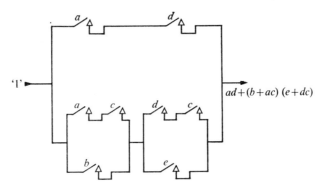

FIG. 12.20. Direct realization of the expression $ad + (b + ac)(e + dc)$.

to achieve the simplest switching circuit. For example, Fig. 12.19 implements the logical relation $a\,d + (b + a\,c)(e + d\,c)$ in a much simpler manner than the direct realization of the expression in Fig. 12.20.

13 *Logic Design*

By J. M. RICHARDS

13.1. Introduction

LOGIC designs are required for a large variety of applications, from the automatic control of a machine tool to the design of a powerful digital computer. Within each application there is usually a wide variety of facilities that might be provided, and a number of different ways in which these could be achieved.

The main task in logic design is to decide what facilities should be provided and how these should be achieved. The best design usually combines convenient operation with minimum cost, though sometimes the speed of operation is also important. This chapter discusses the alternatives open to logic designers, and some of the considerations necessary for good logic design.

13.2. Logic design principles

A novice faced with a complex problem is disposed to consider it as a whole and then wonder how it can be solved. With experience, however, he learns to break the problem down into a number of stages, each of which can be studied in turn. This approach is essential in the logic design for a complex requirement. The task must be analysed into basic functions that can then be designed and integrated into a system to meet the requirement. Analysed in this way, a logic design requires only a few basic functions, each of which may be required many times. The basic functions are: (a) memory, (b) sequencing, (c) gating, (d) decoding, (e) calculation, and (f) amplification.

Let us illustrate those functions by analysing the actions of a man given the task of recording the contents of a six-decade scaler with a binary-coded decimal display. He would first read the most significant decade in the scaler. Since this is displayed in binary-coded decimal form he would next decode it into a decimal digit between 0 and 9, and press the appropriate digit key on the typewriter. After this he must release the key and wait at least the twentieth of a second before driving the next key. When he has typed all six scaler decades in this way he must drive a function key such as 'tabulate', 'space', or 'carriage return' and wait for this function to be completed. In this process he is *remembering* his position in the *sequence*,

293

gating each decade in turn to a *decoder* in his brain and then *amplifying* the signal from his decode to operate the appropriate digit on the keyboard. Thus this example has required every one of the basic functions except *calculation*. The nearest instance of calculation is in the scaler itself, which has counted input pulses to obtain the answer that is being recorded. The following example shows how the same task can be performed by a logic design.

13.2.1. *A read-out system*

Fig. 13.1 is a block diagram of the A.E.R.E. scaler read-out assembly type 3017, which can record the contents of four six-decade scalers on a printer.

Quiescently, the first *scaler* is addressed and its data is presented in binary-coded decimal form on the *data highway*. Then a start pulse sets the

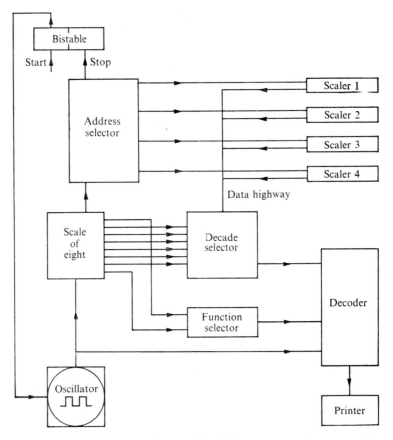

FIG. 13.1. Block diagram of the A.E.R.E. Type 3017 read-out assembly.

start bistable and this frees the *oscillator*. The output from the oscillator drives the *scale-of-eight*, which is a three-bit binary counter and decoder. The second step of the scale-of-eight will gate the most significant decade on the data highway to the decoder. The decoder accepts the four-bit binary coded digit and for the first half of the oscillator cycle it drives the appropriate digit solenoid of the *printer*. Its decode is then suppressed for the second half of the cycle to allow the printer to recover before the next digit is selected.

Thus steps 2 and 7 record all six digits. On the eighth step, a printer function is selected and driven. In this case a 'check-back' from the printer locks the oscillator until the function is complete. The next oscillator cycle resets the scale-of-eight but generates a carry to the *address selector*, which is a two-bit binary counter and decoder. The address selector therefore addresses the second scaler, and data from this scaler is presented to the data highway. Thus the next eight steps record data from this scaler. In this manner the data from each scaler is printed in turn until finally the address selector overflows and resets the start bistable.

We shall now consider how the basic functions of memory, sequencing, gating, decoding, calculation, and amplification are achieved.

13.3. Memory

Memory is required in logic systems for two different purposes. First, there must be active bistables, such as the start bistable in the example above, which hold the current state of the unit. Secondly, data or instructions may be stored in a cheaper storage medium from which they can be read when required. A major factor determining the choice of storage medium is *access time*, i.e. the time taken to extract a *word* of the data stored. In a complex scientific calculation a large number of words of data should be stored with a short access time to each word, and a large core store is desirable. On the other hand, in calculations of pay-rolls the data on each employee can be stored with a long random access time provided the data can be processed sequentially at a high rate. For this kind of application magnetic tape storage is suitable. The most important memory media are listed below.

13.3.1. *Active memories*

All the logic bistables mentioned in Chapter 5 are memory elements. They are used when very rapid access or versatility is required, but they are too expensive for bulk data storage. However, two fairly recent integrated circuit developments increase the applications of active memories:

(a) TTL integrated circuit elements exist which provide sixteen bits of memory that can be addressed one bit at a time. A number of these circuits used together make a rapid access store.

(b) MOST integrated circuit elements are available which provide shift registers of 100 bits or more. While these have a rather slow access time of tens of microseconds, they possess compensating advantages in size and cost over other memories for small stores.

13.3.2. *Magnetic memories*

Magnetic memories are the most popular forms of bulk data storage. They include:

(a) *Magnetic cores.* Each core is a small ring of ferrite material with a 'square' hysteresis loop. Each core contains one bit of information indicated by the direction of saturation. Core stores can provide bulk memories with random access times of about 500 ns, though a further 500 ns is normally required to restore the data that has been read-out. Thus the store cycle time is typically 1 μs.

(b) *Thin film memories.* These rely on similar principles to core stores but achieve a faster access time because the magnetization of thin magnetic films can be switched more rapidly than magnetic cores. Thin film memories are normally used to provide a fast store in large computers. The thin film store of the IBM 360/95 computer has an access time of 67 ns and a cycle time of 120 ns.

(c) *Magnetic tape.* This is a convenient form of storage with short sequential access time (when the data is arranged in the order in which it is required) but long random access time since the whole magnetic-tape reel may have to be rewound. It is widely used for storing computer programs and data.

(d) *Magnetic drums.* These rotate at high speed under a line of recording and replay heads. They provide access in less than one revolution to a considerable amount of data.

(e) *Magnetic discs.* These are rotating discs coated with magnetic material. They are usually scanned by moveable recording and replay heads which can replay any word recorded on the disc in a fraction of a second.

The combination of bulk storage with a reasonable random access time has made the magnetic disc the favoured medium for the backing store of a computer, holding large quantities of program or data that may be required in the main store of the computer from time to time.

296

13.3.3. *Delay-line memories*

In these memories information is stored in the form of pulses on some kind of delay line and is detected and regenerated after every passage through the delay line. The maximum access time in all cases is the delay period. Examples of delay lines are:

ultrasonic pulses in a column of mercury,
magnetostrictive pulses on a nickel–iron wire,
electrical pulses in a coaxial cable, and
light pulses in glass fibres or air.

Interest in these memories has revived recently, partly because of the use of delay lines in colour-television receivers. The most common delay medium involves ultrasonic pulses in glass. One such delay line available gives a 256 μs delay and will store 2048 binary digits at a clock rate of 8 MHz.

13.3.4. *Long-term data storage*

Permanent memory is required for two distinct purposes; the storage of data or results of calculations, and the storage of program for a computer or digital machine. The traditional method of data storage is by printing. Automatic printers range from electric typewriters, through a variety of parallel printers that print a line of data at a time, to printing devices using Xerography. One of these, the Xeronic printer, can print 3000 lines of 128 characters in a minute (Stillwell and Dagnall 1964).

However, the printed word is not easily read by automatic machines and so other methods of storage are used when the information has to be read back into the automatic system. These methods include magnetic tape, magnetic discs, punched paper tape, and punched cards. Punched paper tape is the easiest to produce and read automatically, but punched cards can be sorted manually or automatically and this is often an important advantage.

13.3.5. *Program storage*

Program storage can also be on magnetic tape, magnetic discs, punched paper tape, or punched cards. Punched paper tape enables a long sequence of instructions to be read fairly cheaply, and so is used for tasks such as the numerical control of machine tools. There is also a wide variety of storage media specific for program stores. These include:

Cams. Specially cut cams, which close contracts as required in a timing cycle.
Program drums which rotate under a line of contacts. The required

selection of contacts can be closed at each stage by appropriate inserts in grooves in the drum.

Patch panels in which an arrangement of patch leads can determine the required program.

Plugboards in which pins can be inserted into a matrix of sockets to link a set of horizontal bus-bars to a set of vertical bus-bars. Plugboards are much more versatile if the pins insert diodes between the two bus-bars so that different horizontal bus-bars can be logically linked to the same vertical bus without being connected together.

Static card readers can replace plugboards in many applications and permit the rapid changing of programs. Each program is contained on a punched card, and contacts on the card reader take the place of the plugboard sockets.

While the above methods are probably those most widely used, there are many varieties of program store. In one ingenious system in the IBM Series 360 computers, the capacitance between horizontal and vertical bus-bars is varied by a pattern of holes in a dielectric sheet between them. A signal on one horizontal bus-bar is detected on those vertical bus-bars that are linked by high capacitance, but not on those linked by low capacitance.

Another system under development by ICL recovers data recorded photographically on a glass plate. The system can read out sixty-four bits in parallel, while each plate can retain a very large number of words of a fixed program.

13.4. Sequencing

Sequencing is a requirement in nearly every logic design, since nearly every system has a number of functions that must be carried out in order. The following example illustrates some aspects of this field.

13.4.1. *Typical controller*

The module controller A.E.R.E. type 1857 is the controller of the A.E.R.E. 3016 scaler read-out assembly. It controls a group of scalers, a recording unit that records the scalers on demand, and some sample changer apparatus that makes appropriate changes between successive experiments. It has eight basic control states which are listed in Table 13.1. These states are obtained by decoding three bistables as shown in the Karnaugh plots, which also show the three standard control sequences (see Figure 13.2). In each case the sequence is one step, i.e. every transition involves a change of only one binary, except for the final end-to-waiting

298

TABLE 13.1

Control states of the module controller

State	Function
WAITING	The experiment is quiescent.
READY TO COUNT (C_1)	The scalers can be reset automatically.
COUNTING (C_2)	The experiment is in progress.
READY TO PRINT (P_1)	A print has been demanded but has not yet begun.
PRINTING (P_2)	The scalers are being printed.
READY TO CHANGE (CH_1)	A sample change operation has been demanded but has not yet begun.
CHANGING (CH_2)	A sample change operation is in progress.
END	A control sequence has finished.

transition, during which any state could be decoded briefly. To prevent an incorrect decode, all decodes are suppressed during this transition.

The sequencing logic can often be the most difficult part of a logic design, and it is a part in which the theories of Boolean algebra are of limited use because so much depends on relative timing of different signals. For this

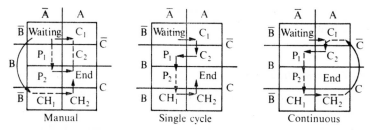

FIG. 13.2. Karnaugh plots of the control sequences of the module controller.

reason sequencing networks should often be checked by constructing a timing diagram. This is a diagram showing the time relationships of the significant waveforms. Fig. 13.3 is a timing diagram of the controller described above.

13.4.2. *Sequencing methods*

In some applications the successive actions are stored as successive words on a sequential program store such as a set of cams, a program drum, or

299

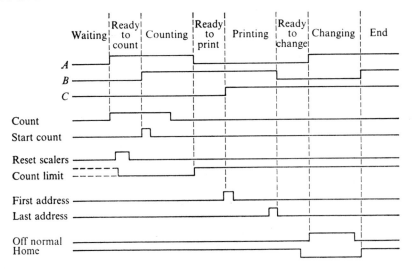

FIG. 13.3. Single-cycle mode '1' state high.
Timing diagram of module controller.

punched paper tape. The timing of successive functions may then be determined by the speed of an electric motor.

An alternative sequencing method is the ring counter, a device that can step round a limited number of positions, each position corresponding to one function in the sequence. This method is common in relay circuitry in which the ring counter is a uniselector. In electronic logic a ring counter is inconvenient because it is difficult to protect from incorrect operation, and so a binary counter and decoder is preferred. This is versatile because signals covering a number of steps can easily be decoded.

13.4.3. *The computer technique*

The computer technique combines a binary counter and decoder with a random access store of instructions. The binary counter is usually incremented after each instruction so that the program is obtained from sequential positions in the store. However, certain instructions, known as jump instructions, can load the binary counter with a new number so that the program sequence jumps to a different position in the store. This organization saves programming since it is often necessary to repeat a sequence of instructions—a program loop—a number of times.

Conditional skip. Conditional skip instructions are usually included in the instruction repertoire. These skip the next instruction in the sequence if a certain condition is satisfied, thereby permitting the instruction sequence to leave a program loop.

Microprogramming. Microprogramming is a method of carrying out complex instructions in the main program. The current instruction in the main program calls up a specific microprogram of instructions to carry it out. This technique is useful in enabling the same instructions to be carried out on different computers in a compatible range, since different hardware facilities provided by the different computers can be taken care of by differences in the microprograms, hence the normal operating programs of the different computers can be identical.

Clocked operation. The start of each function or program step can be derived from the completion of the previous function, but this is often difficult to determine; hence the time allowed for completing the previous function is sometimes fixed by a monostable. Before adopting this approach the designer should consider locking the operation of the system to a master oscillator. The advantages of this clocked technique are that:

(a) the number of timing elements is reduced to a minimum. These elements tend to require large components (capacitors), and do not easily provide precise timing so they may need adjustment;

(b) different functions can be overlapped reliably, without adjustment of a number of delays;

(c) testing and fault finding is eased if a master timing waveform can be run one cycle at a time, so that the unit can be taken through its operations step by step;

(d) the speed of the entire operation can be set up by adjusting one circuit.

Even when a clocked technique has been adopted, there may still be a few requirements for monostable delays, principally when the required delays are much shorter or longer than the oscillator period.

Direct coupling. When monostables are necessary, direct coupled circuits are preferred to capacitively coupled circuits. The difference is not nearly as important as the difference between clocked and unclocked operation, but direct coupling provides the following advantages:

(a) faster recovery and thus less sensitivity to rate of operation,

(b) easier location of faults, since the faulty circuit can often be discovered by d.c. checks.

13.5. Gating

Gating is required whenever a choice has to be made between different signals. In the first example in this chapter data from the selected scaler is gates onto the data highway. Later, when another scaler is selected, the

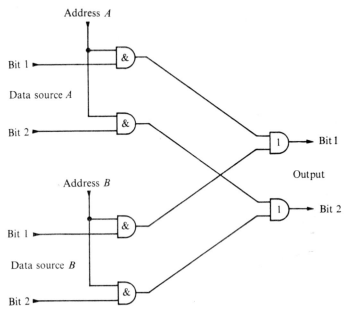

FIG. 13.4. Gating data from two sources.

gates in the first scaler will be closed so that the data from the first scaler can be ignored.

Fig. 13.4 shows the logic diagram of a gating circuit. When address A is in the '1' state, data from data source A is established at the output; when address B is in the '1' state data from source B is established. If neither data source is addressed the output bits remain in the '0' state.

Integrated circuits are available to ease this kind of gating. Fig. 13.5 shows a 'Quad AND/OR Invert' element which can gate one bit of data from four sources.

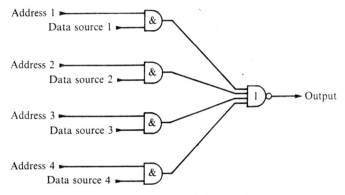

FIG. 13.5. Quad AND/OR invert element.

13.5.1. *Data highways*

Any organized system of data transmission on a number of wires is known as a data highway. A data highway will often permit data from a number of sources to be gated to the same receiver. The data is carried on a number of data lines, one for each bit in the receiver, and each data source feeds these lines through AND gates that can force the line to the '1' state when the source is addressed. The data lines are biased so that in the absence of imposed data they remain in the '0' state. It is important that the gates driving each highway line must have an 'intrinsic OR' capability, i.e. they must drive the line into the '1' state when their output is in the '1' state, but they must not drive the line into the '0' state at any other time because

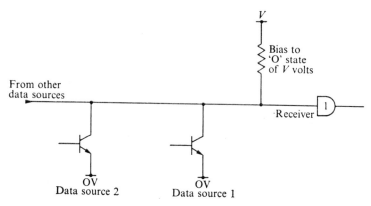

FIG. 13.6. Intrinsic OR connections to a highway line.

another data source having access to the line may be driving it into the '1' state. For this reason data lines are usually driven through diodes, or by the collector of a transistor as shown in Fig. 13.6.

13.5.2. *Data transmission*

Additional factors must be considered when data is to be transmitted between units. The signals used at the two ends of the link must be compatible and there must also be an agreed mode of operation for the logic signals that control and monitor the data transfer. The more widely any standard is adopted the more useful it is, because of the larger selection of instruments that can exchange data conveniently. For this reason each computer manufacturer has fixed his own standards for data transmission so that any of his computers can use any of his peripheral devices such as printers, tape punches, etc.

There have also been efforts to achieve national standards for data transmission. One of these—the British Standard Interface (BS 4421)—can be used to illustrate one particular requirement of any transmission scheme. The transmitter and receiver of data must synchronize their operations so that each can recognize when data has been transferred. The British Standard Interface uses the *hand-shake* method of control, illustrated in Fig. 13.7 and described below.

A signal from the receiver is in the '1' state when it is ready to receive data. The transmitter places the data to be transmitted on the highway and sets a control signal to the '1' state. The receiver then reads the data and switches its control signal to the 'o' state. This transition tells the transmitter that the data has been received so that the transmitter can remove

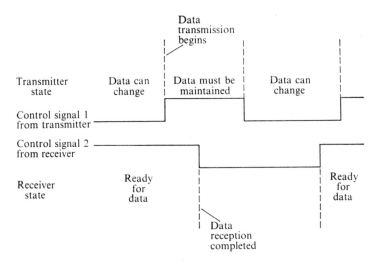

FIG. 13.7. Handshake method of transfer in British standard interface.

the data and switch its control signal to the 'o' state. When the receiver is next able to accept data it re-establishes its control signal in the '1' state and another data transfer cycle can begin. Thus data transmission is accompanied by control signals which ensure that the data is transmitted at a rate within the capability of the system.

Data transmission between units must remain intelligible in spite of noise and signal attenuation, which rule out the use of commercial integrated-circuit logic levels for many applications. One way of overcoming noise interference is to use large logic voltage swings and accept the data through low-pass filters, which exclude the short-duration noise inputs.

A second method is to use *twisted pairs* of wires. Each *twisted pair* acts as a transmission line and carries one signal. In the '1' state the first wire is at a positive and the second at a negative voltage, with a reversal of levels in the 'o' state. This balanced signal propagation produces no net voltage swing on the pair and markedly reduces cross-talk to other pairs. The above effect is enhanced by twisting each pair in the multi-wire cable in such a way that the coupling to each wire of the pair is very similar. The signal is detected by a differential amplifier which rejects common-mode interference. In some environments each balanced pair will be driven and received by transformer coupling which provides very good interference rejection.

The discussion so far has assumed that the data is transmitted in parallel, i.e. that several bits of data are transmitted simultaneously down separate wires and are accompanied by control signals. However, when data has to be sent for long distances, for example over the telephone network, there are great advantages in reducing the number of wires by the use of serial transmission in which the bits of data follow each other in time sequence down the same wire. This mode of operation has been used widely in providing Telex services, but is receiving a new impetus from the development of high-speed serial transmission systems for pulse code modulated telephone traffic.

13.6. Decoding

Decoding is the logical conversion of information from one form to another more appropriate to the current application. For example, binary-coded decimal information from a scaler is decoded to drive one of ten solenoids in the printer; the three control bistables in the module controller must be decoded to indicate the 'counting' state; and in another application the binary-coded address of a word in a core store must be decoded to read out the required word.

Decoding is a common application of logic gates, but one in which the obvious solution is often not the best. There is thus scope for the use of Karnaugh plots and other methods of simplification.

Consider, for example, the decode of a four-bit input code to sixteen outputs. The straightforward approach would be to use sixteen four-input AND gates as in Fig. 13.8.

Fig. 13.9 shows that the same decode can be achieved using twenty-four two-inputs AND gates, equivalent in cost to twelve four-input AND gates. In addition, the method of Fig. 13.8 loads the A input, for example, with nine gate-loads, while in Fig. 13.9 A is only loaded with three gate-loads, thus providing spare fan-out capacity for use elsewhere.

An additional advantage of the circuit in Fig. 13.9 is that intermediate decodes such as $\overline{A}.\overline{B}$ are available. These may save logic circuitry elsewhere in the design.

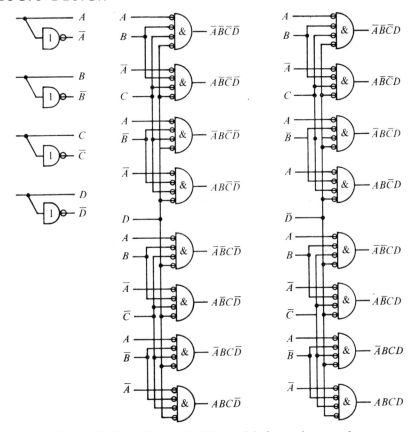

FIG. 13.8. Four-bit decode. The straightforward approach.

13.6.1. *Security of decoding*

When an output of a decoder is used to make some irretrievable change such as clearing a register, it is important that the output is only generated when required. The transition between codes must then be considered carefully. For example, if a serial-carry binary counter is decoded, the transition from step 7 to step 8 is through the following stages:

Binary digits

1	2	4	8	Decode
1	1	1	0	7
0	1	1	0	6
0	0	1	0	4
0	0	0	0	0
0	0	0	1	8

Thus if 4 were used to clear a register, the register would also be cleared between codes 7 and 8. In some cases this does not matter but the possibility

306

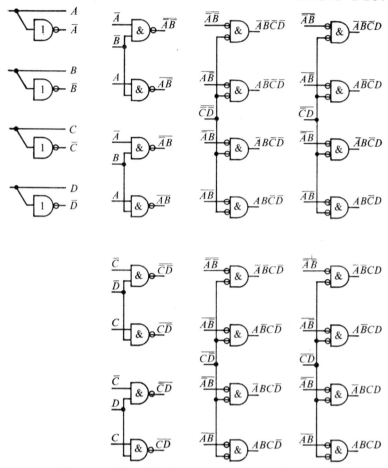

FIG. 13.9. Four-bit decode. Alternative logic circuit.

of trouble should be remembered or avoided. Three ways of avoiding trouble with spurious decodes are given below:

(a) Suppress all outputs from the decoder while the code is changing. The decoder can often be switched off by the clock waveform that initiates the change.

(b) When a serial-carry binary counter is being decoded, use only the odd decodes for irretrievable actions; these are always safe.

(c) Use a one-step code such as the Gray code, in which each transition involves only one bit of input code.

13.7. Calculation

Logic hardware may be required for calculations of many different kinds. These calculations are not limited to the arithmetic operations of addition,

307

subtraction, multiplication, and division, but include parity bit generation, error detection and correction, and more specialized applications such as determining correlation functions between signals, and pattern recognition. The simpler arithmetic calculations are well described in other texts (Walker 1967) while the more complex calculations require a detailed study of the problem.

Integrated circuits are available which provide the logic for full addition, i.e. for addition taking account of the carry from the previous stage. This makes the design of logic to add two numbers in parallel easy (Fig. 13.10)

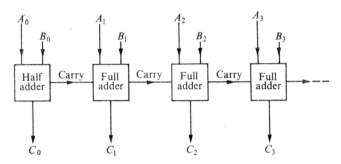

FIG. 13.10. Parallel addition.

but not necessarily cheap. In many problems involving calculation it is worth considering other methods to obtain a solution.

13.7.1. *Counting arithmetic*

One approach makes use of counting registers (scalers) for arithmetic purposes. For example, in one method of binary to binary-coded decimal conversion a binary counting register is loaded with the binary number to be converted. A train of input pulses is then gated into the register and it counts down until it reaches zero; the train is then stopped. The same train of pulses is counted by a binary-coded decimal counting register and at the end of the conversion this register contains the decimal equivalent of the original binary number.

13.7.2. *Serial arithmetic*

This approach can yield considerable savings over parallel arithmetic and introduces only a moderate loss of speed. In a parallel arithmetic circuit the same basic calculating circuit is repeated for each bit so that the total cost is proportional to the number of bits. In serial arithmetic the calculation is performed bit by bit with only one calculating circuit. Thus in serial arithmetic additional facilities in the calculating circuit can be provided at relatively low cost. This method is particularly appropriate when the data is held in shift registers or in a delay-line store.

13.8. Amplification

This is normally provided by the logic elements, or, where more amplification is needed, by power gates in the same logic series. Even when special circuits are required they rarely present much difficulty to the designer.

13.8.1. *Conversion between relay and electronic logic*

The only conversion between different logic systems that requires special care is between relay logic and electronic logic. If a transistor drives a relay and is close to its limit of voltage, it may be damaged by the inductive

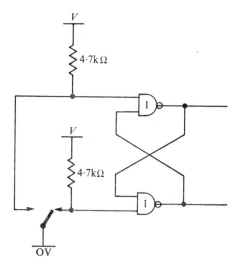

FIG. 13.11. Relay contact squaring circuit using TTL and negative logic convention.

voltage spike from the relay as it turns off. This spike can be suppressed by a diode across the relay coil, but only at the expense of slowing the response of the relay.

When electronic logic is driven from a relay or switch, there may be trouble because of the phenomenon of contact bounce. When a relay switches, the electrical contact may make and break several times as the contacts settle; this succession of signals can cause difficulties in the electronic part of the logic design. Fig. 13.11 shows a relay contact squaring circuit for use with DTL and TTL integrated circuits. This circuit removes contact bounce from the output signals.

13.9. Ergonomics

So far, this chapter has been concerned with the component parts of the logic design. However, a good logic design must also be satisfactory as a whole. This can only be achieved by studying the expected role of the unit. In many cases this study will involve considering the man who must operate the unit, and in these cases as much care should be given to human factors as to other aspects of the design. This aspect of design has been given the title ergonomics.

There is no general theory of ergonomics but the results of the ergonomic considerations can be illustrated by the following examples:

(a) Where accuracy of reading is important a digital display is preferable. Where a quick indication is required, or several parameters must be correlated, analogue displays are better. In a hypothetical two-parameter analysis experiment, data may be printed as a serial list of 100 numbers for the information of the experimenter. The best that he can do with this is to lay it out as a matrix:

262	248	221	195	193	219	237	268	273	263
260	256	247	234	233	242	270	312	275	255
255	266	274	261	268	275	319	287	264	239
221	268	308	294	288	318	283	247	236	227
189	233	292	333	327	300	261	233	222	218
217	195	262	341	365	285	248	316	209	195
196	218	244	331	342	277	241	218	197	194
185	208	243	307	311	268	235	217	188	166
176	197	223	265	272	247	235	220	170	144
165	185	208	225	218	214	208	187	161	132

This matrix takes some time to follow.

Fig. 13.12 displays the same matrix by a contour map. This form of display might be produced on demand on a display oscilloscope attached to a computer. Note how the anomalous result stands out. Simpler useful displays can be produced without access to a computer. The advantage of these displays is that complex information is displayed in analogue form which is the form of data display that the brain can correlate most rapidly.

(b) If a number of meter displays are provided, it aids the detection of faults if the normal operating points on all the meters are similar.

(c) Control panels should provide the information and controls that the user needs. They should not be cluttered with features only required during servicing.

(d) In the layout of control panels, switches and knobs controlling the same feature should be grouped together, even though the circuits they drive may be widely separated. The same direction of rotation on every control should result in a change in the same sense—a clockwise rotation normally increases the controlled parameter. It should also be possible to observe the result of adjusting the control while the control is being adjusted —those with television sets in which important adjustments are at the rear of the set will appreciate this point.

(e) It should be possible to adjust the control to the desired accuracy by easy movements of the hand. However, sweeping the control to the required point should not take too long. If both features cannot be provided by one control, a coarse and fine control arrangement should be considered.

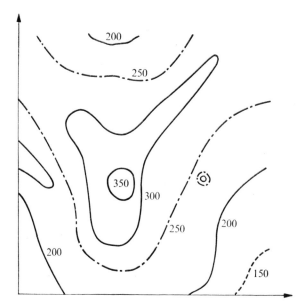

FIG. 13.12. Contour map display of experimental data.

(f) The labelling and layout of the control panels should be carefully studied. Each common control sequence should be run through to ensure that there are no unavoidable difficulties. The proposed layout should then be tried on potential users and their reactions should be observed to see if the possibilities for mistakes and misunderstandings can be reduced.

The final example of this chapter is of an experimental arrangement in which human factors have been taken into account.

13.9.1. *A data-recording system*

This was for an experiment in which cosmic ray showers were being studied. An array of 182 Geiger counters was laid within a triangle with sides 1 mile long. At each point there were two detectors of different sensitivity. After a cosmic ray shower the states of all the Geiger counters were punched out on a card, with data on the date, solar time, siderial time, shower number, pressure, and temperature. Fig. 13.13 shows a card for an event in which every Geiger counter fired.

The pair of Geiger counters at each point were represented by adjacent holes separated from their neighbours, and the card was laid out to give a visual representation of the experimental arrangement. One apex of the triangle had to be shifted as shown, since the card punch drive would be overloaded if it was required to punch more than five holes in a line. Punched cards were selected in this experiment because they can conveniently be sorted during analysis.

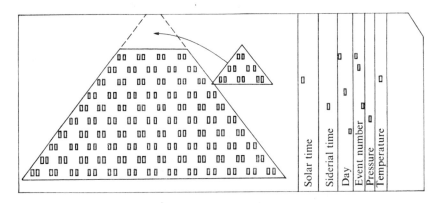

FIG. 13.13. Data card.

13.10. Mechanized and automated experiments

The electronic devices described in the earlier chapters are examples of mechanization of nuclear experiments and are accepted as useful and necessary; so is the mechanization of thinking processes by large memories and computers. Indeed, to take an extreme case, there is no reason why a whole experiment should not be automated, so that the experimenter decides on the experiment writes the computer programmes, sets up the apparatus, and writes the wording of the final report. He then presses a start button and the whole experiment conducts itself, adds to the report the missing data, tables, and graphs, and lights a 'finished' light when this has been despatched to an appropriate scientific journal.

Such an experiment would have a number of advantages over the conventional system:

(a) The experimenter would be forced to give painstaking thought to every aspect of the experiment beforehand, so that no minor (but

312

important) measurement could be neglected, only to be discovered when the report is written.

(b) No well-designed automatic system will make as many errors as a man, and its errors will not be as subtle. A likely human error is to write 25193 for 25913.

(c) The natural but unfortunate tendency of most of us to view with favour the 'good' results and to try to discount the results that do not agree with the others or with our expectations, would be replaced by systematic treatment of every result.

Unfortunately, the idea is impracticable except for standard routine experiments because no one is able to consider every possibility, and the significant events that may indicate an apparatus error or even a new discovery are usually more obvious to the experimenter than to a computer. A computer can nevertheless be a most useful aid to the experimenter. This topic is discussed in more detail in the next chapter.

13.11. Computer-aided design

Computers are also being utilized to assist electronic designers but their use is not yet widespread. They can provide assistance in several different ways:

(a) A computer can be used as a drafting aid so that the designer can generate the required drawings quickly and can modify the drawings conveniently. The computer can keep a record of the modifications and can prepare component lists when these are required.

(b) The computer can be used to calculate the performance of a specified circuit, and to indicate the effects of likely variations in the component parameters. This use is of great importance as an aid to the design of integrated circuit layouts.

(c) The computer can be used to layout a printed circuit for a specified circuit design. The resulting layout is not usually as good as can be achieved by human skill, but this is compensated for by the computer's greater speed and accuracy, and can be overcome by allowing some human intervention.

(d) The computer can be used to generate the detailed logic design from a mathematical statement of the required function. Iveson (1962) has designed a special programming language which can be used in this kind of application.

It does not require great perspicacity to foresee that these applications of computers will become general within a few years.

References

IVESON, K. E. (1962) *A programming language.* Wiley, New York.

STILLWELL, P. F. T. C. and DAGNALL, R. H. (1964) *Electron. Engng* **36**, 756.

WALKER, B. S. (1967) *Introduction to computer engineering.* University of London Press, London.

14 The Use of Digital Computers in Experiments

By J. M. RICHARDS

14.1. Digital computers

DIGITAL computers are complex digital calculating machines. They are able to retain a very large number of intermediate results in their internal *memory*, the computer *store*. While a desk calculator has to be instructed manually at each stage in its operation, a computer holds a *program* of instructions in its internal store. These instructions can be obtained and acted upon by the computer at a much faster rate than is possible with manual input of instructions. Each *word* of computer storage can be used either for an instruction or for data, and so the computer program can change instructions with the same facility that it can operate on data. This allows a sub-program to be written to carry out a task that differs slightly in each application. For example, a sub-program can be written to print out a statement, but the statement to be printed can be selected before beginning the sub-program. This flexibility may appear only a minor benefit from the method of program storage, but without it large-scale computing would not be possible.

The instructions that can be specified in a computer store vary in different types of computer, but all the possible instructions are for fairly simple operations such as addition. Any more complex problem is broken down into these simple instructions, which are carried out at great speed. The time to carry out the basic instructions is limited in practice by the *store cycle time*, which is the time required to read an instruction or data word from the store and then to re-write the same or a modified word in the same store position. If the instruction involves an operation on data in one of the electronic registers that form the active memory of the computer, the whole instruction can usually be carried out in one *store cycle*; but if the instruction requires action on another word contained in the store, another store cycle will be required to obtain this word. The main store of a computer is usually a magnetic core store. Present cycle times are typically 1–6 μs but thin film stores offer the promise of reduced ·times. Developments in integrated circuits ensure that the time for carrying out instructions will normally be shorter than the store cycle time.

Writing the instruction sequences for a computer is known as *programming*. This was originally done by specifying the actual digital words that would be stored in the computer store when the program was operated. This *machine language* is time-consuming to write and it requires a great deal of book-keeping to keep track of the positions in the computer store of the various instructions and data words required. It was soon found possible to leave this book-keeping largely to a computer program known as an assembler, and programs were then written in an *assembler language* in which mnemonics (names which aid one's memory) were used both for the various instructions and for the different words of stored data.

For example, an instruction to add a number A to the present contents of the *accumulator*—which is an active memory register used for calculations—might be written in an assembler language 'ADD A'. This would be assembled into a machine language instruction that might read '2-7413', and 2 indicates the operation 'ADD to the accumulator' while 7413 specifies the location in the computer store that has been assigned for the number A.

An assembler language still depends on the instructions that can be carried out on the particular machine for which is was devised, and so an assembler language program suitable for one computer could not be transferred to a computer of a different type. To overcome this limitation machine-independent languages were developed. Common examples of such *compiler languages* are FORTRAN, ALGOL, and COBOL. Compiler languages are defined in a way that is independent of the computer in use. Thus each machine requires its own compiler program, which accepts a program in the compiler language and translates it into an assembler language program or into a machine language program to suit the machine. The compiled program can then be read into the computer store for operation.

In the example above the FORTRAN statement to carry out the given operation might be $X = X + A$. This may not seem to be a great improvement on 'ADD A', but not only would $X = X + A$ be appropriate for every computer with a FORTRAN compiler but also the statement can be enlarged considerably in complexity, for example

$$Y = (X + A)(B \times C + D)/E$$

specifies the computer operations necessary to calculate

$$\frac{(X + A) \times (B \times C + D)}{E}$$

on the numbers stored in the computer locations corresponding to X, A, B, C, D, and E, and to store the result in the location corresponding to Y. This would take many instructions to program in an assembler language.

Desk calculators operate for our convenience in conventional decimal numbers in which each digit can have one of ten values 0–9 and a digit has ten times the weight of the digit to its right. For example, in decimal,

$$9706 = 9 \times 10^3 + 7 \times 10^2 + 6.$$

Digital computers, on the other hand, nearly always operate in a number system that suits their internal organization. This is the *binary* code in which each digit can take one of only two values, 0 or 1, and a digit has twice the weight of the digit to its right. For example, in binary,

$$1 1 0 1 0 = 1 \times 2^5 + 1 \times 2^4 + 1 \times 2.$$

Each *binary* digi*t*, known conventionally as a *bit*, may therefore be represented by a signal with only two possible states, 0 and 1.

Different types of computer operate with different numerical precision determined by the number of significant digits. The number of bits (binary digits) is usually the same throughout the computer and is called the *word length*. A small computer may have a word length of only twelve bits, which covers the range of decimal numbers from 0 to 4095, while computers intended for long computations may operate with word lengths of forty-eight bits or more.

14.2. The use of computers in nuclear physics

Nuclear physics laboratories make considerable use of large central computers, which are employed in *off-line* computation on experimental data, i.e. in computation some time after the data has been collected.

A typical experiment using such a computer is concerned with the energy spectrum of neutrons in certain conditions. The energy of each neutron is determined by a measurement of the neutron's time-of-flight between two points. From this figure the neutron's velocity and thus its energy can be determined. The time-of-flight data is digitized and analysed into a spectrum by a multi-channel analyser. When sufficient data has been collected the spectrum is recorded and taken to the computer for further analysis. This analysis may involve the subtraction of the contributions of background radiation, a statistical correction for the fact that fast neutrons detected early in the time range can prevent the recording of slower neutrons arriving after them, and perhaps a correction for the energy dependence of the neutron detection efficiency. The output from the computer may be a plot of the corrected spectrum and a list of the significant features expressed in terms of neutron energy; or the computer program may proceed with further analysis of the experiment by comparison of different spectra.

The use of *central computers* plays an increasing part in nuclear experiments since central computers are increasing in size, speed, and convenience of use, and the cost of computing a given job on them is decreasing.

This chapter is concerned, however, with the implications of the development of computers produced for a large range of applications in control and data processing. These *data-processing computers* are versatile and reasonably priced.

The distinction between *central* and *data-processing* computers is primarily one of use, since the same type of machine can usually fulfil either function. Central computers provide a computing service at a central point and feature versatile calculating facilities and arrangements that allow one computing job to follow another in rapid succession. Data-processing computers, on the other hand, are used in close association with an experiment or an industrial process or for a particular task. They feature less complex computing facilities but are more versatile in their input–output arrangements, i.e. their conversation with the outside world.

14.2.1. *Input–output facilities*

A computer communicates with the outside world by the transfer of numerical information to its peripheral devices, which may include a magnetic-tape control unit and a typewriter control unit. This information may consist of

(a) numerical data,
(b) instructions to the receiving device, and
(c) a numerical code for an alphabetic letter, a feature that provides for the communication of words, phrases, and sentences.

Similar methods are used to transfer information in either direction between a computer and a peripheral device. However, there is an important distinction between a passive peripheral device that will accept or deliver information when commanded to by the computer, and an active device that requires the computer to accept or deliver information when the device is ready. The former merely requires an appropriate instruction to the device when the computer is ready for the transfer, but the latter must in some way be able to ask the computer to take action when the device requires a transfer. The active peripheral is common in nuclear physics where, after a nuclear event of interest, the computer is asked to record the resulting data.

Information transfer following a request from the peripheral device can be achieved in three ways:

(a) The computer can be programmed to interrogate its peripheral devices at regular intervals to see if they require attention. When a device requiring attention is found the program jumps into the appropriate routine to service it.

318

(b) The device requiring attention may interrupt the computer by a signal that causes the computer to leave its present program, to store its current position in the program, and to jump to a routine to service the interrupt. In this way the external device obtains a reasonably rapid service but only takes up computer time when necessary.

(c) Many computers have the facility of allowing external devices *direct access* into their store without diverting the computer program from its normal sequence. This facility gives a fast response and enables the computer store to be used like the store of a multi-channel analyser. However, since the computing facilities are not used in the process, the sequencing of different data sources, preliminary checks on the data, and the selection of the correct store location must all be achieved by *hardware*. The hardware is therefore more complex than for the other modes of information transfer.

14.2.2. *Hardware and software*

The existence of data processing computers gives the designer of instrumentation two general approaches. The task may be undertaken by hardware, for example, by the digital circuits discussed in earlier chapters, or alternatively by *software*, i.e. by a computer program in a data processing computer. The hardware equipment will usually be faster in operation but the software solution will often be cheaper and more flexible. Even if the experiment is to make use of a data-processing computer some of the analysis may still be best performed by hardware, if this approach significantly reduces the analysis time.

The decision on the approach to be used for each function in an experimental system is therefore an important part of its design.

14.3. The use of data-processing computers for analysis

A multi-channel analyser has three main functions, namely:

(a) the analysis of the input data into a spectrum by counting the number of times data in each channel is presented to the analyser. For example, the count in channel 276 will be a record of the number of times input data corresponding to the number 276 was received;

(b) the display of the spectrum;

(c) the read-out of the result of the analysis. This read-out is usually via a typewriter or paper-tape punch.

These functions can be taken over by a data-processing computer. For example, the computer may carry out a display program until it is interrupted by a nuclear event. It then leaves the display program and enters a *servicing program* in which it reads the data presented from the experiment, finds the store location appropriate to this data, and then increases the number in this store location by one. The computer then returns to the display program. The servicing program may include checks on the input data to ensure that the event is of interest, and checks on the contents of the incremented store location to detect if it reaches a specified number. Thus the servicing program rarely takes less than ten machine cycles and often many more. The time for one machine cycle, at present typically between 1 and 6 μs, is therefore more important than these fairly short times may suggest, since any routine will require several cycles.

The interrupt facility can also be used to replace a conventional scaler because every event that would normally be counted by incrementing the contents of the scaler can cause an interrupt to a routine that increments the contents of a particular store location. Because of the long *dead time*, i.e. the period before a second event can be counted, and because of the interruption to the computer's other activities, this use would normally be considered only for events that occur at a low rate. The interrupt facility is, however, very suitable for counting time pulses, and these may be counted in more than one way, for example to give the length of the current period and also as a measure of real time. Both these counts may be referred to in the computer program.

Because of the long dead time for programmed acceptance of input data there is often a case for using direct access to the computer store. The input data may be deposited in a part of the store that is regularly examined by the analysis program, or external hardware may perform the analysis if the direct access facility to the computer store allows the hardware to read the content of a specified store location, increment it by one, and replace it in the same location. In this latter case, it is the external hardware that must detect if the data is suitable for storage, or if the content of a store location has reached a pre-set level.

14.3.1. *Computer word length*

In small computers the usual word lengths are 12, 16, 18, or 24 bits. The twelve-bit computers tend to form a class on their own, not only because the short word-length forces the use of a limited instruction code that makes programming more awkward, but also because the full storage at one location of 111,111,111,111 in binary (4095 in decimal) is less than is required for most analysers. Hence two storage locations must often be allocated to each channel of input information. The second location normally records the number of times the first has overflowed. In some cases these disadvantages are outweighed by the advantages of low cost, particularly if the programs to be used are either standard or likely to remain unchanged for long periods. A computer word of sixteen bits is usually sufficient for an analysis, and a choice of word length above this depends on other features of the application.

The number of words of core storage in a data-processing computer is usually 4096, 8192 or 16 384. A store of less than 4000 words would rarely be sufficient for the current programs, data, and working space. If more than 16 384 words of storage are required it is usually possible to allocate parts of the program or data to a cheaper form of storage such as magnetic tape, disc, or drum. Such *backing stores* do not provide as rapid access to the stored words as is possible from the core store. In some experiments the analysis may be stored in the backing store. The computer then does preliminary computation on the data in its core store and continuously updates the information on the backing store.

14.3.2. *Data output*

Small data-processing computers are often provided with paper tape and typewriter input and output devices that will run at ten characters per second. This speed is too slow for applications that require frequent changes of program or considerable output of data. Some consideration in the specification of an experimental system should be given to the use of faster paper-tape readers or punches and to the use of magnetic disc or tape storage that can act both as an extension of the computer storage and as an output medium. In certain cases a direct data link to a central computer can be justified.

14.3.3. *Associative analysis*

Associative storage (Hooton *et al.* 1966) is an idea that is applicable to many tasks in business computing. To take one example, stock numbers are usually allocated in a logical manner but not in a way that directly suits a computer. In a stock-control application reference may be made to the items specified by stock numbers such as 65/13521, 95/2050, and 95/30002, while the vast majority of intervening numbers have never been allocated, and many of the remainder refer to obsolete stock. Computer store is expensive and so it is desirable to limit the allocation of storage locations to information about current stock numbers. It is also important to be able to select the information associated with any stock number rapidly. Storage systems that enable the contents of a storage location to be referred to by some associated *descriptor* are known as associative stores, to distinguish them from the normal sort of store in which the contents of each location are referred to by some fixed number determined by the hardware of the store.

Associative storage is also desirable in the analysis of data from some nuclear physics experiments in which an event is allocated into one of a large number of channels, i.e. in which a very large number of different kinds of events can be described by the apparatus that digitizes the events' parameters. In an experiment involving two related parameters, for example a pulse amplitude and a time-of-flight, each might easily and usefully be digitized into ten bits, i.e. into 1024 channels; thus each event of interest will be described by one of 1024×1024 descriptors. To store events corresponding to every possible descriptor in separate locations would therefore require over a million words of storage. Thus if, for example, only 4096 words of storage are available, the experimenter is faced with the choice of

(a) reducing the accuracy of one or both measurements to store the spectrum in 1024×4 or 64×64 channels, or
(b) limiting the experiment to a small part of the field at one time, for example to study the amplitude spectra associated with the time measurements of 612, 613, 614, and 615.

However, even if the experimenter did obtain a million-word analyser, most of the storage would be wasted if, as seems likely, most of the locations would have very few counts in them. This introduces the possibility of using a smaller associative analyser to collect data only on the events whose descriptors have a significant frequency of occurrence. When a new event is detected by such an analyser its descriptor is compared with all the descriptors at present stored, and if a corresponding location is found the content of this location is incremented.

The operation of comparing an event descriptor with all those already stored can be achieved by various hardware techniques which give an associative or contents addressable store. Alternatively, the descriptors can be stored in a computer and compared in turn by program with the event descriptor. With a suitable arrangement of the descriptors, only a few comparisons are necessary for a programmed search of the whole store. For example, if the descriptors are stored in numerical order the program could initially test against the descriptor half way down the store, and thus decide whether to look in the first or second half of the store. A comparison with a descriptor in the middle of the selected half would then narrow the search down to a quarter of the store. In this way, the program could search through 4096 descriptors with only twelve comparisons.

14.3.4. Hardware versus software analysers

The hardware associative analyser is faster than the programmed computer, but takes a long time to develop, is more expensive, and is not easily altered to include facilities that have not been built into it. A programmed associative analyser on the other hand can be developed fairly quickly and can then be fairly readily modified to suit additional requirements. Further, the computer is available for other work whenever the associative facility is not required. Thus though a hardware analyser can usually be designed for any data analysis that a data processing computer is programmed to do, and can perform the analysis faster and with more convenient controls, a software solution is preferable for the more complicated applications. A realization of the associative analyser is described by Hooton *et al.* (1966).

14.4. The use of data-processing computers for display

The second function of a multi-channel analyser is the display of stored data, but though some analysers have versatile range-setting controls, no analyser can match the versatility of a computer-generated display. However, if the computer generates the display on-line, a large proportion of its time might be spent on this task. For example, if the computer specifies the x and y coordinates of each point to be displayed, a reasonably simple display routine on a small computer may take 50 μs for each point. A 1000-point display would then require 50 ms to generate and the computer would be fully occupied generating such a display twenty times a second. Because of the computing time required for on-line display, computers are sometimes programmed to operate in three distinct modes; data collection, data display, and data read-out. The demand for a display during data collection, however, may lead to the provision of storage specifically for the display. This could be achieved by a storage oscilloscope, or by photography, but both would be rather inflexible for changing data. Fast storage techniques are then preferred and display data is held on a drum, disc, or core store associated with the display. This store may often hold a number of different displays, any of which can be prepared or updated by the computer, so that the experimenter can switch from one display to another for comparison.

The drum store is particularly appropriate to a display based on the television raster. The display screen is scanned by a television raster and successive bits in the store specify whether each point in the raster is black or white. A 512-line raster with 512 bits in each line requires 262 144 bits of storage. A point-by-point display with the same resolution (nine bits in each axis) can display 14 600 points with the same storage. But the raster gives a flicker-free display of any complexity within its resolution, and always uses the same size of store. Furthermore, extra monitor displays can easily be provided. However, point-by-point displays can have the advantages of greater resolution and controlled brightness. The television raster can be given a two-bit brightness control only by allocating two bits to each point. This doubles the storage required.

Computer displays are also used for feeding information into the computer to supplement push button or typewriter inputs. One approach is to provide a keyboard on the cathode-ray tube screen with the keys labelled appropriately by the display. An alternative is to include in the display a point or marker, as in Fig. 14.1, which can be moved to indicate, for example, part

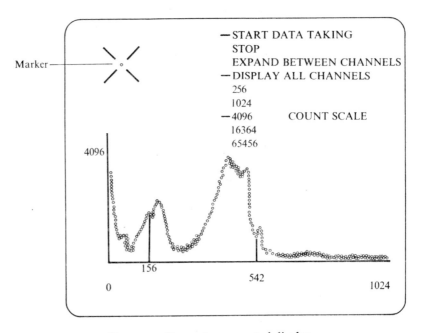

Fig. 14.1. Computer-generated display.

of the display to be printed out, or to select one of several displayed choices. This moveable point or marker can be generated by the computer to follow a *light pen* or photodiode that feeds a signal back to the computer when the point is beneath it. Alternatively the position of the point can be determined by a *track ball*, a ball that projects through a desk top and may be rotated freely about any axis (rotation about the vertical axis has no effect). The light pen is more readily accepted by users as it bears a closer resemblance to writing, but the track ball is easier to program, and its operation does not obscure the display.

An interesting application of computer displays was described by Mollenauer (1964). The computer was programmed so that it would generate the actual display and also a display derived theoretically on the basis of assumptions that could be typed into the computer. Comparison of the two provided a rapid and informative check on the experiment and the assumptions.

14.4.1. *Plots*

An alternative method of display is by the production of plots of the stored data on a graph plotter. This does not give the immediate response of a cathode-ray tube display and cannot be used for control. However, plotted displays have several significant advantages:

(a) a permanent record is made,
(b) greater resolution is possible,
(c) more sophisticated displays can be generated because the display program is not tied to a rapid cycle time, or to a limited number of output points, and
(d) a series of plots can easily be compared.

For these reasons it is, in some cases, more important to provide good plotting facilities than good cathode-ray tube display facilities. Good plotting facilities are normally available in a central computing system so that the connection of a graph plotter to a data processing computer need only be considered if the use of a central computer system would introduce unacceptable delays.

14.5. The use of data-processing computers for control

Computers attached to experiments are not only employed for the analysis and display of the experimental data. Another common use is control of the experiment where computers can again replace and improve on alternative hardware devices. The applications are not exclusive: a computer installed mainly for a control function may also be used for the analysis of the resulting data, while a computer primarily intended for analysis and display may also control its experiment.

In all the experimental sciences, experiments are designed as far as is practicable to isolate one phenomenon for investigation. An experimental approach is chosen that enables the phenomenon to be observed by means of one or more measurable parameters while other relevant parameters are set and constrained.

This same principle also applies to the most complex nuclear physics experiment, where the parameters that are constrained may include the energy, type, and time of occurrence of particles involved in the nuclear event and also mechanical settings in the construction and positioning of the apparatus, while the measured parameter is often the energy of a particle or a time associated with a nuclear event. Both the measured and

325

constrained parameters need to be calibrated and maintained against apparatus drift, while the set values of the constrained parameters are sometimes altered during the experiment. The operations of calibration, maintenance, and alteration of the set points can be delegated to computer control with advantages in speed, convenience, and continuous unattended operation.

14.5.1. *Electronic control by computer*

In nuclear physics experiments several of the parameters that are measured or constrained are characteristics of electronic pulses that represent the parameters of

(a) particle energy,
(b) particle type, or
(c) time of occurrence.

The constrained parameters are first selected by

(a) pulse-amplitude discriminators,
(b) pulse-shape discriminators, and
(c) coincidence units;

and then, if the required conditions are satisfied, the experimental measurement is made and recorded.

If suitable test apparatus is provided, the constrained parameter settings may be checked and the measurement calibrated under computer control. This may be done just to give an alarm if a serious drift is detected, for use when processing the data, or to maintain the apparatus against drift. The latter use requires digital control of the electronic apparatus. Digital control is also required if the computer is to control an experiment that involves alterations to the settings of the electronic apparatus.

Some of the parameters that may require digital control by a computer are

(a) cable interconnections and coincidence channel selection, for setting up test conditions;
(b) discriminator bias levels and delays, for setting constraints;
(c) Amplitude-to-digital converter bias; photomultiplier E.H.T. settings and amplifier gain for controlling measurement apparatus.

One advantage of having the apparatus electronically controlled is that it permits routine tests on the apparatus during the course of the experiment. Such tests would normally be time-consuming and require skilled attention by the experimenter.

326

14.5.2. *Mechanical control by computer*

Certain experiments require a sequence of accurate settings of mechanical parameters. For example, triple-axis neutron spectrometer experiments are designed to determine the period of vibrations of atoms in a crystal by observation of the intensity and location of peaks in the neutron diffraction pattern produced when the crystal is placed in a neutron beam. In Fig. 14.2 neutrons from a reactor strike a crystal that acts as a Laué reflector. The monoenergetic beam of reflected neutrons then strikes the target crystal at the angle θ. The neutrons diffracted through an angle ϕ by the target crystal are analysed by a second Laué reflector, which selects neutrons

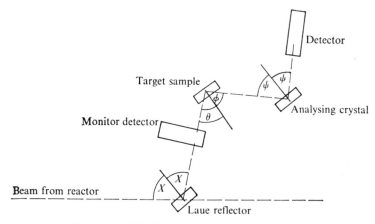

FIG. 14.2. Triple-axis neutron spectrometer.

of one energy, determined by the analysing crystal structure and the angle ψ, and reflects them into the detector. The experiment involves counting the neutrons detected at numerous settings of x, θ, ϕ, and ψ.

Such experiments can be automated so that the counting periods and the intervening alteration of experimental parameters, including the repositioning of neutron shielding, can proceed completely under computer control. Many of the problems are mechanical, for example the accurate measurement of the controlled angles and the accurate positioning of the apparatus in spite of back-lash and friction. The controlling computer can also be used to determine the required positions and to interpret the results of the experiment, but since both these functions require complex calculations they are usually performed on a central computer.

14.6. Computing on data-processing computers

A data-processing computer attached to an experiment is usually occupied by preliminary analysis of the experimental data, by preparation of displays, and by the control of the experiment. It has little spare capacity for computation on the data. Even if sufficient time and storage can be spared on a data-processing computer for scientific computation, a large central computer with its sophisticated compilers and library of standard programs is usually preferred. This is particularly true if, in addition, the central computing facility provides a rapid service. In contrast, a data-processing computer has limited power and there are severe restrictions on any program that is to run in parallel with other computer operations. The computing power of a data-processing computer can be used with advantage, however, when immediate computation of results is most important. One application is in the monitoring of pulse rates from detectors near a reactor. The computer can be programmed not only to check against high and low alarm levels, but also to do more sophisticated checks on the rate of change of readings. In the event of an alarm the computer may display relevant information or print a diagnostic statement.

A second application is in testing, if a rapid answer is required immediately after the completion of a complicated test. One example is in the routine testing of nuclear fuel elements. A rational specification for the distribution of fissile material in these elements requires the mean concentrations to be within certain limits, and deviations from the above mean to be within limits that depend on the area over which the deviation is present.

It is difficult to test to such a specification rapidly and accurately without access to a computer, and so a data-processing computer may be programmed to accept the test data and to compute the rapid accept or reject decisions that are desirable.

14.6.1. *Time sharing*

In doubling the cost of a computer it is possible to increase its computing power by a considerably greater factor. Hence there is a tendency towards large central computers, each serving many users. In data processing there are, nevertheless, considerable advantages in restricting a machine entirely to one application and so this field provides a ready market for small computers. Advantages of scale still apply and thus there is a strong incentive for several experiments in a particular sphere to share a single powerful data-processing computer. This is particularly true if all the experiments are similar and can make use of a large proportion of common program. However, great care must be taken to prevent any unintentional interference between various experiments even by faulty programs. This can be avoided if the structure of the system keeps the programs of separate users apart, a step that prevents the use of a common program. Alternatively, the programs allowed in the machine can be restricted to those that have been extensively tested with all other permitted programs. This latter approach is only suitable for unchanging systems since the developing and testing of new programs becomes time-consuming.

14.7. Programming data-processing computers

The casual user of a central computer will usually write his programs in a compiler language such as FORTRAN. Programs in this language are not far removed from mathematical specifications of the calculation required. They are also machine-independent since they can be converted by compiler programs into the machine language appropriate to the computer being used. The simple compiler programs at present available for small computers have limited application and generate inefficient machine-language programs compared with those that can be developed by a human programmer. Compiler languages cannot normally be used to program data-processing computers if

(a) restricted store size puts a premium on a compact machine program,
(b) the program must operate in minimum time,
(c) data transfer is initiated by an external interrupt signal, and
(d) data transfer is to be made with specific peripherals other than the standard input–output devices and the compiler makes no provision for this.

As a result, much of the programming of data-processing computers is done laboriously in an assembler language that is closely related to the computers' machine language and thus to the structure of the particular computer being used.

A detailed knowledge of the computer is also required because any data processing task involves the connecting of external hardware to the computer. The computer programs are, therefore, part of a system that includes the external devices. Though the programming may well be the major part of the development of the system it is very important that the whole system is worked out in considerable detail before much work is done on individual programs. It is all too easy to rush into details of the construction of small parts of a system only to find that the parts do not fit into an overall scheme and the work has been wasted.

14.7.1. Program interrupt

A program interrupt facility is the flexible way of transferring data between a peripheral and a computer, but it complicates programming, particularly if more than one peripheral can interrupt the computer. In the simplest form of an interrupt facility, an interrupt signal causes the computer to leave its present program and to test all possible sources to find one that is generating an interrupt. This procedure is fairly convenient if the interrupt facility can then be cancelled until the interrupting device has been serviced, but this cannot be done if one device requires rapid service. In such cases, if a lower priority device is being serviced, the source of any interrupt must be discovered. Even if the interrupting device is of low priority its identity must be stored and its interrupt signal cancelled to leave the interrupt line free for use by the high-priority device.

This complexity can be reduced by hardware that accepts interrupts from any source but only interrupts the computer if there is an interrupt of sufficient priority to override that being attended to at present. Such hardware can also provide direct information of the source of the interrupt, thereby obviating the need for a search program.

In any interrupt scheme care must be taken if two programs use the same sub-routine. If the possibility exists that one program might have been interrupted while performing the common sub-routine, the second program must set the sub-routine to its correct quiescent state before entering it, and restore any half-completed calculation after leaving it. Consequently, unless it is tolerable to cancel the interrupt facility while operating any common sub-routine, each servicing routine is usually made complete even though this often involves duplication of programs.

If these difficulties are faced it is possible to construct sophisticated systems using interrupt, but because of the care required in writing and testing, all the programs are usually developed as a package before the inauguration of the system and are then left largely unaltered. This approach is encouraged by the fact that after a computer is installed for this type of application, there are few periods in which the computer is available for program development.

One elegant system has been installed at the National Bureau of Standards in Washington. In this system there are eighty priority interrupt lines and the interrupt principle is extended to the interconnection of software programs. An interrupt can be set in the normal manner by a hardware peripheral to call for a particular program at a particular priority, but an interrupt can also be set under program control. In this way, a program can set an interrupt to call for its successor at appropriate priority or can set several interrupts to arrange a sequence of programs.

14.7.2. *Direct store access*

Because of the difficulties in programming for interrupt operation some systems accept the limitations of using only direct access to the computer store for data input.

In a time-sharing system proposed for Brookhaven Laboratory, data from each experiment is deposited in restricted parts of the core store. A program for this experiment reads the data and updates an analysis of the data on a magnetic-disc backing store. The analysis program for each experiment occupies most of the core store for a short period. It is then interrupted, read rapidly into the backing store, and replaced by the program of the next experiment. The time spent by each program in core store is sufficient to analyse the data deposited by its experiment since the previous period.

A high-speed point-by-point display system for each experiment reads the backing store where data from its own experiment is deposited so that, provided that the data is stored in one of the forms accepted by the display system, the data is displayed without expending computer time.

The system gains considerable flexibility from the fact that users can write their own programs. This is possible because the difficulties of program interrupts are avoided.

References

HOOTON, I. N., BEST, G. C., HICKMAN, S. A., and PRIOR, G. M. (1966) *I.E.E.E. Trans. Nucl. Sci. NS-13*, 553, 559, 566.

MOLLENAUER, J. F. (1964) Display-orientated data simulation and analysis in multiparameter experiments. *Proceedings of conference on automatic acquisition and reduction of nuclear data*, Karlsruhe, 1964, p. 205.

15 Practical Design Aspects of Electronic Equipment

By W. G. L. BROWNRIGG

15.1. Introduction

In this chapter various criteria that should govern the production of an electronic unit are considered, assuming that a final circuit has been achieved. Three main aspects can be considered, namely, components, front panel, and mechanical assembly. Each of these features can be subdivided into a variety of facets whose individual importance may vary from unit to unit. If we take components, these are obviously classified as resistors, capacitors, valves, semiconductors, etc., and care has to be exercised in choosing the right component for reliable operation.

Front panel considerations are usually not so clearly defined. Controls should be arranged in logical order with clear grouping of functions. Legends should be clear and not obstructed by input, output, or cross-coupling connections. In mechanical construction care has to be taken in producing a unit that utilizes the components in the best way for reliability with correct electrical functioning, i.e. no cross-talk or interference due to poor lay-out and with a reasonable degree of accessibility for maintenance.

15.2. Components

There is an increasing tendency when thinking of components to visualize some form of complete circuit, either as one of the commercially available 'single-chip' circuits or logic bricks, or as some form of 'plug-in'/'wire-in' assembled board such as the Harwell 5000 series. However this type of component will be considered under mechanical construction and the more conventional concept of components is dealt with here.

15.2.1. *Resistors*

During the last decade substantial advances have been made in resistor design. At A.E.R.E. metal oxide resistors have replaced cracked carbon resistors and show promise of being more reliable. Improvements have been made, however, in cracked carbon resistors during their quarter-century life and they now have higher stability and reliability than 10 years ago.

The main advantages of metal-oxide over carbon are:

(a) Reduced size for equal wattage.
(b) Higher temperature operation.

TABLE 15.1

Comparison of electrolytic capacitors

(Extracted from *STC Component News*, Vol. 6, No. 4, 1964)

Characteristic	Aluminium electrolytic	Electrolytic		
		Tantalum foil	Wet tantalum anode	Dry tantalum anode
Capacitance range	0·5 to 150 000 µF	0·2 to 200 µF	0·2 to 1250 µF	0·1 to 330 µF
Capacitance tolerance:				
Standard (%)	−50, +100, +150 / −10	±20	+10, +20, +75 / −10	±20
Minimum (%)	±25%	±10%	±15%	±5%
Form	Polar or non-polar	Polar or non-polar	Polar	Polar
Operating voltage:				
d.c.	2·5 to 500	6 to 150	3 to 150	6 to 50
a.c. 50 Hz operating	40 to 320 int.	Limited	Limited	Limited
Power factor: 50 Hz	6 to 35% dependent on voltage	12% (varies with C and V)	12% (varies with C and V)	8% (varies with C and V)
Leakage current at 25°C / 85°C	0·006 mA/µF V / 4 × 25°C value	0·02 µA/µF V / 10 × 25°C value	0·02 µA/µF V / 10 × 25°C value	0·02 µA/µF V / 10 × 25°C value
Temperature:				
Operating range (°C)	−40 to +85°C	−55 to +125°C	−55 to +125°C	−55 to +125°C
Capacitance change (%)	−60% at −40°C	−20% at −40°C	−50% at −55°C	−8% at −55°C
Capacitance change (%)	+20% at +70°C	+9% at +85°C	+20% at +125°C	+7% at +125°C
Stability:				
Capacitance change with temperature and life	Large	Medium ±15%	Large ±25%	Small ±8%
Size:				
Varies as	CV approx. Small	CV approx. Very small	CV approx. Very small	CV approx. Very small
For equivalent CV rating				
Cost	Very low	High	Moderate	Moderate
Specification:				
U.S. MIL	62	3965	3965	26655
Brit. DEF	5134	5134	5134	5134

Characteristic	Metallized paper	Metallized polyester	Paper	Polyester
Capacitance range	0·01 to 20 pF	0·01 to 20 pF	0·001 to 200 pF	0·01 to 20 pF
Capacitance tolerance:				
Standard %	±20	±20	±20	±20
Minimum %	±5	±2	±2	±1
Operating voltage:				
d.c.	50 to 600	50 to 600	50 to 200 000	50 to 1000
a.c. 60 Hz	25 to 250	25 to 250	50 to 75 000	Seldom used
Dissipation factor:				
% at 60 Hz	0·4 to 0·6	0·2 to 0·3	0·2 to 0·5	0·3
% at 1000 Hz	0·6 to 0·8	0·4 to 0·5	0·2 to 0·5	0·5
% at MHz (low capacitance values)	Relatively high	Relatively high	Higher; varies with type	Relatively high
Power factor (at 20°C)				
Insulation resistance:				
(*MΩ μF at 25°C)	600 to 1200*	5000 to 50 000*	3000 to 20 000*	50 000*
Comparison 85°C 25°C	1·60	1·40	1·100	1·25
Temperature:				
Operating range °C	−55 to +125	−55 to +125	−40 to +100	−55 to +150
Stability:				
Capacitance change with temperature ageing	Medium	Medium	Medium	Medium
Size:				
Varies as	CV^2	CV^2	CV^2	CV^2
For equivalent CV rating	Small	Small	Medium Small	Small
Specification:				
Brit. DEF	5136	5138	5131	5138

(c) Smaller total excursion in value. This is the first type of resistor where total excursion is stated. Total excursion of a resistor is the change in resistance due to the combined effect of changes in temperature and voltage and takes into account the initial selection tolerance of the resistor.

The main disadvantages of metal-oxide resistors are:

(a) Not really suitable for replacing grade-1 cracked carbon resistors over a wide resistance range due to the production problems in producing metal-oxide resistors of very high value.

(b) Heavy termination wires. These are necessary to allow generated heat to escape from the resistor, but they present problems in assembling the resistors since the wires are of similar diameter to mounting pins now used.

15.2.2. *Capacitors*

The use of low-voltage lines for transistors has made the electrolytic capacitor more necessary than it was in the past. For decoupling such lines the solid

334

15.2

capacitors

News, Vol. 6, No. 4, 1964)

| Polystyrene | | Silvered mica | | | |
Unprotected	Metal cased	Wax dipped	Resin dipped	Resin moulded	Metal cased
10 to 500 000 pF	5000 pF to 2 μF	4 to 750 000 pF	4 to 750 000 pF	4 to 750 000 pF	100 pF to 1 μF
±10 ±0·5	±10 ±0·5	±10 ±0·25	±10 ±0·25	±10 ±0·25	±10 ±0·25
50 to 750 Seldom used	50 to 2·5 kV 50 to 350	125 to 750 Seldom used	125 to 750 Seldom used	125 to 750 Seldom used	125 to 2·5 kV R.F. voltage varies with current and frequency
<0·0005		<0·001			
100 000 MΩ	>100 000 MΩ	>3000 MΩ	>50 000 MΩ	>50 000 MΩ	>50 000 M
−40 to +70	−40 to +85	−25 to +70	−40 to +85	−55 to +100	−55 to +85

tantalum capacitor of, say, 10–50 μF at 15 V-working is commonly used. These capacitors are also suitable for general use. Their advantages over early electrolytics are longer life (no 'drying up'), superior high-frequency performance, and smaller size.

The following are the most frequently used dielectrics:

(a) Polyester dielectric. This is a general purpose cheap capacitor.
(b) Miniature lacquer dielectric. Properties: very small size, high insulation, special purpose, expensive.
(c) Polystyrene foil. Properties: very low loss, free from hysteresis, highly stable, very low leakage.
(d) Silver mica. Properties: precision capacitor very low loss, highly stable, good temperature coefficient, fairly expensive.
(e) Metallized paper. Properties: going out of favour, although some makes are very good.
(f) Hi K ceramic. Properties: poor dielectric, wide tolerance, poor stability, but very small size and good H.F. performance. Tables 15.1 and 15.2 show the comparison of capacitors produced by S.T.C.

335

15.2.3. *Valves*

The use of valves at A.E.R.E. is now very limited and restricted to conditions that transistors cannot yet readily meet, for example electrometer valves (where currents of the order of 10^{-12} A or less are being detected although field effect transistors may soon perform this work), very low noise amplifiers, large voltage excursions, etc.

15.2.4. *Transistors and diodes*

Transistors are now commonly used in a wide variety of equipments. The advantages of the silicon types of transistor over the germanium types are:

(a) Silicon transistors can operate at higher temperatures.

(b) The leakage current I_{CO} of silicon transistors is several orders of magnitude less than that of a germanium transistor and can usually be ignored, even in d.c. coupled circuits.

(c) Silicon transistors are capable of handling higher voltages than germanium transistors.

(d) Silicon allows production techniques which accentuate these features and also allow high-frequency operation.

More recent advances in the semiconductor field are junction and metal-oxide field effect transistors (FETs and MOSTs) and integrated circuits (ICs).

The FET and MOST have low noise and high input impedance. The IC is replacing logic elements of assembled discrete components with an immense saving in space, power, and cost. Progress is also being made in the analogue field where a reasonable selection of operational amplifiers is available.

15.2.5. *Potentiometers*

With the introduction of the 'Helipot' type of 3- and 10-turn potentiometer it has become possible to achieve higher resolution, greater control in setting up, and better re-positioning accuracy. Normally these potentiometers have a coarse control with a vernier dial refinement for accuracy and some of these have locking mechanisms that prevent accidental movement of the knob. The knob dial potentiometer is a miniature version of the 'Helipot' with the potentiometer assembly mounted inside the knob.

Another type of potentiometer commonly used is the 'Trimpot' type. Its rectangular shape and small size makes it attractive for many uses, especially so in replacing potentiometers of awkward shape on printed wiring boards. It has a smaller wattage rating than the standard potentiometer but because of the lower power dissipation in transistor circuits this is not a serious obstacle to its use.

15.3. Front panels

The front panel is probably one of the most important aspects of a unit and it is therefore worth-while spending some considerable effort on its design. Care should be taken in logical grouping of controls that are related to each other. All controls governing the shape of an output pulse, for example rise time, duration, amplitude, etc., should be grouped logically. Too often the various controls affecting the shape of an output pulse are badly placed relative to each other with, for example, the coarse and fine knobs separated by a control affecting another parameter of the unit. External cabling to

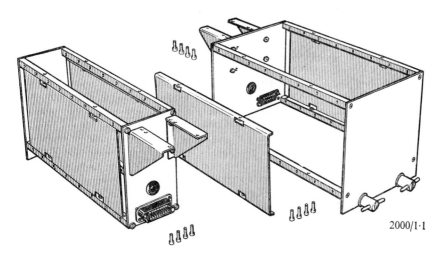

2000/1·1

FIG. 15.1. Perspective view of frame units with transformer mounting assemblies—2000 series module.

the front panel must be arranged so that connectors do not obscure legends unnecessarily or make it very difficult to connect cables to their mating parts.

Instrument type numbers or catalogues are normally shown on the front panel. The type numbers are grouped to give some idea of the equipment series. These numbers are listed below.

1000 series: This includes the normal 19-in front panel black-box type of equipment, the health physics 19-in front panel cream-box type of equipment, and other portable equipment (non-unitized equipment).

2000 series: Unitized equipment (see Fig. 15.1 for mechanical details).

3000 series: System comprising some 2000 series units.

4000 series: Reactor control equipment, U.K.A.E.A. Winfrith.

5000 series: Board assembly sub-units.

5500–5599: Plug-in standard board assemblies (see Fig. 15.3 for mechanical details).

5600–5699: Wired-in standard board assemblies (see Fig. 15.3 for mechanical details).

6000 series: Reserved for board assemblies.

7000 series: Small modular units (see Fig. 15.2 for mechanical details).

8000 series: U.K.A.E.A. Culham equipment (all types).

9000 series: U.K.A.E.A. Aldermaston (all types).

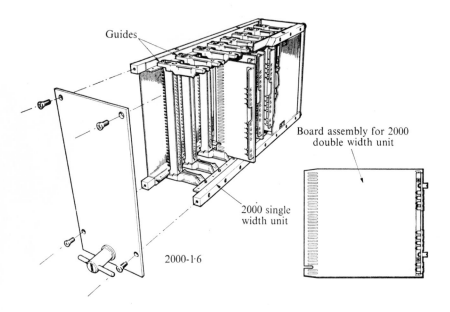

FIG. 15.3. 5000 Series guiding system. Guide arrangement for plug-in board assemblies.

The positioning of sockets should also be considered in the grouping suggested above. In the case of 2000 series units a few rules laid down in the *Specification and Guide* 2000 (1966) limit the positions of input and output sockets. This guide also gives information on the size of engraving and the positioning of certain information, provides a glossary of approved terms and abbreviations, and also a list of symbols for use on front panels. This is to make the individual 2000 series units conform to an overall system of style and nomenclature.

Fig. 15.2. 7000 Series module, frame and extension unit.

15.4. Mechanical assembly

15.4.1. 1000 *Series*

In general, during the past decade the construction of units has undergone some radical changes. Ten years ago the conventional 19-in front panel unit was the standard in the 1000 series equipment. In this the chassis usually extended over the total inside area of the unit containing holes for mounting valves and transformers, etc. Each unit had a variety of ventilation louvres, its own power pack, and its own front panel height. This method of construction tended to bring about poor reliability due to the high wattage dissipated in the box (inherent in valve equipment): in fact some units had cooling systems added at a later stage to improve reliability.

15.4.2. 2000 *Series*

About ten years ago the Electronics and Applied Physics Division of A.E.R.E. introduced a modular system of plug-in units known as the 2000 series (Fig. 15.1), and centred around the continued use of valves. In this the chassis consists of four tie-bars with a front panel and a rear panel. The useful height of the chassis is 6 inches and the width $3\frac{3}{8}$ inches. Five of these units can be plugged into a shelf unit whose width is that of the old 19-in front panel units. The method of using this frame is to mount the components on synthetic resin-bonded fabric boards and bolt those to two vertically placed tie-bars. Side covers are provided with each chassis and they help to create a chimney effect of ventilation allowing free passage of air through the unit to cool the components, with the air inlet and outlet from the rack controlled by the shelf unit. The side covers also provide some protection for the inside of the unit during transport, i.e. exposed components are not damaged by careless stacking of units.

The advantages of this system are listed below:

(a) No individual power supplies.
(b) Maintenance by replacement.
(c) Higher reliability due to cooler operation.
(d) Greater ease of maintenance because of greater accessibility.
(e) Flexibility in building systems or in changing systems.
(f) Lower cost for complex systems.

The disadvantages are as follows:

(a) The difficulty of power supplies. This involves the question of unitized or bulk power supplies and impedance effects of both wiring and shelf unit. Fig. 15.4 shows this effect with power units varying slightly in design. The power unit is in the adjacent position on the

339

shelf to the one on which the measurements are being made. If some decoupling (about 50 μF) is added to the line, the impedance curve is virtually level at 0·2 to 10^6 MHz, thereafter following the general inductive line; an improvement can also be shown when the power unit and the test position are remote from each other.

(b) Loss of volume: only 15 per cent of rack volume is capable of being used by components.

(c) Greater cost for simple systems.

(d) Loss of front panel area for controls: 33 per cent of area of equivalent 19-in unit available due to positioning of tie-bars.

This is not an ideal system but it is a vast improvement on the earlier chassis form.

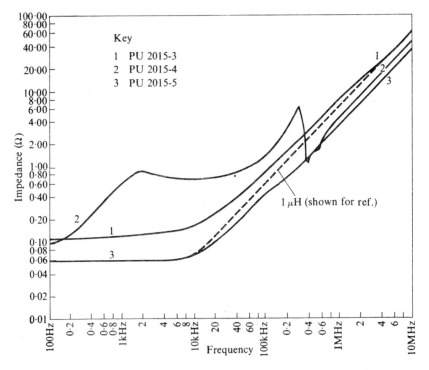

FIG. 15.4. Plot of impedance versus frequency—2015 power supply unit.

15.5. NIMS modules

The United States Atomic Energy Commission Committee on Nuclear Instrument Modules was founded in 1964 to draw up specifications for standard modules to assure mechanical and electrical interchangeability—a standard was laid down in July 1964. The above standard, known as

NIMS format (Nuclear Instrumentation Module Standard) allows twelve single-width modules to be accommodated in a standard 19-in bin. All multiples of the single-width module can exist and can be inserted in any combination. Two module heights are permitted, $5\frac{1}{4}$ inches and $8\frac{3}{4}$ inches. Adaptors exist to allow the use of a $5\frac{1}{4}$-in panel with panels of $8\frac{3}{4}$ inches. The power supplies have been tightly specified having such features as ±0·5 per cent voltage stability at any time during any 24-hour period, ±0·5 per cent voltage drift over 6 months, 3 mV peak-to-peak noise or ripple, etc. The features that may cause undue system failure are as follows:

(a) There is no overall systems-control and independent firms may design circuits in the NIMS format and have little regard to compatibility with units designed by other firms. Sample signal coupling levels laid down are: logic '1', 4 to +12 V; logic '0', +1 to −2 V. Analogue levels are: Class I, 0 to +1 V; Class II, 0 to +10 V; and Class III, 0 to +100 V.

(b) This is the first attempt at an American unitized system and many problems have inevitably arisen even though a careful study has been made of other national unitized systems. Most of these problems have been solved.

(c) Due to the rapid advances of technology the NIMS format single-width module appears too large for use with integrated circuits.

(d) Cooling or ventilation relies on convection only. This is inadequate for the power dissipation that can arise when using integrated circuits. Some forced air-flow is necessary to carry away the generated heat before the components become hot enough to encourage the chimney effect to operate.

15.6. General considerations

The various considerations involved in deciding on a unitized system are

(a) mechanical style,
(b) mechanical size of chassis,
(c) mechanical size of front panel,
(d) application: digital, analogue, or both,
(e) number of electrical couplings required,
(f) power lines required,
(g) allocation of lines, and
(h) coupling signals and levels.

Dealing with the various items in the above order:

(a) The general mechanical style is reasonably well agreed. The module should be a printed wiring card capable of being inserted into an edge connector and having a front panel mounted on some sort of bracket. The bin should be capable of accepting the plug-in cards and would be of a size to be held in a standard Post Office type rack.

(b) The mechanical size of the chassis, i.e. the printed wiring card size, would be constrained as follows. The height would be controlled by the size of edge connector used. The length should be controlled by the space required at the rear of the rack or bin for wiring, cooling power supplies, etc.

(c) The height of the front panel should be closely linked with the height of the chassis and the width should be governed by the size of control components, for example the largest control component is envisaged as being a 'Knobpot' of $\frac{3}{4}$-in diameter, so the width would be approx. $\frac{3}{4}$ inches plus some tolerance amount.

(d) The argument in favour of using such a modular construction for purely digital functions or as interface equipment between the standard A.E.R.E. 2000 series and computers is that the width of the front panel can be reduced until it is only slightly wider than the relevant dimension of the uniting edge connector. As the bulk of the components used would be integrated circuits in general, and probably dual-in-line case in particular, this would allow sufficient clearance over components; the edge connector is 0·34 inches and the overall height of a dual-in-line is 0·31 inches. This would give a maximum number of 33 modules/19-in bin.

(e) The argument in favour of using such a modular system for analogue work is that the existing 2000 series module is too large. Only one side or part of it is used for carrying the circuit and it is only occasionally that such circuits can be duplicated in units and used simultaneously. However, if the front panel of the new modular system was increased in width to allow a variety of small manual controls, for example knobs and pots, this would provide a useful medium for an analogue modular system.

(f) Complete systems are likely to contain both analogue and digital components for many years to come. It is feasible therefore to consider a common modular construction for both the digital and the analogue units.

(g) The assessment and collation of commands necessary to control interface, analogue, and computer units required to transfer data along the data highway assumes a knowledge of the number of bits per word together with the commands accompanying the words. Similarly, if a unit has acquired information that should be passed along the data highway for processing it must be capable of generating a signal demanding attention.

(h) Power supplies for either, or both, digital and analogue units have to be provided.

(j) All the signals and power lines must be allocated a position in the data highway and in the uniting plugs and sockets.

342

(k) The types of pulses used for digital information have to be defined and the input and output pulses have to be specified for both sides of the interface units whether coupling to analogue units, printers, 2000 series, or computers, etc. Levels and pulses have also to be defined for analogue units. Some compatibility must exist between signals and impedance for digital and analogue modules.

Many of the above points have still to be resolved. The problem is made even more difficult by the problem of considering mechanical and electrical compatibility with existing modular systems. (One mechanical proposal is shown in Fig. 15.2.)

15.7. Miniaturization

The introduction of transistors allowed the first step in miniaturization to get under way by an immediate space-saving in the size of a transistor over the size of a valve. The second step was achieved because less power was required and therefore smaller wattage components could be used.

The next logical step is the introduction of integrated circuits where a bistable or a gate can be obtained in an envelope whose volume is comparable with that of a small signal transistor. This step is being implemented.

The other major change in construction followed the use of printed wiring. The difference between printed wiring and printed circuits is that:

(a) Printed wiring is the formation of conductors of copper only, produced by some means, on to an insulated base and bonded to this base.

(b) Printed circuit is the formation of conductors and some components by some means as above. Little use has been made of printed circuits at Harwell but in one case the delay-line inductors and conductor paths for a distributed amplifier used this technique.

In the early stages of printed wiring, components were assembled on to the etched board, which was then mounted in some sort of frame on the chassis. Later the idea of using the board as a plug was introduced and a wide range of edge connectors were designed commercially, all serving the same purpose but having many disadvantages such as poor contacts, tearing-off of the copper pads forming the plug section, and using too much of the useful area of a board. Subsequently families of edge connectors have been designed commercially and these overcome or minimize many of the above disadvantages.

At A.E.R.E. some of the requirements for printed wiring standards have recently been modified because of improved control and techniques by the manufacturers. The system of producing the board as a plug section is now commonly used. The method of assembly of a 2000 series chassis employing plug-in boards (or cards) is that the edge connector is mounted vertically by some means to the two left-hand tie-bars of the chassis and the socket orientated with the numbered pins towards the rear of the unit. Cards are plugged in from right to left. The method for mounting the edge connector is shown in Fig. 15.3. A strip of angle suitably drilled is mounted on the vertical inner faces of the tie-bars as shown and the guides are bolted to these angles in pairs. The edge connector is located by the pairs of guides providing automatic alignment of the guide and the entry slot for the card in the socket.

Conditions in printed wiring have reached the stage when there is a reasonable yield of products employing newer techniques. The natural improvement of printed wiring is to produce connections from one side to the other of a double copper-clad laminate during the manufacturing process of the board. Several commercial processes are now available for the plated-through process. One standard technique using double copper-clad laminate is to drill all the necessary holes before any chemical work is carried out on the board. The walls of the holes are sensitized with a chemical and copper is then deposited electrochemically. The required copper pattern and the holes are now gold-plated and using the gold as an etch-resist the excess copper is removed.

Another process starts with only the insulating base laminate and any copper is plated to the surface simultaneously with plating through the holes. The advantage of the first system is that in all probability the adhesion of the copper to the base board is better but in using the second technique the copper throughout is of a soft electro-deposited type and there is less tendency for imperfections at the junction of the two types of copper.

By using the through-plating technique it is possible to make connections to a variety of layers of copper and this technique is known as multi-layer. With the multi-layer system numerous conductor paths and earth planes can be accommodated in a thickness of material no greater than the $\frac{1}{16}$-in thickness used for standard printed boards.

15.8. Location and mounting of components

The location of components is governed chiefly by the requirements of *in situ* servicing, which will usually be done by means of extension cards. This leads to three possible methods. For board assemblies where the frequency is less than 10 MHz all components should be mounted on the lettered face. This type of mounting allows access to all components from the front of the unit. For low frequency boards of high component density it may be permitted to relocate signal inactive diodes and other inactive components on the numbered face. The method of mounting diodes on the numbered face may also be allowed where fast circuit considerations make this advisable. Where earth planes are used in fast circuits of 10 ns upwards special consideration has to be given to component lay-out and servicing problems.

Components should be mounted on the opposite side of the board to the copper conductors. The leads should be threaded through the required holes and bent at approximately 45°, soldered to the copper, and cropped. Special attention should be given to the mounting of transistors, keeping their electrodes clear of other components for ease of servicing and identification of the parts of the circuit. It has been found that the best way to mount transistors is in holes in the board, with the electrodes fanning naturally to the rest of the circuit.

Transistors should be marked with their circuit references (J10, J15) on the board and all electrodes should be clearly indicated (E, B, C). Diodes and electrolytic capacitors should be marked with their circuit references and should have their polarities indicated. Inductors and large components should have their circuit reference marked on the board. Similar consideration should be given to wire-in cards, remembering that these cards are mounted vertically and along the length of the 2000 series unit. The components should therefore face outwards for ease of servicing.

In assemblies of high packing density with conventional wiring it is advisable to use, where necessary, wiring whose insulated sleeving can withstand high temperatures. PTFE sleeving or insulation is suitable for this purpose. The reason for this is that some wiring may appear on either side of the card, and when replacing a component there is a danger of melting the insulation of a wire on the other side of the card lying across the heated pin.

15.9. New constructional techniques

15.9.1. *Welding*

The so-called welding technique is more closely associated with braizing if a metallurgical examination is carried out on the specimen. The technique commonly used for micro-circuits is to use a split-head electrode welder and carry out the operation with both electrodes on the same side of the work (see Fig. 15.5). If a component lead is to be welded to a substrate conductor, it is placed in position and the welder is then lowered onto the lead position. A controlled amount of energy is released and the two metals fuse together. What happens is that initially the component lead carries most of the current and as it is heated its resistance rises and the substrate conductor then carries its share of the current. The temperature of both metals rise until

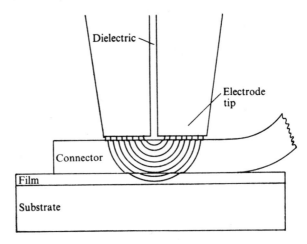

FIG. 15.5. Parallel gap welder.

the metals intermingle. Later welders have resistive feedback, which interrupts the flow of charge into the join when fusion occurs.

15.9.2. *Wire wrap*

This is a technique introduced by the Bell Telephone Laboratories some years ago to overcome a constructional difficulty (Fig. 15.6). A small relay had been designed with a large number of ways on it and the connection terminations were packed closely together making it virtually impossible

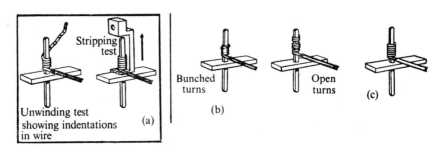

FIG. 15.6. Wire-wrapped joints. (a) Test of the joints. (b) Unacceptable joints. (c) Correct joint.

to solder wires to them when using existing techniques. The solution adopted was to wrap wire round each termination. It was reasoned that if the wrapping post (the termination) was correctly shaped and if the wire was held under sufficient pressure then a cold weld could be effected between the wrapping part and the wire. To facilitate this, square-section wrapping

346

posts improve welding probability by increased pressure on the corners. Tests showed that high reliability could be achieved using wrapped joints.

An alternative wrap was devised when a second wire was used to wrap the component or lead wire to the wrapping post. This avoided undue strain on the junction of the component lead and body. The claimed advantages over soldering are

(1) high reliability of connection,
(2) less skill required,
(3) avoidance of heat close to components.

The main disadvantages are that if a component is removed it can only be re-used by soldering it into circuit, due to the damage to the leads. Also there is a possibility of pieces of wire being dropped into units.

Present trends indicate that many advances in circuitry will be made in the next decade, resulting in changed concepts of construction. Integrated circuits already perform most of the standard circuit functions previously achieved with discrete circuits. Integrated circuits using MOSTs are capable of operating at 10 MHz. Large scale integration (LSI) is now being used for custom-built complex logic blocks. Thin-film circuits are also being used for custom-built circuits.

There have been improvements also in the machines used for unit construction for such purposes as welding, flow or wave soldering, reflow soldering, and wire wrapping.

Acknowledgements

The author would like to thank Standard Telephone and Cables Ltd. for permission to publish Tables 15.1 and 15.2 and also *Electronic Equipment News* for permission to publish Fig. 15.6 from an article by R. A. Leather of Ferranti Ltd. on wrapped joints and connections.

16 *Reliability*

By L. A. KILBEY

16.1. Introduction

THIS discussion is biased towards electronic equipment. To avoid misconception later, let it be said at the outset that well-designed electronic equipment is normally superior, in terms of the average failure rate, to other types of equipment. Furthermore, this is attained without routine maintenance comparable to, say, lubricants in mechanical devices. Another important and characteristic difference between mechanical and modern transistorized electronic equipment is 'wear'.

To illustrate the point, compare an average motor car engine with a transistor amplifier having the same number of parts as the car engine. In the engine most parts associated with movement will suffer significant wear, and the amount of this wear can be readily measured in such parts after say 60 000 miles travel, which at the average car speed of 30 miles per hour is only 2000 hours use. In a transistor amplifier an average working life of at least twenty times that of the car engine would be expected before measured degradation of a few parts occurs, and this degradation is unlikely to be significant. An even more important difference than the time to wear out is the effort (man-hours) required to restore each of these two items to the limits specified by the design specification; for the mechanical and electronic devices considered the ratio would be at least 10:1. A similar situation exists between the electronic and the electromechanical devices that usually comprise a major portion of modern automatic data processing (adp) systems. This is confirmed by experience, which shows that with a particular transistor adp system used, about 75 per cent of the total failures were due to the mechanical or electromechanical items.

Reliability will now be discussed on as broad a front as possible to obtain a reasonable perspective of the subject. The main topics are

    (a) the need for improved reliability,
    (b) an introduction to reliability,
    (c) physics of reliability, and
    (d) application of reliability concepts.

16.2. The need for improved reliability

If, as has been stated, electronic devices are comparatively more reliable than many other devices, why is the subject nevertheless given so much prominence in modern electronic equipment and in complex systems in particular ? The answer lies basically in the application needs that stem from the political, scientific, and economic pressures. A few of the more obvious examples for purposes of illustration are as follows.

(1) *Satellites for peaceful or military purposes.* After the final command to launch the device is given, a failure in the device cannot be repaired in the workshop nor can a technician be sent to carry out the repair *in situ.* The total investment in the device project in terms of human effort and money is extremely high, and in military cases national prestige and/or survival is deemed to be at risk. Hence the chance of success for such systems must be of a corresponding high order, otherwise the devices may not warrant the investment from national resources, i.e. the mean time to failure (mttf) of this non-repairable device needs to be much greater than its working life.

(2) *The transatlantic telephone cable or a large automated sheet steel plant.* In both of these the economic losses that could result directly from a complete failure of the system can be extremely high, say £10 000 per hour in the steel plant. When a failure occurs it is necessary to contend with the human reaction time involved in logical diagnosis of the fault, the time needed for repair, and the time necessary for the functional checking of the system before its recommissioning. In such cases an important factor to be taken into account, when considering the economic viability of the project, is that of assessed reliability of the system. The reliability assessment requires a forecasting activity that is based upon understanding the mechanisms of failure, equipment design considerations, and the use of statistics applied in both these areas. Furthermore, the same basic data and information are needed, together with other information, to assess the likely effect of maintenance activities on the performance of the required functions of the system, i.e. maintainability.

(3) *Laboratory electronic instruments such as those used extensively in research and experimental groups.* Here the need for reliability improvement stems from a number of factors, some of which conflict with reliability goals. The main factors in this area are

- (i) the increasing number of functions required from electronic instrumentation, and
- (ii) the need for faster response time, higher accuracy, and resolution.

The two factors above, which have evolved out of new technology developed during the past decade or so, have (so far) been resolved by the availability, from large scale production, of relatively new equipment and parts. The transistor and the integrated circuit are good examples. The

24

remaining factors have largely been dealt with by the improvement in reliability techniques and available data in relation to electronic equipment. For the future, the application of electronic instrumentation to processes and operations of industry, commerce, engineering, communication, health services, etc. will undoubtedly demand further advances in reliability engineering. The foreseen economic gains will then be more nearly realized rather than penalized by the 'overall' cost of maintaining the functional performance at the required level for operational effectiveness, i.e. competitive for given quality of product or service.

16.3. Introduction to reliability

The study of electronic system reliability arises from the need to improve future designs and to predict from previous experience how effective these will be operationally. Interest in the reliability of past designs may provide a fascinating historical account but has little engineering merit, unless the engineer attempts to examine and analyse the data accumulated from past experience and applies the findings to optimize new designs. The complexities entailed in identifying, normalizing, and compiling this experience constitute a relatively new challenge to the orderly utilization of data that appear in prodigious quantities and subtle forms. To meet this challenge new disciplines of engineering judgement and mathematical analysis have to be organized and proven in the practical 'workshop' of engineering design and operational application. Therefore, the problems of reliability appertaining to electronic systems present all the ramifications of an operations research endeavour. Despite the complexity of the task, a direct and significant contribution can be made to the problem by deriving engineering constants and parameters suitable for predicting the reliability of new system designs. One approach is to summarize the critical aspects of the problem on an engineering plan, and then to present a framework of analyses suitable for a broad application as an expedient procedure for engineering guidance (Bazovsky 1961). Further, before discussing any of the details involved it is essential to recognize the issues at stake when component parts are viewed as the 'key stone' of system reliability. The 'role' of the component parts become critical at the time of selection for a given circuit application. The reason is that previous to the time of selection a great deal of human effort and money has been invested in the genetic characteristics of the component part. After the time of component part selection, various environmental factors associated with the specific application enter as stress factors. These may markedly affect the reliability of the part and therefore of the system in which it is used.

In any component type the general flow of engineering investment is of itself a complex matter entailing the accumulation of basic knowledge, both pure and applied, and of the materials and processes used, together with the economic factors affecting cost. Therefore it is essential from a system designer's point of view to accept, at any given time, the component part resources as having a relatively high degree of inertia arising from the causes mentioned. These points are illustrated in Fig. 16.1.

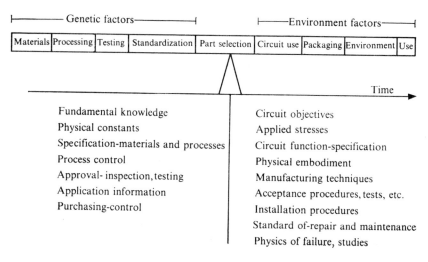

FIG. 16.1. Factors affecting part reliability.

In the normal process of designing electronic equipment, systems are broken down into units which are then subdivided into circuits which are divided further into component parts. In each step, some degree of design commitment is involved. In practice there is a strong tendency to freeze decisions once made, since any changes often propagate their effects throughout the system. During the project planning stage the objectives are formulated for the different programme phases, i.e. applied research, development, production, etc. These objectives are set to meet the total project target, and include the function requirements of the anticipated or defined application. At the development phase it is common practice to assign engineering goals, for example signal characteristics, logic philosophy, accuracy, response time, etc. During the development stage the degree of achievement of these is verified by some form of test schedule. Normally, the target for reliability should be assigned at the start of the development programme but it is often omitted, although it is an important parameter in the determination of the required performance of the equipment or system under various conditions of operation; for example in automatic protective systems such as those used for nuclear reactors (Green and Bourne 1965).

Omission of the reliability target in the schedule of design requirements for an item can be costly to manufacturer and user since either one, or both, may need to spend unbudgeted money and time at the production stage for redesign and/or redevelopment to achieve an acceptable operational performance for the required or intended application. As neither the producer nor the customer desire to incur unforeseen expenditure at the production stage the question is: what is the reason for this omission? It is suggested that it is the necessity to support any risk assignment by a valid and practical routine analysis and proof. With the techniques so far developed, the experimental proof of attainment is necessarily costly. This is due to

(a) large number of samples,
(b) the time of evaluation for acceptable confidence levels, and
(c) because accelerated life tests usually have only a limited use.

These sources of information can be classified by the terms, sample testing, prediction, and field experience. The British Standard document, BS 4200, Part 2 (1967) gives definitions of the variants of reliability terms. The basic information required must be obtained from testing the particular item and even the small sections of a complex system require a long period of time for a large number of units for test, leading to an excessively burdened test programme, which in some cases may not be practical. For example, assuming an exponential distribution and allowing only single failure, about 4×10^8 unit test hours would be required to prove, at 90 per cent confidence level, a failure rate of 0·001 per cent per thousand hours. For certain component parts this is a desirable target figure. On the other hand, a knowledge of the process mechanisms involved in the change of a priority electronic property of a part, for example ΔR, ΔC, etc., might make it possible for reliability statements to cater for a wider variety of conditions than is now possible, since a knowledge of the mechanical, physical, chemical, and other reactions involved would improve prediction techniques. This though leads to the consideration of the physic of failure, or reliability physics.

16.4. Reliability physics

Reliability is defined as the probability of success, i.e. the ability of an item to perform a required function under stated conditions for a stated period of time. It follows from the definition that the converse, unreliability, is related to the chance of an item failing when it is required to perform a function in the required manner. The subject of reliability is concerned with the concept of probability and the type of probability distributions that may be used to represent the behaviour of the system. Feller gives a good review of the mathematics for the theory of probability (Feller 1957).

The reliability parameter is concerned with the frequency of failure, which is related to the physics and mathematics of reliability engineering. The mathematical science of reliability is considered by some engineers to be ahead of practice since techniques and methods have been developed for assigning numerical form to the reliability of a system or assemblies of component parts. These methods take into account the configuration of circuits, redundancy, failure sequence effects, and the abort modes of the system. The accuracy of prediction depends upon a number of factors, particularly on the validity of the *assessed reliability characteristic of the individual component part*. Admittedly, a worth-while advantage derives from the application of these techniques but normally there remains an area of uncertainty, the *component part reliability characteristics*, upon which reliability prediction, analysis, and synthesis, rests. In the final analysis the calculations depend upon our understanding of the physics of component, sub-system and system failures. Some of the topics related to reliability physics will now be briefly considered.

16.4.1. *Failure criteria*

The general definition for failure is given as 'the termination of the ability of an item to perform its required function'. Failure may also be classified according to cause, degree, suddenness, etc., or a combination of these terms, for example catastrophic failure (one that is both sudden and complete) or degradation failure (one that is both gradual and partial). An item, a component part or equipment, fails when stress combinations or individual stresses exceed the strength of the item, as defined by the item's inherent strength and the criteria by which it is determined. By definition, failure criteria are the tolerances on physical, geometric, or material properties which, if exceeded, will cause the performance of the item to be degraded outside defined performance limits for the specific function. Alternatively, the criteria may be of a catastrophic nature, such that when exceeded the item performance characteristic is moved to its extreme value which may be either 'zero' or 'infinity' condition. These concepts are reflected in the British Standard document BS 4200 (1967) for terminology. The word 'function' in the definition of 'failure' covers all those aspects of the item that have the possibility to terminate its performance in terms of a continuum of some parameter, usually time. Therefore, it is important when making reliability assessment or prediction, that all the appropriate

353

characteristics are considered together with their relevant parameters, and that the boundary values for these characteristics are realistically formulated for the given situation. Such boundary values represent the limits within which the item is expected to give the required operational performance.

Some formal postulates for the definition of failure criteria have been put forward and these relate failure to flaws in *objects* and flow of energy.

Some aspects of silicon rectifier failures occurring in certain instruments have been examined using the flaw concept. A selection of the photo micrographs obtained during this work are illustrated in Figs. 16.2–16.4. In practice, it is generally accepted that all obtainable objects contain flaws. This leads to another well-known concept, the largest flow concept, i.e. the strength in any object, is determined by the magnitude of the largest flaw. This relationship of strength is postulated as follows. 'The largest flaw changes with time due to the flow of energy, the change occurring as a rate reaction. But strength is a function of the largest flaw, therefore strength changes with time in accordance with the reaction equations.' From the definitions already given we can develop the area of reliability physics into the region of engineering technology that is concerned with the application of these principles to equipment specifications. In these specifications it is essential, in the interests of improved reliability, for engineers to apply such definitions as function, stress, strength, and failure rate and, in addition, to apply the relation of these definitions to the rate equations. Let us now consider briefly those definitions that we need to take account of in equipment specifications.

16.4.2. *Application stresses*

Application failure stresses are electrical, chemical, and physical. They are internal to a design although these stresses may originate externally. Such stresses are applied for primary function performance and relate to the basic physics of a design, and they can be classified as to their operating state as shown in Table 16.1.

16.4.3. *Environmental stresses*

Environmental failure stresses may be external or internal to the design. Such stresses result from the environment in which a function is being performed. They can be classified by their nature as naturally occurring or induced. A selection of these stresses is given in Table 16.2.

It is important to distinguish between a *stress* and a *failure stress*. The failure stress cannot be defined until it is known to approach the design

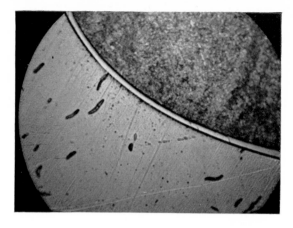

Fig. 16.2. Section of normal junction. Note smooth parallel fronts.

Fig. 16.3. Section of junction—excessive alloying.

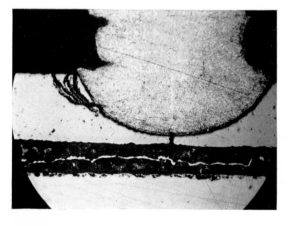

Fig. 16.4. Section of junction—effect of external voltage overload.

TABLE 16.1

Application failure stresses

Failure stress	Relates to:
1. Thermal	the existing temperature distribution and heat flow pattern
2. Chemical	the composition of the chemical reactions that occur
3. Electromagnetic	the electric and magnetic conditions, either steady or fluctuating
4. Rheological	the properties of fluids and the visco-elastic materials that affect their transport behaviour
5. Elastic	the part behaviour properties with respect to deformation under steady or cyclic forces
6. Strain	the deformation produced by the internal forces to which the part is subjected

TABLE 16.2

Environmental stresses

Failure stress	Related to any environment
1. High temperature	that causes heat input to a part so that its temperature approaches the design range maximum
2. Low temperature	that causes heat output from a part so that its temperature approaches the design range minimum
3. Changing temperature	that causes heat in, or out, flow at a rate that approaches the design maximum
4. High pressure Low pressure Changing pressure	similar to above
5. Ionizing radiation	that produces charged and uncharged particles capable of producing further ionization
6. Electromagnetic fields	in which stresses result from subjecting an object to steady or alternating electric or magnetic fields or electromagnetic radiation

limit and for this reason failure stresses are based on history and are defined by knowledge gained in the past, with extrapolation or interpolation from one stress condition to another, i.e. extrapolated reliability.

16.4.4. *Failure mechanisms*

Taking our previous definitions of failure, any study of the mechanisms of failure is a study of the dynamics of energy storage mechanisms. These dynamics consider two types of agents, chemical and physical. Unfortunately, any attempted classification of the two agents poses a problem, since some chemical agents often act physically, and vice versa; sometimes both act in the two capacities simultaneously.

When considering failure mechanisms, the concept of matter in motion versus matter at rest, or the dynamic versus the static, must not be forgotten. From our previous discussion we can say that a failure mechanism is related to failure criteria and failure stress, but without a definition or assumption of these factors we cannot define the associated failure mechanism. Some of the more common mechanisms of failure are tabulated in Table 16.3.

TABLE 16.3

Some common failure mechanisms

Arcing	Diffusion	Noise
Corrosion	Erosion	Radiation damage
Contamination	Fatigue	Secondary currents
Corona	Leakage	Voltage overload
Current overload	Migration	Wear
Dielectric breakdown		

16.4.5. *Failure modes*

Although failure modes are often peculiar to a particular equipment, task, and environment, there are incipient modes of failure related to the largest flaw concept and particular component usage. The failure mode is defined as 'the way in which a strength capability may be exceeded in terms of geometric or material properties'. An alternative definition of failure mode is as 'the manner in which the stress (i.e. energy that an object must accept) exceeds strength'.

16.4.6. *Failure rates*

In applying the scientific method to our subject, measurement must play an important role in reliability physics. Observations must be quantative as well as qualitative, and not all the reliability measurements can be made directly. One important measure of reliability, or unreliability, is the failure rate parameter. Failure rate is a statistical connotation which is related to the number of failures that would be expected to occur over a large number of cycles, or a long period of time, or a large number of trials. It is, therefore, a frequency function, i.e. reliability is the rate at which failures occur.

The term 'failure rate' is commonly used in electronic engineering where it is often thought of as having a time scale only, while in fact it can have many bases. Observations covering several types of simple and complex electronic items or systems have yielded data that suggest that most systems conform on average to a fairly standard type of 'life characteristic' curve.

This failure rate curve can be divided into three sections. First, an early failure period that has a 'normal distribution'; this represents a failure pattern that arises from the production, testing, and installation stages. The

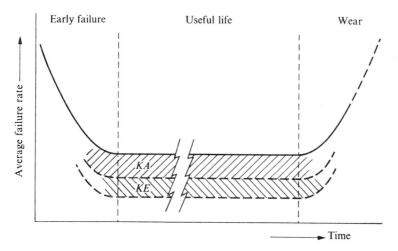

FIG. 16.5. Typical failure-rate curve.

second period, which is an 'exponential distribution', is referred to as the useful life or constant failure rate period since it remains constant at some particular value. The third and final period, which is also a 'normal distribution', arises from the effects of ageing, i.e. when an item begins to wear out. Applying the postulate of failure criteria to field observations of failure rates, the following correlations can be made to the traditional failure rate curve of Fig. 16.5.

16.4.7. *Generic failure rate*

All equipments have an inherent failure rate that may be attributed to the largest flaw concept. Using this concept, generic failure rate can be defined as 'the number of failures per unit of time, cycles, or trials that occur under *ideal* internal and external stress conditions'. Departure from the ideal stress condition increases the number of failures per unit of time, etc. and this is illustrated in Fig. 16.5 by modifiers for the application and environmental stresses, i.e. by KA and KE.

Observed 'initial' or 'early' period. In all groups of manufactured items there is a certain number of items that are defective, and from field observations of equipments the case is seen in which there is a decreasing number of failures during the early period of use or testing. The antecedent of these failures is basically a production quality control problem.

Observed 'normal' operating period. The early period is normally followed by a period of almost constant failure rate, often referred to as the *useful life period*. Under different operating environments or different internal design stresses the failure rate during the normal operating period will have different values above the generic level. It can be measured in terms of failure rate modifiers, and the area in which these operate is shown in Fig. 16.5, where KA is a modifier related to the application failure stress increase or decrease, and KE is a modifier related to the environment failure stress increase or decrease.

Observed wear-out period. The last section of the curve is referred to as the *wear-out period* in which the failure rate is increasing with time, etc. This is a time variation in 'strength' normally termed 'deterioration', which is caused by diffusion, corrosion, migration, etc. It reflects an increase in failure rate that is of a cumulative or additive nature, and is caused by both internal and external stress levels.

16.5. Application of reliability concepts

From the performance viewpoint, the predicted reliability for a given class of item is largely determined by the actions taken during the three principal phases in the chain of events leading to the production of the item, i.e. applied research, development, and design engineering. This 'inherent' reliability can be adversely affected by actions during the production cycle, and the accuracy of the prediction will be strongly influenced by the effectiveness of the quality control methods employed throughout the production cycle. Also, during the life span of the item or equipment there is the effect of the quality and effectiveness of the maintenance, fault diagnosis and repair work actions to which an item is subjected. This also affects the degree of divergence between 'predicted' and 'observed' reliability. It is the designer's task to ensure that suitable and adequate testing techniques and programmes are available to the production and maintenance organizations involved, so that a proper assessment can be made on the various performance factors which involves the mathematics of probability and the type of probability distributions relevant to the behaviour of the equipment.

To accomplish a valid failure rate characteristic it is necessary to have available complete specifications and test information on the characteristics of the item. When the process of stress analysis and reliability is completed for all the significant individual parts, the resulting statistical summary will form the basis for the appraisal of a given equipment design. Data accumulation and evaluation routines are necessary for failure rate information, and these routines involve the design information, the probability allocations, and the weighting factors affecting the component parts (and the equipment) due to the special circumstances of their application. The following procedures involve the interpretation of the statistical data to give 'assessed reliability' for the specific equipment design when used in the particular operational environment. The overall procedure is illustrated in outline in Fig. 16.6. This procedure, when properly applied by a competent assessor,

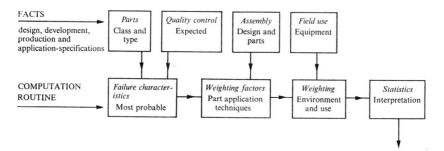

FIG. 16.6. Flow diagram, reliability assessment and/or prediction.

can give a realistic indication of the reliability parameter for the given circumstances, and this numerical result would be available prior to the availability of the production units. Thus, it would permit advanced planning for the operational mission, for example logistics and strategies. Many of the 'professional' equipment designs, as distinct from 'domestic' designs, are usually not produced in batches in the statistical sense. Therefore, they are not so amenable to the more usual sampling techniques. The procedure outlined is often the only one that can be justified on economic grounds when the quantity of the unit production for a given design is small, say one to twenty units.

16.5.1. *Probability distribution*

A probability distribution describes the way in which the probability is dependent on some variant that may be *continuous* or *discrete*. The mathematical expression of the relationship is referred to as the probability function. Some of the types of probability distributions that may be relevant

TABLE 16.4

Some probability distributions—general form

---

1. *Binomial*

$$p(r) = \binom{n}{r} p^r (1-p)^{(n-r)}.$$

$p(r)$ is the probability of r successes out of n trials. $\binom{n}{r}$ signifies all possible combinations of n things taken r at a time, i.e.

$$\binom{n}{r} = [n!/\{r!/(n-r)!\}].$$

The mean is given by np and the variance by $np(1-p)$.

2. *Poisson*

$$p(r) = (a^r/r!) \exp(-a).$$

$p(r)$ is the probability of exactly r events taking place in n trials when the chance of an individual event is p. a is the total number of events n multiplied by the probability p of each event. Therefore a can be described as the average number of occurrences of the event. The mean is given by a and the variance is a.

3. *Exponential*

$$f(x) = (1/\lambda) \exp(-x/\lambda) \qquad \text{for} \quad x \geqslant 0.$$

λ represents mean time to failure and x represents the continuous chance variable. The mean is given by λ and the variance by λ^2.

4. *Normal*

$$f(x) = [1/\{\sigma(2\pi)^{1/2}\}] \exp(-(x-\mu)^2/2\sigma^2) \qquad \text{for} \quad -\infty < x < \infty.$$

μ, *the mean*, is the average of all the values of the variable when each value is weighted by its associated probability of occurrence. σ^2, *the variance*, is the average of all the squared deviations forming the mean where each such deviation is weighted by its associated probability. x is the continuous chance variable.

---

for representing the reliability behaviour of electronic equipment are summarized in Table 16.4. The two main classes of probability distributions are:

(i) those that are concerned with the chance of discrete events, or combinations of discrete events taking place; and

(ii) those that are concerned with likely variations of chance events with respect to time.

The binomial and the Poisson distributions are examples of the first class and the normal and exponential distributions examples of the second class.

16.5.2. *Discrete variables*

Often when a unit of a given population and equipment design fails, it is 'immediately' replaced (or repaired) to restore the lost function(s) to a specified degree. In these circumstances, when a unit does not remain permanently in the failed state, there is a chance that the particular unit may fail again during the time interval of interest to the operational mission. In such cases, the chance of repeated failures is best described by the Poisson probability function.

In many practical cases involving electronic parts, sub-assemblies of these parts, etc., where discrete events take place within the bounds of a known number of tests or trials, the probabilities of occurrence of combination of such events can be obtained by using the binomial probability function. In such cases, it is assumed that the performance of an electronic item has two possibilities, success or failure, in so far as the operational mission objective is concerned.

The binomial probability function approximates to the Poisson probability function. If $nP = A$, then

$$P(s) = (a^s/s!)\exp(-a), \tag{16.1}$$

$P(s)$ being the Poisson probability function. The binomial probability function also approximates to the normal distribution since the values for the 'mean' and 'variance' of the binomial are given by $\bar{m} = nP$ and $\sigma^2 = nP(1 - P)$. The larger the value of n the closer the approximation. It can be shown that, if $nP^{3/2} > 1\cdot07$, the error in using the normal instead of the binomial distribution never exceeds 5 per cent (Raff 1956). The Poisson distribution is the more useful of the two because it permits of ready calculations of a probability value when only one parameter, the value of the mean, $\bar{m}$ is known.

16.5.3. *Continuous variables*

Only two of a number of the available standard distributional forms will be considered. For electronic equipment, the failure pattern is often expressed in the form of an average failure rate for the particular design. A typical common failure rate characteristic pattern has already been given in Fig. 16.5. Many equipments are designed to operate in the 'useful life' portion of this characteristic curve, where the failure rate is relatively constant. In such cases the exponential distribution is appropriate. If, the density function for the exponential distribution is expressed in terms of the time variable, then

$$f(t) = (1/\lambda)\exp(-t/\lambda). \tag{16.2}$$

The probability of failure, in any time interval, is given by the integration of this density function between the time limits of interest. If $Pf(t)$ is the probability of failure (unreliability) in the time interval o to t,

$$Pf(t) = \int_0^t (1/\lambda) \exp(-t/\lambda) \, dt \tag{16.3}$$

$$= 1 - \exp(-t/\lambda) \tag{16.3a}$$

and the probability of success (reliability), $Ps(t)$, is given by

$$Ps(t) = 1 - Pf(t) \tag{16.4}$$

$$= \exp(-t/\lambda). \tag{16.4a}$$

Now, assume the equipment being considered has a failure rate represented by some function, say $\theta(t)$, then the probability of failure in a small time interval dt between t and $(t + dt)$ is given by

$$Pf(t + dt) = \theta(t) \, dt \tag{16.5}$$

and the probability of success in the same time interval is

$$Ps(t + dt) = 1 - \theta(t) \, dt. \tag{16.6}$$

When $\theta(t)$, has a constant value, say $(1/\lambda)$, the exponential function of eqn (16.2) applies. Thus the exponential distribution represents a constant failure rate, and is applicable to the average performance of items working in the 'useful life' portion of the typical failure pattern considered. Under these conditions $\theta(t)$ equals $(1/\lambda)$, a constant. The reciprocal of the constant failure rate λ is referred to as 'the mean time between failures' or mtbf.

The normal probability distribution is now briefly considered in relation to reliability. This function is useful when considering the likely effects of failures due to 'wear' of parts. It is also useful for the inaccuracies that can arise from the deviations that exist between the actual and the true characteristic values of the output of an equipment, i.e. the errors (see Fig. 16.7). Usually such errors can be considered as 'systematic' or 'non-systematic'; the first follows some known law, the second does not. Errors of the first type, when recognized and revealed, may be reduced by a suitable modification to the equipment design and/or the manufacturing processes, providing the cost of so doing is acceptable. If this is not so, the error values may in principle be calculated, and/or measured. They could then be allowed for directly in the operational decisions, but the practical difficulty in such cases is that of establishing the fixed levels on which they are based. Errors of the second type have the element of chance and so are random errors. The measure of the degree of scatter that such errors produce is referred to by the term 'precision'.

Whilst the probability functions and distributions considered can render a mathematically exact value of the relevant probability, in many practical cases the majority of the parameters of distributions (or even the distribution itself) are not known with precision. This brings in the concept of accuracy, or confidence, into the predicted probability. It is desirable that this confidence be defined in some quantitative manner. This involves a number of statistical techniques which are not discussed here but they are fully covered in standard texts (Kendall and Stuart 1961).

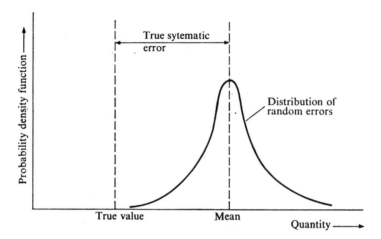

FIG. 16.7. Error diagram.

16.5.4. *Field experience*

The application of the techniques considered to instrumentation systems in the field requires some knowledge of the type of faults that can occur and a knowledge of the physical processes involved in the different failure mechanisms. Also needed is a knowledge of the appropriate probability distributions that represent the phenomena in a given environment. Given sufficient statistical evidence from past operational experience of identical or similar equipment, the unreliability can be determined along with the other more conventional performance characteristics of a design. However, this approach has its difficulties because of the fast moving frontiers of electronic technology and the continuing increase in complexity of the instrumentation designs. The motivating force behind this situation stems from the potential advantages to be gained in many fields from the application of advanced and complex electronic instrumentation. To realize the potential advantages economically these new and complex systems need to have a high degree of reliability.

In attempting to reconcile in practice the opposing factors of complexity and reliability, efforts at the Atomic Energy Research Establishment have been towards the capability of designing low failure nucleonic instrumentation. This has involved studies in a number of allied work areas, for example:

(i) The generation and use of component parts failure rates under the particular conditions of interest to A.E.R.E. designed equipments (historical data).

(ii) The study of parts, equipments, and system failure data, and the mechanisms of failure.

(iii) The study of reliability prediction methods (mathematical models).

(iv) Developing new and improved methods and techniques for measurement, testing, and control of design parameters, through the location of the sensitive parameters affecting their performance.

In the past, most of the effort available for reliability work within the Electronics and Applied Physics Division of A.E.R.E., was directed towards the improvement of the parts used, and to the design of low failure rate

TABLE 16.5

Comparison of some passive component failure rates in A.E.R.E. designed equipment

(Failure rate in per cent/2000 hours, which corresponds to the operating time in 1 year)

	Thermionic valves			Transistors	
	1953–5	1956–8	1959–61	1961–3	1964–6
Resistors					
Cracked carbon	0·73	0·5	0·4	0·07	—
Composition carbon	0·25	0·15	0·1	0·01	—
Wire-wound	0·3	0·27	0·2	0·02	—
Metal oxide	—	—	—	0·001	0·0015
Capacitors					
Paper	0·22	0·17	0·13	0·03	—
Electrolytic	0·95	1·0	1·11	0·16	0·09
Plastic	—	—	—	0·02	0·009

equipments, through the availability of additional information and knowledge from studies of the failure mechanisms. To illustrate the practical outcome of this work, the failure rates of those parts used in large numbers (many thousands, in A.E.R.E. designs) are given in Table 16.5 for a period spanning about 10 years. Likewise, Table 16.6 lists equipment failure rates.

In practice, the larger and more complex equipment designs that had been in operational use long enough to mature operationally had an observed reliability numerical result better than that predicted from calculation. It is suggested that often the statistical failure rate, following seemingly random laws, is the result of using a large number of mixed parts with mixed ages. It would appear from more detailed studies of the failure mechanisms involved that an effect is preceded by some basic ascertainable cause.

TABLE 16.6

A.E.R.E.-designed equipment. Change in equipment failure rate

	Period (years)					
	1949–51	1952–4	1955–7	1958–60	1961–3	1964–6
Failure rate per 100 component parts, per annum (~2000 hours)	1·0 Datum	0·5	0·38	0·33	0·2	0·06
	Category A. Equipment having parts population between 100 and 499					
All designs with a population, $N > 10$						
			1·0 Datum	0·6	0·2	0·09
			Category B. Equipment having parts population between 500 and 5000			

In a research and development establishment such as A.E.R.E., the experimental nature of the work means that the users' requirements in terms of the equipment design performance are often in a state of change. Nevertheless, A.E.R.E.—designed equipments for quantity production (of several hundreds or more) are for economic reasons designed to suit as wide a variety as possible of users. Such a situation makes if difficult to accept the general definition for reliability without a slight modification. This is because of the particular demands of the experimental work applications, which make it extremely difficult to assign to any given performance characteristic of a particular equipment an allied probability of occurrence for any given parameter values. Fortunately, it is found that many of the A.E.R.E. designs do approximate, on the average, to a simple pattern where the equipment behaviour can be described by reference to the variation in its performance characteristic values within the 'specification limits'. These limits, generally define an upper and lower boundary within which the variations in performance characteristic parameters are taken as acceptable, i.e. for correct operational functioning of all the equipments of a given design for all the likely applications. It follows that when a performance variable strays beyond the 'specified limits', the functioning capability of the equipment is categorized as 'not acceptable', i.e. failure.

In the particular circumstances and working of the experimental type applications now being considered, the 'specification limits' may, at any given time, be unrealistic for reliability purposes in view of a user's system requirements. Hence the 'specification limits are not 'a priori' those that have been written into the appropriate document for the equipment design. For reliability purposes it may be necessary to formulate 'envelope' values for accuracy, resolution, response, etc., i.e. characteristics for a particular design application. Such values should represent the limits within which the design would normally be expected to operate correctly for the user's application, assuming no significant component part fails. In some instances it may be possible to assign some statistical confidence to these envelope values. This leads to the probability of failure (unreliability) of an item being taken as 'the chance of the performance characteristic lying outside the specified envelope boundaries for correct operation (functioning) for the specific mission task'.

To be effective, reliability engineering demands mixed strategies, selection being according to the particular task, and the circumstances of the design, development, production, and application activities. In an establishment where a large number, say hundreds, of particular equipments have been operating under the relevant conditions for a long period of time, a suitable strategy to adopt could be the following: the continuing acquisition and recording of data concerning the performance of the particular equipments, coupled with a detailed testing programme capable of showing whether an equipment is, in all aspects of its life environment and use, normal or not normal.

Such information can provide a realistic indication of the envelope boundary values, and these limits may then form the basis for estimating the reliability performance of similar equipments (Raiken 1963). In this approach, progress clearly depends on the detail of the test data, and on the available expertize to distinguish between normal and not normal. Progress also depends on the capability of the test/measuring instrumentation available, and of the data processing/analytical techniques. In addition to these, an improved knowledge of reliability physics, failure mechanisms, etc. is needed to improve further the reliability prediction techniques for any given objective. When extensive information of the type mentioned above it not available for a variety of reasons, for example due to the fact that the item design is new, or part recorded history of similar items is lacking, then it may be necessary to seek out information from sample testing by the producer of the item. Alternatively it may be necessary to commission such reliability test procedures, remembering that it is very difficult to carry out exhaustive tests on the reliability of given equipments. This is due to a number of factors among which are cost, available time, space, effort, etc.

16.5.5. *Assurance*

Certain customers for electronic equipment feel that some form of testing is required to demonstrate that the required reliability value has been attained in the production units. They would argue that the reliability parameter should be tested *in production*, along with the other characteristic parameters of the equipment. Those making this demand often fail to recognize that the reliability parameter can only be demonstrated in terms of 'time', and this requires long testing programmes for production units. For example, even a moderate mtbf of 500 hours would require several months of testing, even if carried out 24 hours per day. Such a period of testing during the production phase of a unit for a single performance parameter invalidates an important requirement for any production test, namely that the test period for any single performance parameter should be relatively short compared to the rate of producing the units.

The burden on production departments for a meaningful reliability test often requires a disproportionate amount of time, money, working space, and effort, in relation to the other production tests. Therefore the cost of the reliability test is high, and the customer should justify such a test on economic or safety criteria, i.e. in relation to the specific risks of his particular project application. It should be noted that such testing as an accept/reject criteria can be damaging financially to both the producer and the customer.

A way around the situation outlined above is to implement a limited programme of sample testing, from which some useful information can be obtained. Such programmes involve making measurements of failure rate of a small sample of the particular item over a limited period. From the results obtained an apparent failure rate may be predicted with a particular degree of confidence. The degree of confidence is related to the size of the sample and to the test duration. In practice this method is very similar to that used generally for statistical quality control. Further information on this latter topic is readily available (Conner and Wells 1961).

When such sampling techniques are used for estimating the reliability parameter of a given equipment design, it is advisable to monitor the performance of the equipments in actual use, partly to confirm the initial

367

prediction and partly to improve the degree of the confidence limits on the estimate of failure rate with increasing tests.

In the past it was occasionally necessary to give some quantitative indication of the expected reliability of a given equipment design prior to the availability of sampling test results. Such assessments are now more frequently required, particularly in those cases where an equipment has been designed to achieve a high order of reliability. This arises out of the particular situation because any economic sampling test may have to proceed for a long time (years) before sufficient confidence could be placed in the experimental results. Empirical approaches have been put forward to suit this situation (Kaufmann and Kaufman 1960). For such cases it is generally necessary to use some prediction technique based on the available design information. It is here that a knowledge of the available theoretical techniques is needed to carry out the required calculation. Such procedures are often feasible because in many new equipment designs a majority of the component parts used are well tried and of proven performance. In addition information on component parts failure rates is more readily available than similar information on equipment designs. Over the past decade or so, a number of establishments have published figures on parts failure rates.

16.6. Present trends

Currently, reliability engineering is more concerned with the possible interactions, intended or otherwise, between the equipments and the computer processing programmes. Many new systems now employ one or more computers linked with controllers, etc., as part of the instrumentation. All these items are interconnected in some way at any particular point in time to provide monitoring, processing, control, and information functions. These conditions tend to emphasize the undesirable effects that can result from electrical interference phenomena. Such effects, when coupled to the possible interactions with programmed instructions and their execution at any point in time, result in a new set of problems for the reliability analyst. Many of these new and complex system designs are only produced in small quantity. To apply production tests in such cases to establish the reliability parameter is not practical or economic. For this reason the emphasis of reliability work falls on

(i) the parts or circuit elements, and
(ii) the actual configuration and arrangements of the elemental parts in relation to the units of function, for the particular system design.

The standardization and/or rationalization of the constituent parts for use in different designs enhances the possibility of more reliable large scale production from established production lines. The parts manufacturer, by the use of the appropriate techniques and test measurements, could make available quantitative reliability assessments with some agreed level of confidence. This would then permit the use of established statistical techniques for a similar assessment of the equipment reliability by the producer of the equipment. The additional cost for this work should not, in these circumstances, be necessarily prohibitive in relation to other costs of producing the items.

In future it is likely that an increasing number of manufacturers of parts will provide, in a standard form, quantitative reliability data on their particular products. Then it will be possible for most designers and producers of electronic equipment to provide at an acceptable cost reliability assessments using prediction techniques with limited sample testing.

The further development of electronic components and instrumentation depends on the contributions to knowledge in many fields of science and technology, physics and chemistry in particular, to increase our understanding of the materials used and of the failure mechanisms in relation to the pattern of the applied stresses. Relevant topics are states of order in ferromagnetic materials, small deviations from statistic disorder in polar dielectrics, and the inclusion of disturbance centres in semiconductors. Further work in reliability physics should aim to clarify the influence that these factors have on the characteristic parameters of any given electronic part and/or equipment unit. By this means the complex instrumentation systems now being conceived are more likely to perform their designed-for functions throughout their envisaged life span; and this with a degree of reliability that will satisfy the necessary economic criteria to justify application of these complex electronic systems to all areas of endeavour where they may benefit mankind.

References

BAZOVSKY, I. (1961) *Reliability theory and practice*. Prentice-Hall, New Jersey.

British Standard document 4200 (1967). Terminology. British Standards Institution, London.

CONNER, W. S. and WELLS, W. T. (1961). Predicting failures in a guidance set. *Proceedings of the seventh national symposium on reliability and quality control in electronics*. I.R.E., New York.

FELLER, W. (1957). *An introduction to probability theory and its applications*. Wiley, New York.

GREEN, A. E. and BOURNE, A. J. (1965) Reliability considerations for automatic protective systems. *Nucl. Engng* **10**, No. 111.

KAUFMANN, M. I. and KAUFMAN, R. A. (1960). Predicting reliability. *Mach. Des.* 1960.

KENDALL, M. G. and STUART, A. (1961). *The advanced theory of statistics*, Vols. 1 and 2. Griffin.

RAFF, M. S. (1956) On approximating the point binomial. *J. Am. stat. Ass.* **51**, No. 274.

RAIKEN, A. L. (1963) The problem of forecasting the reliability of instruments. *Proborostroenie* No. 6, June 1963.

Index

NOTE: Authors of chapters are listed separately on p. xix

371